STP 1249

Assignment of the Glass Transition

Rickey J. Seyler, Editor

ASTM Publication Code Number (PCN):
04-012490-50

ASTM
1916 Race Street
Philadelphia, PA 19103

Printed in the U.S.A.

Library of Congress Cataloging-in-Publication Data

Assignment of the glass transition / Rickey J. Seyler, editor.
 (STP ; 1249)
 "ASTM publication code number (PCN): 04-012490-50."
 Includes bibliographical references and index.
 ISBN (invalid) 0-8081-1995-X
 I. Seyler, Rickey J., 1950– . II. Series: ASTM special technical publication ; 1249.
 IN PROCESS
 641.8'23—dc20
 94-27245
 CIP

Copyright © 1994 AMERICAN SOCIETY FOR TESTING AND MATERIALS, Philadelphia, PA. All rights reserved. This material may not be reproduced or copied, in whole or in part, in any printed, mechanical, electronic, film, or other distribution and storage media, without the written consent of the publisher.

Photocopy Rights

Authorization to photocopy items for internal or personal use, or the internal or personal use of specific clients, is granted by the AMERICAN SOCIETY FOR TESTING AND MATERIALS for users registered with the Copyright Clearance Center (CCC) Transactional Reporting Service, provided that the base fee of $2.50 per copy, plus $0.50 per page is paid directly to CCC, 222 Rosewood Dr., Danvers, MA 01923; phone: (508) 750-8400; fax: (508) 750-4744. For those organizations that have been granted a photocopy license by CCC, a separate system of payment has been arranged. The fee code for users of the Transactional Reporting Service is 0-8031-1995-X/94 $2.50 + .50.

Peer Review Policy

Each paper published in this volume was evaluated by three peer reviewers. The authors addressed all of the reviewers' comments to the satisfaction of both the technical editor(s) and the ASTM Committee on Publications.

The quality of the papers in this publication reflects not only the obvious efforts of the authors and the technical editor(s), but also the work of these peer reviewers. The ASTM Committee on Publications acknowledges with appreciation their dedication and contribution to time and effort on behalf of ASTM.

Printed in Baltimore, MD
September 1994

Foreword

This publication, *Assignment of the Glass Transition,* contains papers presented at the symposium of the same name, held in Atlanta, GA on 4-5 March 1993. The symposium was sponsored by ASTM Committee E-37 on Thermal Methods in celebration of its 20th anniversary with the cooperation of the Plastics Analysis Division of the Society of Plastics Engineers (SPE-PAD) and the North American Thermal Analysis Society (NATAS). Rickey J. Seyler, Eastman Kodak Company, served as chairman of the symposium and is editor of the resulting publication.

Contents

Overview   1

Audience Discussion I: Dealing with Condensed Moisture—R.P. TYE   4

Audience Discussion II: Measurement of Temperatures and the Precision of the Assigned Glass Transition Temperature—R.P. TYE   6

THEORY AND OVERVIEW

Opening Discussion—R.J. SEYLER   13

The Nature of the Glass Transition and Its Determination by Thermal Analysis—B. WUNDERLICH   17

Phenomenology of the Structural Relaxation Process and the Glass Transition—C.T. MOYNIHAN   32

Glass Transition Measurements by DSC—H.E. BAIR   50

Assignment of Glass Transition Temperatures Using Thermomechanical Analysis—C.M. EARNEST   75

The Application of Dynamic Mechanical Methods to T_g Determination in Polymers: An Overview—R.P. CHARTOFF, P.T. WEISSMAN, AND A. SIRCAR   88

Assignment of the Glass Transition Temperature Using Dielectric Analysis: A Review—S.A. BIDSTRUP AND D.R. DAY   108
    Discussion   118

Calorimetric Studies on Glasses and Glass Transition Phenomena—S.-S. CHANG   120
    Discussion   136

Analysis of DSC Thermal Curves for Assigning a Characteristic Glass Transition Temperature, Dependent on Either the Type or Thermal History of the Polymer—J. R. SAFFELL   137

INSTRUMENTAL TECHNIQUES

Sensing Glass Transitions in Thin Polymer Films on Acoustic Wave Microsensors—J. W. GRATE   153

Plasticization of Polystyrene by High Pressure Gases: A Calorimetric Study—M. L. O'NEILL AND Y. P. HANDA   165

Glass Transition in Polymers: Comparison of Results from DSC, TMA, and
TOA Measurements—H. G. WIEDEMANN, G. WIDMANN, AND G. BAYER   174

MATERIALS

Glass Transition and Heat Capacities of Inorganic Glasses: Diminishing
Change in the Heat Capacity at T_g for $x\text{Na}_2\text{S} + (1-x)\text{B}_2\text{S}_3$
Glasses—J. KINCS, J. CHO, D. BLOYER, AND S. W. MARTIN   185

Glass Transition of a Liquid Crystal Polymer—B. CASSEL AND A. T. RIGA   202
   Discussion   212

Glass Transition(s) of Ionomers—R. A. WEISS   214

Measurement of the Glass Transition Temperature of Elastomer Systems—
A. K. SIRCAR AND R. P. CHARTOFF   226

Assigning the Glass Transition Temperature in Oriented Poly(ethylene
terephthalate)—M. J. MOSCATO AND R. J. SEYLER   239

APPLICATIONS

The Glass Transition Temperature of Glassy Polymers Using Dynamic
Mechanical Analysis—E. L. RODRIGUEZ   255

Measuring the Glass Transition Temperature of EPDM Roofing Materials:
Comparison of DMA, TMA, and DSC Techniques—R. M. PAROLI AND
J. PENN   269

Evaluation of Upper Use Temperature of Toughened Epoxy Composites—
J. L. JANKOWSKY, D. G. WONG, M. F. DiBERARDINO, AND R. C. COCHRAN   277

Glass Transition Measurements on Automotive Coatings by DSC, DMA, and
TMA—M. K. GUPTA   293
   Discussion   301

Closing Discussion: Highlights and the Challenges that Remain—R. J. SEYLER   302

Author Index   305

Subject Index   307

Overview

This Special Technical Publication (STP) represents a compilation of presentations from an international symposium addressing the Assignment of the Glass Transition which was held 4-5 March 1993 in Atlanta, GA. The symposium and this subsequent publication reexamine an age-old phenomenon in amorphous and semicrystalline materials, the glass transition. Perspectives from the physics, materials science, engineering, and manufacturing communities for both organic and inorganic materials were shared in the two-day symposium. Five invited lectures and eighteen papers by authors from four countries were presented in four sessions including Theory and Overview, Instrumental Techniques, Materials, and Applications.

The symposium was the fourth in a continuing series on the use of thermal analysis in materials science sponsored by ASTM Committee E–37 on Thermal Measurements and marked the 20th anniversary of this committee. The symposium also enjoyed cooperative sponsorship from the Plastics Analysis Division of the Society of Plastics Engineers (SPE-PAD) and the North American Thermal Analysis Society (NATAS). I, R. J. Seyler (Eastman Kodak), served as the symposium organizer with the support of committee members Charles M. Earnest (Berry College), R. Bruce Cassel (Perkin-Elmer), and Alan T. Riga (Lubrizol).

This book follows the same organizational structure as the symposium and includes 20 of the 23 technical presentations. At the behest of numerous participants I have agreed to include my technical introduction and conference summary remarks in this publication. These summary remarks represent my personal observations of the salient points expressed throughout the symposium and some of the challenges that remain with regard to glass transition assignments. An enthusiastic and interactive audience participation throughout the symposium contributed measurably to the overall success of this symposium. Two audience discussion points by Ronald P. Tye (Ulvac Sinku Riko) were directed at a number of authors. Since these points were recurring throughout the presentations and discussions, it was decided that closure would be handled *en masse* at the beginning of the text. I encourage the reader to regard these points with due consideration prior to reading the text.

Assignment of the Glass Transition was a particularly appropriate subject matter for an ASTM E–37 technical symposium. The committee has a standard test method using differential scanning calorimetry (DSC) for obtaining T_g (E 1356) and has just approved a second standard test method using thermomechanical analysis (TMA) (E 1545). It is also in the process of developing additional standard test methods for the glass transition using dynamic mechanical analysis (DMA) and tensile mode TMA, as well as considering a dielectric analysis approach.

The **Theory and Overview** section includes the invited lectures which provide an excellent review of the glass transition and its measurement using thermal analysis techniques. Wunderlich in the opening paper discusses the thermodynamic aspects of the glass transition with a primarily organic polymer viewpoint. The structural relaxation process concept offered by Moynihan has been more openly embraced by the inorganic glass and physics communities. When taken together, these two papers offer a unique, perhaps "modern," perspective of the glass transition. One notable feature of the structural relaxation process concept is the use of a "fictive temperature" to define the glass transition temperature. Chang and Saffell both further the cause for use of a fictive temperature from calorimetry or

2 ASSIGNMENT OF THE GLASS TRANSITION

scanning calorimetry data. The paper by Chang is a particularly well prepared treatise on calorimetry observations of the glass transition. Ultimately, one must measure a glass transition temperature, which is no trivial matter. The lectures by Bair, Earnest, Bidstrup and Day, and Chartoff, et al. review the glass transition measurement and associated difficulties for each of the thermal analytical techniques DSC, TMA, dielectric analysis, and DMA, respectively. The reader is strongly encouraged to become familiar with the numerous extrinsic factors cited by these authors and how they impact the respective measurements to assign a glass transition temperature for a material.

A variety of properties undergo significant changes in the glass transition interval. The **Instrumental Techniques** section examines several additional measurement procedures or compares the utility of several alternatives. Assessment of thin film (organic, inorganic, and metallic) properties has been identified as a serious analytical need both within ASTM E–37 and the scientific community at large. Grate establishes a viable option for thin films using acoustic wave sensors. Wiedemann compares the utility of optical measurements both as a complement to and as an alternative to the more traditional DSC and TMA procedures. The concept of an extrinsic parameter impacting measured values is reinforced in the paper by O'Neill and Handa where gas pressure as a component of the specimen "atmosphere" causes different temperature assignments.

The **Materials** section offers a sampling of the classes of materials that can exhibit a glass transition. Despite layman references to melting of glass, inorganic glasses are generally recognized as amorphous materials and therefore undergo a glass transition. Martin shows how the chemical composition of an inorganic glass can both shift the temperature interval of the glass transition and modify its appearance and ability to be measured. The paper by Cassel and Riga (liquid crystal) and the one by Weiss (ionomers) examine two classes of polymers in which the existence of a glass transition remains somewhat controversial. The elastomers discussed by Sircar remind us that the glass transition can occur at subambient temperatures and that material properties are significantly different above and below the glass transition. Moscato and Seyler use poly(ethylene terephthalate) as a model semicrystalline material to show how both crystallinity and process history can greatly affect the glass transition. They further suggest that these effects can be so unique as to result in directionally different glass transition temperatures within the same specimen.

The final section, **Applications,** offers the reader some practical experiences in assigning glass transition temperatures. Rodriguez provides an excellent complement to the Chartoff lecture by sharing practical experiences using DMA for three commercial glassy polymers. Assessment of differences and utility of EPDM roofing materials were addressed by Paroli and Penn. Jankowsky et al. demonstrate the value of glass transition measurements to define application limits with toughened epoxy composites examples. In the final paper, Gupta brings a "whole laboratory" perspective to glass transition measurements by examining automotive coatings with DSC, DMA, and TMA methods.

It was the intent of Committee ASTM E–37 in sponsoring this symposium and the subsequent STP to review the current understanding of the glass transition and the measurement practices being applied to its identification. This would serve as a technical resource for the various task groups in subcommittee E–37.01 which are engaged in preparing or revising standard test methods to ensure their utility, relevance, and quality. I believe all of this and more has been accomplished here. One point in particular has become pervasive in E–37 thinking: We are assigning a reproducible temperature from explicit experimental protocols to represent the glass transition interval, not measuring an unequivocal glass transition temperature. This conversion in thinking is critical for successful communication between laboratories measuring the glass transition phenomenon in materials and is being integrated into all of our standard test methods. It is hoped that this text will be as useful a resource to others as well.

The undertaking of any such technical activity as the symposium and this STP cannot succeed without a tremendous amount of support and participation. A very special thanks to my symposium committee Drs. Earnest, Riga, and Cassel and the ASTM staff including Margie Lawlor, Dorothy Savini, Rita Hippensteel, and Therese Pravitz. Many thanks also to the lecturers, presenters, and volunteer reviewers who have made this a quality technical accomplishment.

Rickey J. Seyler, Ph.D.
Eastman Kodak Company
Rochester, NY 14650;
Symposium chairman and editor

Ronald P. Tye[1]

Audience Discussion I: Dealing with Condensed Moisture

A variety of authors discussed measurements of different thermoanalytical methods that are or can be used for identification of the glass transition regime. For a number of material types and especially for polymer-based materials, the glass transition region is below 273K and certainly often below the dew point at normal temperate conditions. In many cases it is in the cryogenic temperature regime.

In the discussions on particular materials the significant effects of moisture on the glass transition of such materials was often highlighted. Several of the authors discussing instrumentation also provided necessary requirements for equipment design and for undertaking evaluation of appropriate properties.

However, little or nothing was said by any of these authors regarding the necessity of maintaining a dry specimen environment to avoid such effects. In fact, one author mentioned only that the specimen was cooled directly by immersion in liquid nitrogen. Once the environment temperature reaches a value below the dew point for the conditions of the laboratory, any moisture in the environment will start to condense on the specimen and its immediate surroundings.

We were made well aware of the effects of moisture on a particular material. In calorimetry, for example, the heat capacity of water is very high compared to that of most materials and small amounts can have a significant impact on results. Apart from one author who mentioned the environment purely from visual operational purposes, and not as a necessary item to control, no one else mentioned this very important parameter. However, condensed moisture within the system can also affect control, response time, and other operational parameters of an apparatus.

Thus, all authors who discussed apparatus requirements, measurement techniques, and results of tests on material where low temperatures were involved should indicate what steps, if any, were or should be taken to ensure that moisture effects are or can be eliminated. It would seem that as a minimum any test apparatus used for measurements on moisture-sensitive materials below the dew point should have the capability of being evacuated and back-filled with a "dry" gas and that all tests were or should be carried out in such an environment.

Closure Statements

[EDITOR'S NOTE: I have attempted to compile the closure comments with appropriate acknowledgment of the responding authors along with my own closure position in this group response to this Audience Discussion. Comments attributed to authors but not highlighted by quotations or offset text are my interpretations of their responses.]

This comment by R. P. Tye is indeed a very important consideration when conducting measurements to assign a glass transition temperature. Unless you are operating in complete drybox conditions, moisture is always present which is especially problematic for those conducting subambient temperature measurements. The presence of moisture and its poten-

[1] Consultant, Ulvac Sinku Riko Inc., Andover, MA 01810.

tial impact is included in the list of recurring points given in the "Closing Discussion: Highlights and the Challenges that Remain" by R. J. Seyler. C. M. Earnest has incorporated a section on effects of moisture with TMA in response to Mr. Tye's comment. S. A. Bidstrup comments that elimination of moisture is difficult with dielectric measurements. Good results have been obtained by drying a specimen mounted on a sensor in a vacuum oven and maintaining that dry vacuum environment during cooling and testing. A. K. Sircar also revised his manuscript on elastomers recognizing the moisture complications at subambient temperatures and further comments:

> The question raised by R. P. Tye about the possibility of moisture condensation on the sample while determining low temperature T_g is relevant for elastomers and has been mentioned in our paper. Tye suggests the test apparatus be evacuated and backfilled with dry gas. However, this will not eliminate condensation of moisture while transferring the sample to the DSC after cooling in liquid nitrogen (as required for the samples which undergo rapid crystallization and discussed in this paper).

One can extend the point by Sircar for any of the techniques employed to observe the glass transition. For certain dynamic mechanical experiments, R. P. Chartoff indicates it may be necessary to freeze the specimen to gain sufficient hardness before clamping into the mechanical driver and to obtain accurate specimen dimensions. These circumstances put both the instrument hardware and the specimen at risk to moisture condensation during loading unless the entire procedure is conducted in a glove box environment. He further notes, however, that moisture diffusion into the rigid glass specimen should be slow relative to the timeframe of an experiment. If proper drying of the specimen before freezing is accomplished, this surface moisture effect should be minimal.

J. W. Grate required the use of a dry nitrogen atmosphere with his quartz transducers when measuring very thin films. His closure comments note that although the moisture does not directly affect the transducer, it modifies the base signal which will change exponentially to a new value with desorption of the moisture. In such thin films even a surface layer of water is critical because it represents a greater percentage of the total specimen than for those of the more traditional thermal analysis techniques.

Despite concerns regarding moisture effects in glass transition measurements, the presence of moisture is a virtual reality for scientists and engineers when considering applications of materials. H. E. Bair comments: "Sometimes it is necessary or desirable to study the effects of moisture on the glass transition." He notes that in many cases the difference in the assigned T_g from first and second heats in a DSC experiment is predominantly due to moisture. Faster heating rates and hermetic encapsulation of the specimen are offered to improve the ability to assess the water effect on T_g with DSC. S. A. Bidstrup and D. R. Day mention the observance of water through permittivity and loss factor signals in dielectric measurements which can be used under controlled conditions to sense changes in moisture concentration or to determine moisture diffusion coefficients.

Ronald P. Tye[1]

Audience Discussion II: Measurement of Temperatures and the Precision of the Assigned Glass Transition Temperature

The major conclusion of this meeting was a general agreement that the glass transition is a phenomenon that occurs over a temperature range. The range can cover an interval of varying limits from a few to many tens of degrees Celsius, highly dependent on the behavior and history of a particular material or material type. Furthermore, the definition of the range is affected by numerous experimental parameters dependent upon the particular technique used to study the phenomenon. It is clear that there is no single measured glass transition temperature for a material and that the onset and end of the regime cannot be clearly defined in terms of a sharp transition. A better practice would be to use the term "assigned glass transition temperature."

However, in a number of presentations, particularly in papers devoted to results of experimental studies on polymer materials, several authors still referred to the measurement of a glass transition temperature. Furthermore this temperature, however determined experimentally, together with the temperatures of onset and completion were often quoted to tenths of degrees Celsius and in a few cases to hundredths of degrees.

In a majority of cases such values were derived from slopes of tangents to curves or from subjective evaluation of inflection points in curves. In general, authors indicated that small differences in choice of where the tangent was drawn or inflection point determined could affect the resultant assigned temperature significantly, often by degrees. Yet particular temperatures were still quoted to the above levels of precision.

A second point in this discussion of an obvious inconsistency related to thermoanalytical techniques is the measurement of temperature itself. The many factors involved in temperature measurement and their effects on thermal analysis instrumentation are well known and have been summarized recently [1]. In general, the temperature of the specimen is not measured directly by a temperature sensor, usually a thermocouple. It is usually inferred from the measured temperature of the surrounding environment or from that of another object in close proximity to the specimen. Furthermore, in most analyses the specimen is being heated or cooled at controlled rates which can vary over a broad range from 1°C to 20°C and higher per minute.

The effects of this latter parameter on the glass transition and other phenomena are well known and were discussed by several authors. The solution chosen to overcome the difficulties of actual temperature measurement is to calibrate the thermoanalysis system by measuring the melting point(s) of a known material(s) having a well-defined melting point(s) under the same heating/cooling rate used for measurement on the actual specimen. Under these conditions it is assumed that the specimen temperature can then be defined by suitable interpolation.

Unfortunately the developers and users of thermoanalytical instrumentation, particularly the systems that use only a computer and software for the final analysis, fall into a trap due

[1]Consultant, Ulvac Sinku Riko Inc., Andover, MA 01810.

to a misunderstanding or misuse of known melting points for calibration. The melting point of a reference material is a thermodynamic property determined to a high precision of measurement under well-defined steady-state conditions. These require special apparatus, a very high level of temperature control, and a temperature measurement system capable of measuring temperature to a level higher than that required to define the melting point. As an example, to define a melting point to a hundredth of a degree Celsius it is necessary to define temperature differences to thousandths of a degree.

In the review mentioned earlier [1] the conclusion was reached that for a combination of the following major factors:

- use of one uncalibrated sensor
- choice of range(s) for which linearity in voltage with temperature is assumed
- measurement and read-out circuitry
- "remote" evaluation of specimen temperature
- heating rate effects on sensor size and response
- heating rate cause of nonuniformities of temperature within the apparatus
- differences in thermal diffusivity between specimen and reference

The temperature definition of a thermoanalytical procedure is no better than 0.5°C and probably much worse.

Thus the essential point is that the melting point of the reference material(s) that is assigned in the analysis software is some two orders of magnitude greater than the level to which it can be measured in practice with the instrument. Thus any derived temperature to tenths or hundredths of a degree Celsius based on input is purely a computer generated artifact having no physical basis.

It is recommended therefore that this issue be recognized and that in the future the temperature calibration of thermoanalytical measurement systems be based on inputs of melting point(s) of a reference material(s) at a more practical and verifiable level based on the temperature measurement instrumentation that is utilized.

Closure Statements

[EDITOR'S NOTE: I have attempted to compile the closure comments with appropriate acknowledgment of the responding authors along with my own closure position in this group response to this Audience Discussion. Comments attributed to authors but not highlighted by quotations or offset text are my interpretations of their responses.]

R. P. Tye is correct in stating that the glass transition occurs over a temperature range which is strongly influenced by numerous extrinsic factors. He is also correct in stating that there currently exists no unequivocal glass transition temperature defined for any material, although a number of scientists have suggested that the fictive temperature derived from DSC measurements may be suitable to uniquely define T_g. However, he misses an important point in his second paragraph when he takes issue with people reporting a glass transition temperature. The scientific and engineering communities have a practical need for a singular temperature to identify the glass transition. This is necessary to readily communicate and tabulate the glass transition for various materials, as well as to develop relationships with properties, composition, process performance, etc. What is practiced is to observe a property change associated with the glass transition as a function of temperature or frequency with an established analytical procedure and to assign a temperature from the observed transition interval as T_g according to a given protocol. Such assigned temperatures are commonly referred to as the glass transition temperature. They are single values that have been assigned to represent the glass transition; they are not unique identifiers. As such I am inclined to accept Tye's position that values for the glass transition temperature be reported to whole degrees Celsius for general applications. H. E. Bair comments "glass transition temperatures of whole degrees Celsius is practical and adequate for 95% of the time." J. W. Grate expressed a similar concurrence.

J. R. Saffell takes a somewhat different position stating:

> I agree that an attempt to define the glass transition to 0.01°C resolution is misleading. However, I believe it is worthwhile to assign values to 0.1°C resolution but with quoted repeatability of ± 1 or ± 2°C. My paper includes histograms showing repeatability of fictive, onset, and peak temperatures of DSC thermal curves.

M. J. Moscato commented that his tabulated values are the average of replicate assignments.

In reviewing the various closure comments for this discussion point, I was reminded of several old DSC experiments I conducted on some specialty polymers of low molecular weight. Observable performance differences followed subtle differences in the assigned values of T_g when recorded as a series under very controlled measurement conditions involving a single DSC on the same day. Sample to sample differences of 0.3 to 0.6°C trended across a half dozen samples and could be repeated on a different day despite greater differences in the absolute values of the temperatures between sets of measurements. When I shared this example with Bair, Grate, and R. M. Paroli, all three agreed that under special circumstances like these it is both appropriate and necessary to report T_g with the greater resolution.

I wish to make a final point with the readers regarding measurement reporting. Values of T_g are derived from measurements of temperature. Measurements are not exact. They have uncertainty and are proximate values, not true values. From a statistical perspective, there is information content in additional digits of a measurement though it diminishes rapidly toward statistical noise with each additional digit. ASTM Committee E–37 has adopted a policy for handling and reporting measured or calculated values from its standard test methods, the essence of which is:

> All available decimal places of a measured or calculated value will be retained through the final calculation. This final value shall be rounded off to the decimal place equivalent of 2 significant places in the standard deviation. If replicate measurements were not conducted to establish the uncertainty, the Interlaboratory Test 95% repeatability limit (r) of the method may be used as an estimate.

Using this guideline and the generally accepted uncertainty of ± 1° to ± 4°C for T_g assignments with thermal analysis techniques, it is acceptable to report values of T_g to 0.1°C.

[EDITOR'S NOTE: For a background to this ASTM E–37 policy readers are encouraged to read (1) Delury, D. B., *NBS Special Publication 300*, Vol. 1, 1969, pp. 392-401; (2) Anderson, R. L., *Practical Statistics for Analytical Chemists*, Van Norstrand Reinhold Co., NY, 1987; (3) Eisenhart, C. et al., *NBS Special Publication 644*, "Expression of the Uncertainties of Final Measurements Results: Reprints," January 1983.]

Tye raises a second major point in his discussion; that of measurement of temperature itself in thermoanalytical instrumentation and the use of melting point references. Saffell offered these comments:

> He (Tye) pointed out that measuring of the temperature indirectly in such techniques as TMA and DMA may lead to errors due to temperature variations within the measuring cell. This is true, but not

true for DSC techniques whereby we are measuring the sample and temperature which is in direct contact with the specimen. My paper outlines the technique for using fictive temperature to calibrate out thermal gradients between the thermocouple and the specimen at heating rates up to 100°C/min. This is not a new technique; it was pioneered by Petrie and Flynn. The use of a highly accurate melting point with high resolution but neglecting the differences of thermal conductivity, specimen mass and specimen-pan thermal resistivity, must be avoided.

He (Tye) recommends that the melting point of a reference material is a more verifiable method for calibrating thermal analytical equipment. There is no equivalent alternative in DMA and TMA, but the use of the fictive temperature in DSC offers a viable alternative that more realistically replicates the specimen thermal environment than using the melting point of crystals.

Tye should be commended for his efforts (both here and elsewhere) to make us realize what temperature it is we are actually measuring in thermoanalytical instrumentation and for the necessity of temperature calibration. One cannot over-emphasize the need to calibrate the temperature axis in thermoanalytical experiments under conditions as equivalent to the actual measurements protocol as possible. I also strongly recommend use of a two-point temperature calibration which brackets the temperatures of interest rather than a one-point calibration which many instrument software packages offer.

I am, however, uncomfortable with Tye's position regarding the extent of our ability to measure temperature in thermal equipment. Temperatures can be measured to a greater extent of accuracy and certainty than Tye proposes, especially at slower heating rates. The issue for me is to what extent does the measured temperature approximate the "true specimen temperature" in a thermoanalytical experiment. This then becomes more an issue of bias. Thus what I believe Tye is actually arguing is how well we can determine the specimen temperature, not how well we can measure a temperature in the measurement cell of a thermoanalytical instrument.

Reference

[1] Tye, R. P., "The Contribution of Precise Temperature Measurement and Control to Thermal Analysis," *Proceedings 21st NATAS Conference*, Atlanta, GA, 1992, pp. 125-132. See also *Thermochemica Acta*, Vol. 226, 1993, pp. 343-351.

Theory and Overview

Rickey J. Seyler[1]

Opening Discussion

This STP represents the ASTM symposium, Assignment of the Glass Transition, which celebrated the 20th anniversary of ASTM Technical Committee E–37's efforts at generating consensus standards for thermal measurements. Subcommittee E37.01 on Test Methods and Recommended Practices has 18 currently active task groups addressing a host of subjects including vapor pressure, temperature scale calibration, oxidative stability, heat flow, volatility, mass loss, and the glass transition. The two task groups of Subcommittee E37.03 on Nomenclature and Definitions continually review technical terms appropriate for use in thermal measurements. ASTM E–37 is responsible for 18 active standards including E 1356, Test Method for Determining Glass Transition Temperatures by Differential Scanning Calorimetry or Differential Thermal Analysis. A second procedure for determining the glass transition using thermomechanical analysis awaits Society approval.[2]

Two additional ASTM standards under the jurisdiction of Technical Committee D–20 on Plastics allow for observation of glass transition temperatures in polymeric materials. They are: D 3418, Test Method for Transition Temperatures of Polymers by Thermal Analysis and D 4065, Practice for Determining and Reporting Dynamic Mechanical Properties of Plastics. Definitions for the glass transition and glass transition temperature may be found in five ASTM documents addressing terminology from Committees D–20, E–37, and also Committee F–17 on Plastic Piping Systems.

Glass transition and glass transition temperature are both used pervasively in the materials community. Underlying this usage is the implication that the glass transition is adequately understood and T_g is an unequivocal value. Survey a random selection of materials scientists and the majority, if not all, will be comfortable in their knowledge of the glass transition. Provide these same individuals with an amorphous specimen and ask them to report the glass transition temperature, and it will be accomplished with equal conviction of certainty. Review the reported values and it is likely that single temperatures of significantly varying values will have been provided. Why?

Let us begin our examination of the glass transition with a definition. A general definition is provided in ASTM E 1142, Terminology Relating to Thermophysical Properties.

glass transition—the reversible change in an amorphous material or in amorphous regions of a partially crystalline material, from (or to) a viscous or rubbery condition to (or from) a hard and relatively brittle one.

The remaining four documents including D 83, D 4092, E 375, and F 412 simply substitute "polymer" for "material" in this definition. A discussion is appended to these definitions:

---

[1]Eastman Kodak Company, Rochester, NY 14650-2158; chairman of the symposium and editor of this STP.
[2]Project TM-01-20-4 was approved as E 1545, Test Method for Test for Glass Transition Temperatures by Thermomechanical Analysis, in June 1993.

ASSIGNMENT OF THE GLASS TRANSITION

The glass transition generally occurs over a relatively narrow temperature region and is similar to the solidification of a liquid to a glassy state [; it is not a phase transition]. Not only do hardness and brittleness undergo rapid changes in this temperature region, but other properties, such as, coefficient of thermal expansion and specific heat capacity, also change rapidly. This phenomenon sometimes is referred to as a second order transition, rubber transition, or rubbery transition. When more than one amorphous transition occurs in a material, the one associated with segmental motions of the backbone molecular chain, or accompanied by the largest change in properties is usually considered to be the glass transition.

This discussion is included in all of the ASTM definitions with the exception that "polymer" is substituted for "material" and the parenthetic expression in brackets is added to the definitions from D 20 and F 17.

Defining glass transition temperature within ASTM has not achieved quite the same extent of consensus. E 1142 provides:

glass transition temperature—a temperature chosen to represent the temperature range over which the glass transition takes place.

DISCUSSION—The glass transition temperature can be determined readily by observing the temperature region at which a significant change takes place in some specific electrical, mechanical, thermal, or other physical property. Moreover, the observed temperature can vary significantly depending on the property chosen for observation and on details of the experimental technique (for example, heating rate, frequency of test.) Therefore, the observed T_g should be considered valid only for that particular technique and set of test conditions.

D 4092 and the other documents offer a somewhat different definition:

glass transition temperature—the approximate midpoint of the temperature range over which the glass transition takes place.

NOTE—The glass transition temperature can be determined readily only by observing the temperature at which a significant change takes place in a specific electrical, mechanical, or other physical property. Moreover, the observed temperature can vary significantly, depending on the specific property chosen for observation and on details of the experimental technique (for example, rate of heating, frequency). Therefore, the observed T_g should be considered only an estimate. The most reliable estimates are normally obtained from the loss peak observed in dynamic mechanical tests or from dilatometric data.

This brief exercise in definitions has very pointedly identified many of the issues to be addressed at this symposium. The glass transition occurs over a temperature interval for which no single temperature is unique, but merely representative. Additionally, given the exact same material, the reported value of T_g may differ if generated by a different measurement protocol. Hence the subject of this symposium, *assignment* of the glass transition, for we do not measure THE glass transition temperature but rather make measurements to observe the glass transition and then assign a temperature, T_g, to mark its occurrence.

Returning to the ASTM standard test methods for a moment, both E 1356 and D 3418 use differential scanning calorimetry (DSC) or differential thermal analysis (DTA) to observe

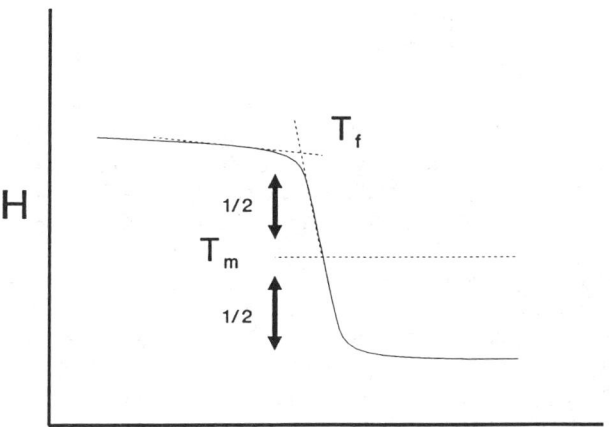

FIG 1.

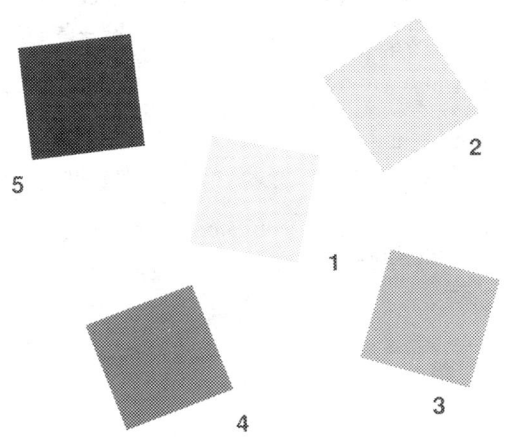

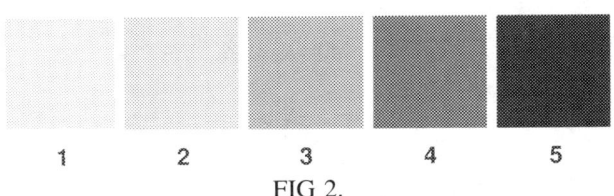

FIG 2.

the glass transition as an enthalpic step change in the baseline (Fig. 1). E 1356 allows use of either the extrapolated onset, T_f, or the midpoint of the step change, T_m, for T_g. D 3418 indicates a preference for T_m to represent T_g as is widely practiced in the polymer community but does allow for use of T_f as an alternative. D 4065 is a general purpose dynamic mechanical analysis (DMA) method which uses the peak value of the loss modulus to represent the transition temperature regardless of its type. Since several decades of frequency could be used in this method and frequency shifts of 6° to 8°C per decade is possible, significantly different T_g's may result from its use. The DMA method under consideration by E37.01.14 is very similar to D 4065 but prefers to represent T_g with the extrapolated onset of the drop in storage modulus. Display of the storage modulus on either a logarithmic or linear scale is allowed but slightly different temperatures will result. E37.01.20 has prepared a standard test method for ASTM approval that uses the compression mode of thermomechanical analysis (TMA) for identifying the glass transition. T_g is assigned as the extrapolated onset to the change in thermal expansivity or alternatively, as the extrapolated onset to softening under loading. Other measurements such as dielectric analysis are also suitable. E37.01.18 is currently laying the foundation for a possible future test method using dielectric measurements to observe the glass transition.

[An audience participation exercise, Glass Transition ? Yellow, was conducted using five tiles of different yellow hues. There is nothing sacred about the choice of yellow and is therefore repeated here using gray tiles of differing density. Each of the tiles was randomly selected and shown to the audience individually with the question, what color is this? The general response for each contained yellow in its color description. The tiles were then redisplayed together as a progression from reddish-yellow to greenish-yellow.]

Consider the tiles in Fig. 2a. What is the color of each? Each is a different shade of gray. If the tiles are now ordered in a progression from light to dark as in Fig. 2b, we have a transition that is a visual corollary to the glass transition. If we were to relate a level of heat capacity for each unit of density in these tiles, one could use the step change curve of Fig. 1 to depict this gray scale. T_f would be the point between tiles 1 and 2 and T_m would relate to tile 3.

Having "seen" a glass transition, all are entreated to seek an enhanced understanding of this phenomenon—the glass transition.

Bernhard Wunderlich[1]

The Nature of the Glass Transition and Its Determination by Thermal Analysis*

REFERENCE: Wunderlich, B., "**The Nature of the Glass Transition and Its Determination by Thermal Analysis,**" *Assignment of the Glass Transition, ASTM STP 1249,* R. J. Seyler, Ed., American Society for Testing and Materials, Philadelphia, 1994, pp. 17–31.

ABSTRACT: In this paper operational definitions of the glass transition and the pertinent condensed phases are given. This is followed by a description of the basic observations of glass transitions via thermal analysis using calorimetry, dilatometry, as well as thermomechanical, dielectric, and dynamic mechanical analysis. The time dependence of the glass transition causes hysteresis phenomena and physical aging. Strain effects on the observation of the glass transition are identified. Complications resulting in symmetric and asymmetric broadening of the glass transition are linked to nanometer and micrometer scale phase separations. Finally, a rigid amorphous phase is identified in some semicrystalline polymers.

KEYWORDS: glass transition, operational definition, solid, liquid, mesophase, thermal analysis, hysteresis, physical aging, microphase separation, nanophase separation, glass transition broadening, rigid amorphous phase, mobile amorphous phase, strain

Operational Definitions

The glass transition is one of the two transitions that characterize the change of a solid to a liquid, the other being the melting transition. To define the term "glass transition," it is best to specify an experiment that results in a clear distinction whether the object under scrutiny is a glass (below the transition temperature) or a liquid (above the transition temperature), i.e., to give an operational definition [1]. This attempt raises, however, an old and common problem. Our language was created long before the scientific principles of many of the phenomena to be described had been understood, and changes in the meaning of a once established word, based on scientific evidence, are hardly ever possible. Common language usage requires a "solid" to be: "A substance that offers some resistance to pressure and is not easily changed in shape: distinguished from gas and liquid" [2]. Operationally this definition could only be used if the pressure and degree of change in shape were quantified, something that is impossible since some well established liquids, such as the melt of high-molecular-mass poly(tetrafluoroethylene), are harder to deform than the corresponding crystal, usually thought of as being a solid. A better, operational definition is to require that: "A solid is a condensed phase below the glass transition temperature, if amorphous, or below the melting transition temperature, if crystalline." To discuss the nature of this glass transition and the thermal analysis techniques for its measurement is the purpose of this article.

---

[1]Department of Chemistry, The University of Tennessee, Knoxville, TN 37996-1600, and Division of Chemistry, Oak Ridge National Laboratory, Oak Ridge, TN 37831-6197.
*The submitted manuscript has been authored by a contractor of the U.S. government under Contract DE-AC05-84OR21400. Accordingly, the U.S. government retains a nonexclusive, royalty-free license to publish or reproduce the published form of this contribution, or allow others to do so, for U.S. government purposes.

Although the just-given definition of a solid seems simple, some amplification of the structure and molecular motion in the possible condensed phases is helpful. Classically one recognizes the three condensed phases that are shaded in the schematic diagram of Fig. 1. The glass is an amorphous solid and possesses the random structure of its corresponding liquid state. The crystal, in contrast, is an ordered solid, characterized by its long-range perfection in structure. The melt is a liquid phase, disordered (amorphous) and usually mobile and easily deformed (but not always, as mentioned above). It may share relatively perfect short-range order with the corresponding crystal by having only fractionally smaller coordination numbers for the nearest neighbors (quasicrystalline melt) [3].

The additional six phases shown in Fig. 1 are so-called mesophases or intermediate phases. Their degree of order is in between those of the liquid and crystal. Well-known are the liquid crystals with some degree of orientational order [4]. Elongated or disc-like portions of the molecules (the mesogens) are preferentially aligned parallel to each other, but otherwise, the liquid crystals have the disorder and mobility of a liquid. The plastic crystals have usually a close-packed cubic crystal structure. The molecules that commonly display plastic crystalline phases, such as $C{=}O$, CH_4, $Si(CH_3)_4$, and C_{60}, are close to spherical or cylindrical and are rotationally disordered and mobile in the mesophase [5]. Conformationally disordered crystals, called condis crystals for short, are disordered and mobile with respect to some or all of the conformational isomers available through bond rotation, and are thus only found in flexible molecules [6,7]. The overall molecules in themselves are fixed with respect to position and orientation.

The observed transitions between the various crystals and the melt are indicated on the right side of Fig. 1. The degree of disorder, in terms of entropy change ΔS, commonly found on losing conformational, orientational, and positional order is also indicated in the figure. The numerical limits have been established by comparing experimental data on hundreds of well characterized transitions.

To understand the upper three mesophases, designated as glasses, one must characterize not only their structures, but also their motion, as was already done for the corresponding pair melt/glass. The pairs of phases, linked on the left, have the same structure, but the glass is missing the typical liquid-like molecular motion. This motion is of large amplitude and can, depending on the molecular structure, be translational, rotational, and conformational.

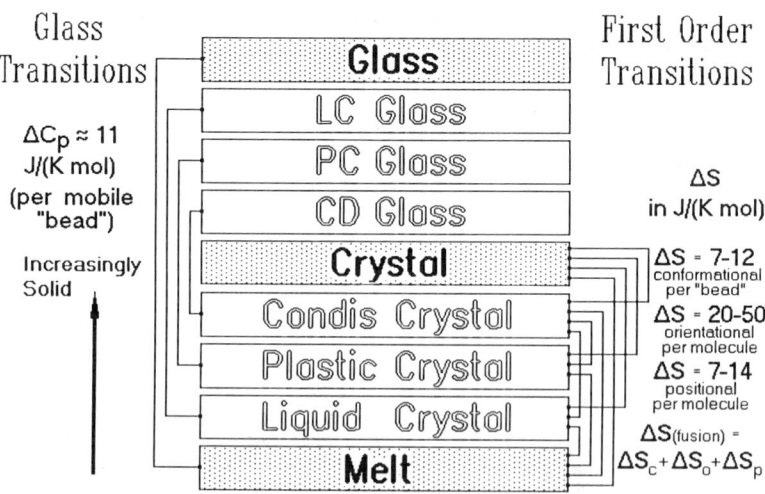

FIG. 1—*The condensed phases.*

The amplitude is large relative to that of vibrations that typically make up the major thermal molecular motion of all condensed phases. The timescale of the large-amplitude molecular motion is in the picosecond (10^{-12} s) range, while the macroscopic response of the corresponding material in the vicinity of the glass transition temperature is slowed to a timescale of perhaps seconds to milliseconds. The large difference in timescale of molecular motion and macroscopic response is based on the cooperative nature of the overall motion in the vicinity of the glass transition.

The schematic diagram of Fig. 2 illustrates in a two-dimensional (2-D) plot the correlation of order (ordinate) and motion (abscissa) to the various transitions. The three mesophase glasses of Fig. 1 correspond thus in structure to the mesophases, but lack the appropriate large-amplitude motion. All five upper phases in Fig. 1 are, according to the definition given above, solids. They are called glasses because of incomplete order and frozen-in large-amplitude motion.

The transitions marked on the left side of Fig. 1 and also indicated in Fig. 2 are glass transitions, taking place at temperatures T_g. Study of many glasses revealed for mobile units (beads) of the size of a (rigid) cluster of perhaps one to six atoms an increase of heat capacity of about 11 J/(K mol) at the glass transition temperature [8,9]. Another empirical rule is that the product of the expansivity-difference between liquid and glass, $\Delta\alpha$, and the glass transition temperature T_g is typically 0.11 (dimensionless free volume fraction) [10]. Similarly, the viscosity of a glass is typically 10^{-12} Pa s (at least for a molecular mass below 50 000) [11] and many liquids become brittle at T_g [12].

Based on the just-discussed phase properties, it is possible to propose the following operational definition: "The glass transition temperature is reached on vitrification or devitrification of half of the sample." The experiment can involve measurement of any extensional property that changes proportionally to the transition between the two phases in question. Note that, especially for inorganic glasses, the term "devitrification" is sometimes, erroneously, used for "crystallization" from the glassy state, a process that is not indicated in Fig. 1. Before any crystallization is possible, the glass must first be able to carry out large-amplitude motion. At this moment the glassy character is, however, lost, i.e., the sample has become, at least locally, a liquid. Crystallization from the glassy state is thus a two-step process, devitrification followed by crystallization. The experimental separation of the two steps may, however, be difficult. Another frequently found misrepresentation is the statement that a glass is a supercooled liquid. Obviously a glass is a solid, not a liquid. The range of existence of a supercooled liquid reaches from the melting temperature, T_m, to only the glass transition temperature, T_g, and not below.

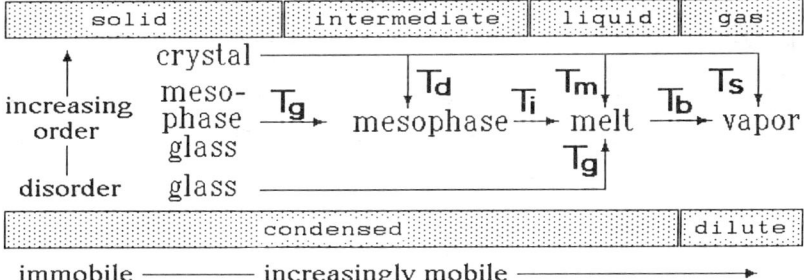

FIG. 2—*Transitions in one-component systems. Dictionary definition of a transition: "A passing from one condition to another."*

Characterization of the Glass Transition

The thermal analysis experiments that can most easily be used to determine the glass transition are calorimetry (differential scanning calorimetry, DSC), dilatometry, and mechanical analyses (thermomechanical analysis, TMA, and dynamic mechanical analysis, DMA) and dielectric measurements. A large collection of experimentally determined glass transition temperatures can be found in Ref *13*. Calorimetry, represented in Fig. 3, is most easily used for a quantitative assessment of the glass transition [9]. Half of the transition is found at T_g. Besides drawing the heat capacity of a 50% glassy and 50% liquid sample and finding the intercept with the measured heat capacity of the sample, T_g is usually close to the point of inflection which, in turn, is often not much different from the temperature of intersection of the extrapolated enthalpies of the liquid and the glass. For full characterization, T_b and T_e, the beginning and end of the glass transition range, should also be given. Furthermore, T_1 and T_2 are useful, identified as the intersections of the tangent at T_g with the extrapolated glass and liquid heat capacities, respectively. The change in heat capacity at T_g, ΔC_p, is also a required identifier. It is a measure of how much of the sample has participated in the transition, information of major importance if the analyzed material contains more than one phase. Finally, all five data must be recorded at a fixed cooling rate. Measurement on cooling has the advantage of starting from the liquid state, which is usually in equilibrium, to go reproducibly to the nonequilibrium glass. Measurement on heating, in contrast, begins with a nonequilibrium state that must be first characterized, as will be discussed in the next section. Fortunately, cooling and heating a sample at the same rates in succession through the glass transition leads to almost the same heat capacity curves for many polymers (to be sure, the to-be-analyzed material needs to be checked for its approximate reversibility, I have been told, some inorganic glasses will not behave in this fashion). After such prior cooling, the easier to measure DSC heating curve can be used to approximate the glass transition parameters at the given (cooling!) rate. If the sample cannot be changed to a convenient thermal history by cooling through the glass transition region, it remains possible to integrate the heat capacity through the transition and extrapolate the solid enthalpy to high temperature and the liquid enthalpy to low temperature and find the glass transition (characteristic of the prior cooling rate) at the intersection of the two enthalpy curves at the "fictive" temperature, as defined by Tool in 1946 [*14*].

Figure 4 illustrates dilatometry for a glass transition determination, using the example of

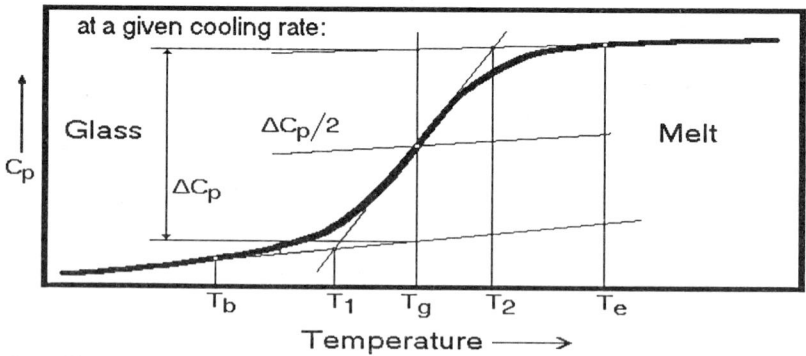

FIG. 3—*Characterization of the glass transition. The transition should be measured on cooling at a specified cooling rate. There should be no ΔH nor ΔS. ΔC_p is often ≈ 11 J/ (K mole of mobile unit).*

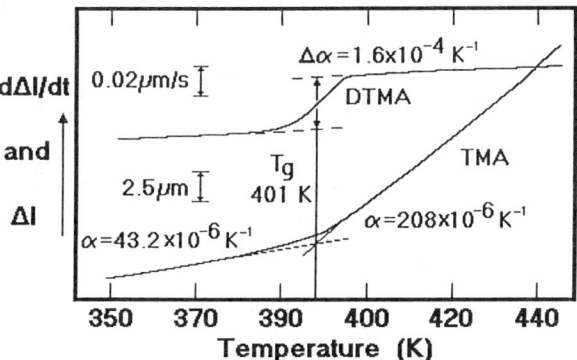

FIG. 4—*TMA of an epoxy printed circuit board (load 0; HR 5 K/min; sample 1.44 mm).*

linear expansion as measured by TMA with minimal force on the sample. Usually the integral measurement of length and the expansivity are recorded simultaneously, so that the temperature versus length curves can be extrapolated to their intersection at T_g. Naturally, this extrapolated T_g is characteristic of the thermal history of the sample, independent of the heating rate. Analogous, multiple-parameter sets as in the heat capacity measurement can be recorded for the expansivity α for the more detailed characterization of the sample. Most often, TMA is not carried out with sufficient precision to yield a full analysis of the glass transition. Samples with low viscosity in the liquid state must be treated with special sample holders by TMA and are better measured by dilatometry [9]. Measurements by penetrometer are more qualitative and cannot be directly related to the glass transition definition given above.

The measurement of the glass transition by DMA is illustrated in Fig. 5. The example chosen is an analysis of poly(vinyl chloride), measured at constant frequency [15]. The glass transition can easily be recognized at 354 K by the precipitous drop in the shear modulus G as the sample changes from the solid state to, in this case, the rubbery liquid. It is not possible to identify half-devitrification as in calorimetry and dilatometry, so one usually chooses the peak temperature of $\tan\delta$ ($= G''/G'$) where G' is the real part (storage modulus)

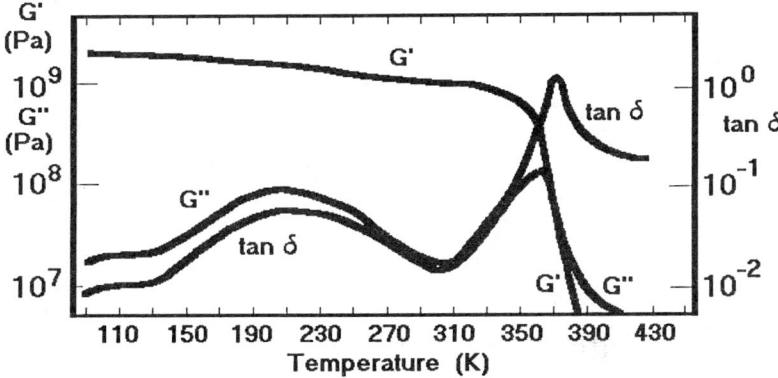

FIG. 5—*DMA of unplasticized poly(vinyl chloride)* $(CH_2—CHCl—)_x$. *Glass transition temperature 354 K.*

and G'' the imaginary part (loss modulus), with $G = G' + iG''$. The measured glass transition temperatures correspond often to those evaluated by calorimetry and dilatometry, when correcting for the different experimental timescales (for precise comparisons, the proper time-temperature correlation should be established [16]). The extrapolated onset-temperature of the decrease of the storage modulus G' in the glass transition range is an important and often used practical indication of the upper temperature limit of the solid phase (at the given timescale); there is, however, no exact correlation to the above-defined T_g.

Varying the frequency of DMA as well as the temperature, it is possible to establish activation energies and relaxation time spectra [16]. These data, although important for the macroscopic description of the glass transition, are difficult to link to the underlying, cooperative molecular motion. Additional maxima in tanδ may occur if a local motion is coupling with the chosen type of deformation. Frequently, these latter peaks have only a minor or no corresponding effect in calorimetry or dilatometry.

The polarization of a material, measured via its relative permittivity (dielectric constant), undergoes a change at the glass transition that can be treated similar to the DMA results. A complex permittivity $\epsilon^* = \epsilon' + i\epsilon''$ is defined, where ϵ' and ϵ'' are the in-phase and out-of-phase permittivities, respectively. Analogous to DMA, ϵ'' represents the loss factor, or dielectric loss, and a loss tangent, $\tan\delta = \epsilon''/\epsilon'$, shows a maximum at T_g that corresponds at similar timescale often to the calorimetric and dilatometric glass transition. The main contributor to the polarization is the dipolar or orientation polarization, resulting from reorientation of permanent dipoles in the alternating electric field. Only in case that these dipoles become mobile at the glass transition, as is often observed, is there a good correspondence of dielectric and mechanical analyses.

Time Dependence of the Glass Transition

A multitude of different glasses of the same chemical structure, but with different physical properties, can be produced by changing the thermal history, for example, by cooling at different rates or annealing below T_g. The latter is also called physical aging. Figure 6 illustrates two such different glasses with plots of their free enthalpies G. Although the liquid is always a stable phase ($G_{\text{liquid}} < G_{\text{glass}}$), the liquid state cannot be realized at the chosen cooling rates, enabling the transition to the less stable glasses. Experiments on polystyrene, for example, show a change in glass transition temperature from 365 K at a cooling rate of 1 K/h to 380 K at a cooling rate of 1 K/s [17]. Even for a cooling rate of 1 K/

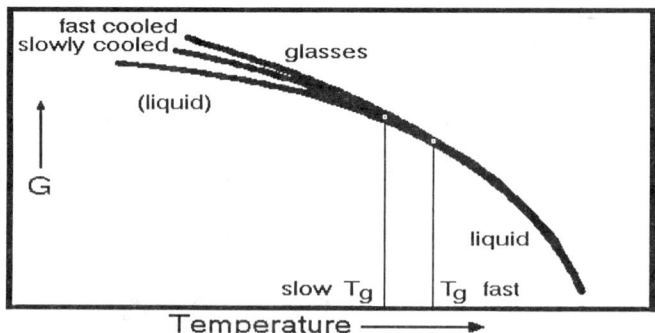

FIG. 6—*Gibbs energy diagram of the glass transition. The liquid (melt) always has the lower G; kinetic reasons cause the change to the higher G of the glass at T_g on cooling.*

year the observed glass transition is not expected to decrease below 351 K, and for a cooling rate of one kelvin per century, 344 K is extrapolated for the glass transition temperature. Glass transition measurements with much shorter timescales are available through DMA or dielectric analysis, but again, a cooling rate of 1000 K/s would not raise the glass transition above 395 K. Because of this relatively small variation of the transition temperature with cooling rate, the specification of timescale is often neglected [13].

Schematically the changes in volume V, enthalpy H, and entropy S as well as the derivatives heat capacity C_p and thermal expansivity α on cooling and heating are shown in the top graph of Fig. 7. Coupling equal heating and cooling rates (path 1) leads to close to "normal" glass transition behavior as was shown in Fig. 3 and permits measurement on heating as well as on cooling. The thermodynamic quantities change close to reversibly between the values characteristic for the glass and the melt at a temperature determined by the timescale of measurement. On coupling slow cooling with fast heating (path 2), the motion of an equilibrium melt cannot be realized at the temperature where the system froze on cooling. As a result, one obtains a superheated glass that drifts quickly toward equilibrium, as soon as the changing timescale of molecular motion on heating permits. In C_p or α this behavior is characterized by a maximum. Since the integrals over C_p and α over a cyclic path must be zero (the integral functions H and V are functions of state), the corresponding dotted areas of Fig. 7 must be equal. This behavior of glasses is called the hysteresis behavior and can be used to obtain quantitative information on the thermal history. The bottom portion of Fig. 7 illustrates a series of differential thermal analysis traces on polystyrene cooled at various rates, but heated at the same heating rate [17]. The change from a glass transition coupled with an endothermic hysteresis peak to a glass transition preceded by a shallow exotherm can be seen. The "normal" glass transition is observed, as mentioned, only when cooling and heating rates are approximately equal. Time-dependent glass transition temperatures measured by thermal mechanical analysis are described, for

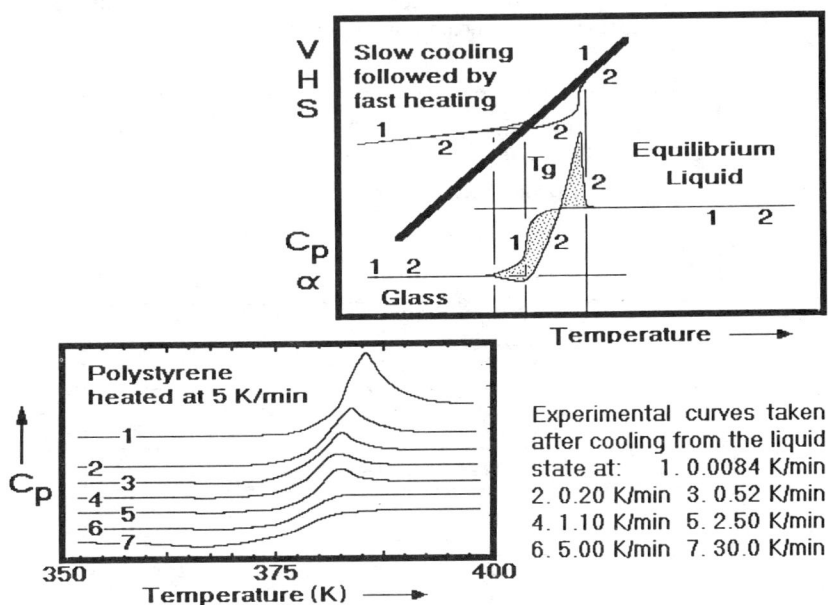

FIG. 7—*Hysteresis at the glass transition.*

example, in Ref *18*. Dynamic mechanical and dielectric analyses have not been used for the study of hysteresis behavior.

Heat capacity or expansion coefficient measurements with a hysteresis peak do not immediately reveal the position of the glass transition, since the sample behavior is governed by both heating rate and thermal history. In these cases the detailed shape of the thermal analysis curve is ignored and one extrapolates the integral curves of liquid and glassy enthalpy or volume to the point of intersection, as discussed with the description of Fig. 3 [*14*]. To distinguish the hysteresis endotherm from a first-order transition (i.e., an equilibrium transition accompanied by an enthalpy and entropy change) one should study the heating-rate dependence of the endotherm. Faster heating rates increase the hysteresis endotherms, while a first-order transition endotherm is, naturally, heating-rate-independent. In both cases, the data must first be corrected for possible instrument lags.

A number of quantitative, phenomenological descriptions, parameterizing the time-dependence of the glass transition, has been developed and are discussed in more detail in the paper by C. T. Moynihan, which follows this presentation. An effort to base such a description on the "hole theory of liquids" by Frenkel and Eyring indicated that the cooperative nature of the glass transition needed to be considered [*17*], a problem still not resolved.

Strain Effects on the Glass Transition

In case strain of any kind is frozen into the glass, it is released in the vicinity of the glass transition and can be measured by the heat, volume, or length effects. Most commonly, strain release leads to an exotherm in calorimetry and a dimensional change in TMA. Rubber elastic strains are of special interest since their release is coupled with an endotherm (entropy decrease on coiling of stretched polymer molecules) and shrinkage of the sample. An example of the analysis of strained samples is shown in Fig. 8 for polystyrene cooled under various pressures [*19*]. After release of the pressure at room temperature, the strained glassy samples (pressure densified) are heated in a DSC experiment to the liquid state (upper

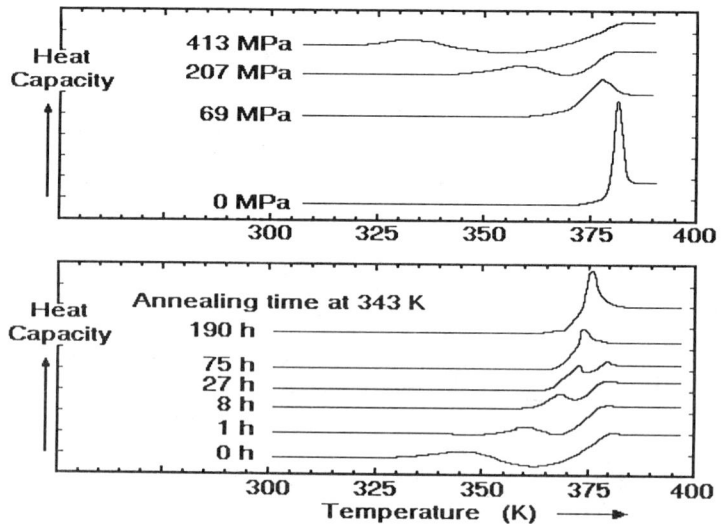

FIG. 8—*DTA of polystyrene cooled under high pressure.*

curves). The atmospheric pressure (0.1 MPa ≈ 0) sample shows a typical hysteresis curve (largely strain-free). It can be seen from the curves taken after cooling at increasing pressure that several strain effects operate during the glass transition. Only the temperature of the end of the glass transition region, T_e, remains fixed. At this temperature an overall strain free conformational equilibrium is reached. The frozen strains start relaxing at increasingly lower temperatures for higher experimental pressures. Experiments of this type indicate that much information can be deduced from a quantitative analysis of the glass transition region.

For strained fibers and films, thermomechanical analysis is particularly suited to study the shrinkage in the glass transition region [20]. Surface strains in small beads of polystyrene could be shown to be analyzable by calorimetry [21]. A method of thermal analysis that relies on the disappearance of birefringence of strained samples that relax at the glass transition was developed by Kovacs and Hobbs (thermo-optical analysis) [22].

The strain-containing glasses of Fig. 8 can change their perfection through annealing below the glass transition temperature to a state of lower free enthalpy. The lower graph of Fig. 8 shows such annealing sequence on a pressure-cooled glass. Volume and enthalpy relaxation do not go parallel in these experiments, another indication that several ordering parameters are needed for a full description of the glass transition [19].

Glass Transitions in Phase-Separated Samples

The composition and molecular mass dependence of the glass transition has received considerable attention, but is of interest to the present discussion only as it changes the nature of the glass transition. The change of the glass transition temperature on composition (copolymerization) has been described, for example, in Ref 23, while that of solutions (blends) in Ref 24. The molecular mass dependence was given first in Ref 25. The changes in the glass transition temperature have been linked to the variation in free volume, chain stiffness, interaction energy, and changes in entropy.

Special effects are observed as soon as the composition of the sample becomes inhomogeneous. Although the change of the glass transition temperature with composition of copolymers and solutions can be described with similar empirical equations, solutions involving macromolecules have always a symmetrically broadened glass transition region. Most likely, this broadening is caused by the inability of the molecules to randomize the structure of their molecular backbone on a nanometer scale. Polystyrene and poly(α-methyl styrene), for example, have a glass transition range of $T_2 - T_1$ and $T_e - T_b$ of 5 to 8 and 25 to 30 K, respectively. Such values are typical for many homopolymers. In a 50/50 solution of the two polymers, possible for sufficiently low molecular masses of the two components, the glass transition range increases to 33 and 75 K, respectively [26]. Similar broadenings are found for solutions of polymers with low molecular mass components (as in plasticizing).

A macroscopically phase-separated system has two glass transitions, as is shown in Fig. 9. For complete separation, as in the case of the 10^7 molecular mass poly(α-methyl styrene), both glass transition ranges are quantitatively those of the homopolymers (in T_g, ΔC_p, and sharpness). As the molecular mass decreases, the poly(α-methyl styrene)-rich phase deceases in T_g and $T_e - T_b$ becomes larger, an indication that a certain amount of polystyrene remains in the poly(α-methyl styrene) [26]. As the phase size approaches micrometer dimensions, as in block-copolymers of the same two components, the two glass transitions broaden asymmetrically, due to the increase in interfacial material [27]. Figure 10 contains plots of the heat capacities of such block-copolymers. The composition (in mol%) and structure is indicated next to the curves [successive curves are displaced by 40 J/(K mol)].

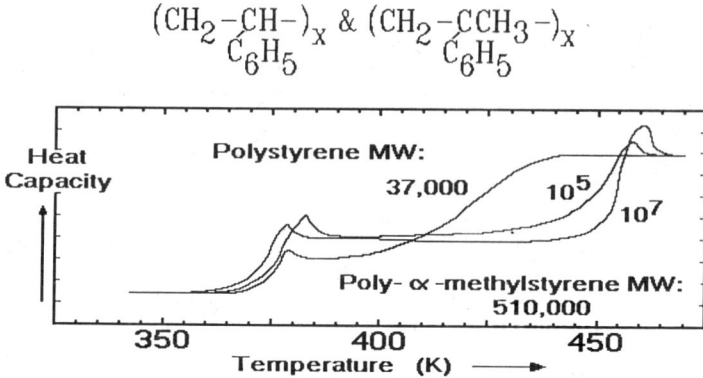

FIG. 9—*Change of breadth of glass transition in copolymers. Thermal analysis curves of 50 mass% blends of polystyrene and poly(α-methylstyrene) at a heating rate of 10 K/min.*

For the two upper block-copolymers the asymmetric broadening of the two glass transitions is so severe that it almost looks like a single transition.

Rigid Amorphous Polymers

In semicrystalline polymers the crystals are separated into microphases. At sufficiently low temperatures, the crystallites are surrounded by and connected via tie-molecules with glassy material. Three observations can be made in such nonequilibrium structures: (1) the glass transition changes its temperature, (2) the ΔC_p decreases from the value expected from the crystallinity of the sample (presence of a rigid amorphous fraction), and (3) the glass transition can become broad.

The increase in the glass transition is shown in Fig. 11, where data are given for poly(oxy-1,4-phenyleneoxy-1,4-phenylenecarbonyl-1,4-phenylene) (PEEK). The glass transition

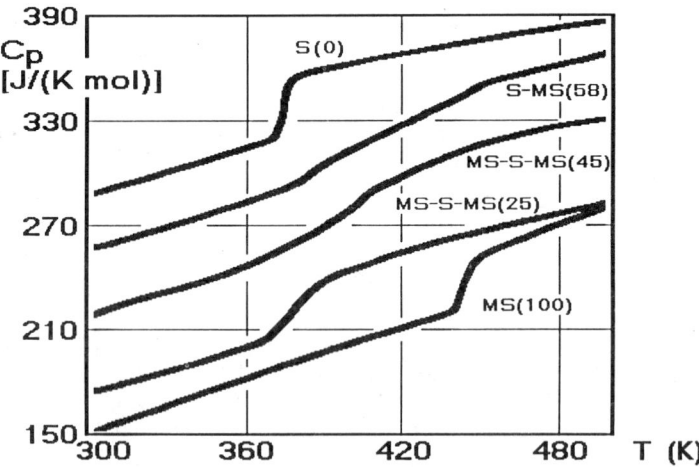

FIG. 10—*Heat capacities of di- and tri-block copolymers of styrene (S) and α-methylstyrene (MS) through T_g.*

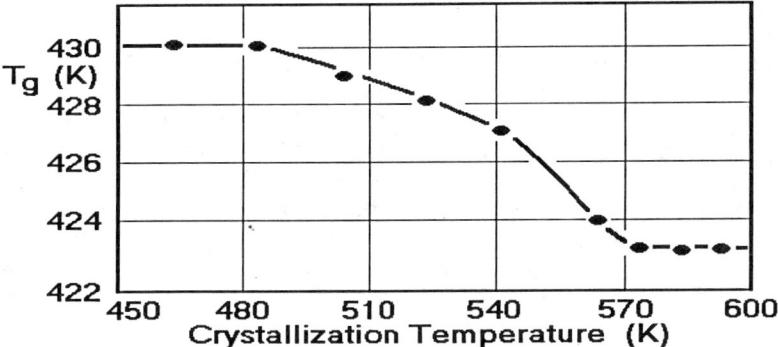

FIG. 11—*Change of the glass transition temperature of PEEK as function of crystallization temperature.*

temperature of quenched, fully amorphous PEEK is 419 K. On fast crystallization at a low temperature, T_c, the glass transition is not only broadened, but also raised to 430 K. As higher crystallization temperatures are chosen, the interface is less strained and T_g decreases, as shown in the graph. In addition to the shift and broadening of the glass transition, a quantitative analysis of ΔC_p shows that there is a rigid amorphous fraction [28].

Since it was indicated in Fig. 3 that ΔC_p is a characteristic constant for a given material, a decrease in ΔC_p must be an indication that some of the amorphous part is hindered to such a degree that it is rigid, i.e., it possesses the lower heat capacity of the glass instead of the higher heat capacity of the liquid.

Figure 12 illustrates the first quantitative analysis of the rigid amorphous fraction as carried out on polyoxymethylene [29]. The left figure shows the great discrepancy that results if one computes the heat capacity using the crystallinity measured from the heat of fusion ($w_c = 67\%$; $w_a = 33\%$, the heat capacity should thus be the dotted curve). Only by

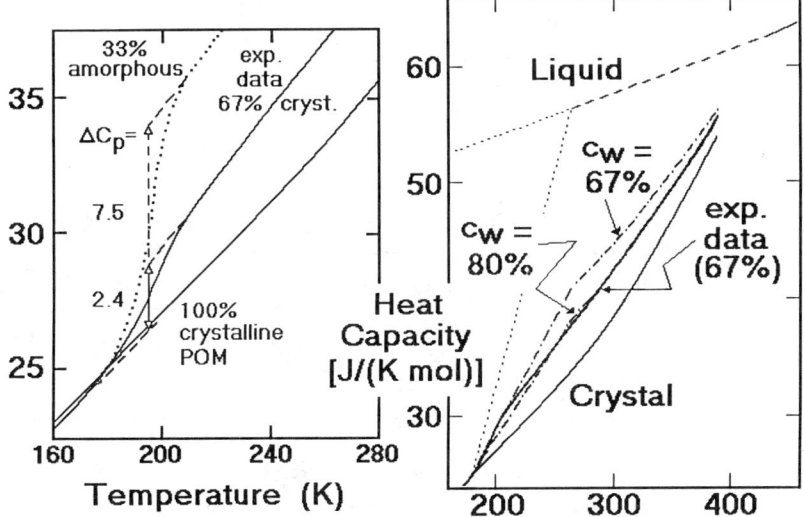

FIG. 12—*Heat capacity of semicrystalline poly(oxymethylene).*

assuming that the solid fraction of the sample is 80% can experiment and measurement be made to match (see the right plot). The resolution of this problem is the assumption that the 13% difference in solid content is caused by a rigid amorphous fraction.

Figure 13 illustrates in its top graph a glass transition analysis for the change of the rigid amorphous fraction with cooling rate for semicrystalline poly(thio-1,4-phenylene), [poly(phenylene sulfide) or PPS] [*30*]. The abscissa of the plot is given in terms of the natural logarithm. On fast cooling, the crystals are poor and have a very large surface area. These seem to be conditions to produce a large fraction of rigid amorphous polymer. On slow cooling (and also on annealing), the rigid amorphous fraction decreases.

For each polymer the condition of formation and the amount of rigid amorphous fraction seems to be different. For poly(oxy-2,6-dimethyl-1,4-phenylene) and poly(butylene terephthalate), for example, the rigid amorphous fraction can be practically 100% of the amorphous fraction, while for more mobile molecules, such as poly(ethylene oxide), the interaction between crystal and amorphous fraction seems to produce only a moderate upward shift in T_g without effect on ΔC_p.

The bottom of Fig. 13 illustrates an even more complicated situation. The macromolecule is the random copolymer poly(oxybenzoate-*co*-oxynaphthoate), a stiff-chain material, designed for high-performance fibers and composites. The graph shows a rather small endotherm and, when compared with the limiting heat capacities of the solid and liquid, an extremely broad glass transition range ($T_2 - T_1 > 100$ K). Without independent knowledge of the limiting heat capacities, the analysis of this material would not have been possible. Other observations of the effect of the crystallinity on the glass transition is a reduction in the hysteresis effect, first observed on poly(ethylene terephthalate) [*31*].

Semicrystalline polymers show also a complicated DMA picture. The glass transition results only in a partial softening because of the high level of physical "crosslinking" due to the crystals (to be distinguished from the more permanent, chemical crosslinking). Only above the melting temperature is the fully liquid state reached. A typical torsion pendulum result is shown in Fig. 14 for linear polyethylenes of different crystallinity. Three regions of

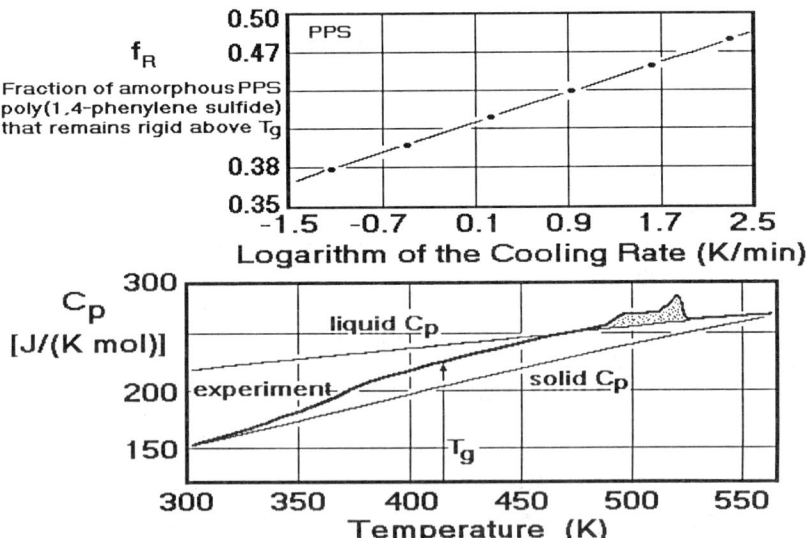

FIG. 13—*Glass transition analyses.*

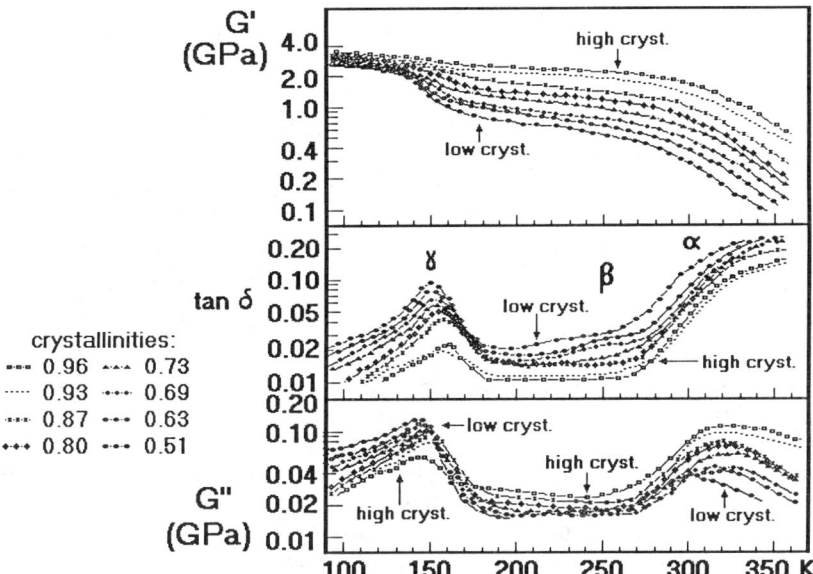

FIG. 14—*DMA of polyethylenes measured at 1 Hz.*

maxima in G'' and tanδ can be found below the equilibrium melting temperature of 414.6 K. Customarily these relaxations are called α, β, and γ. A detailed interpretation of such DMA data has been reviewed by Boyd [*32*]. The α-transition was linked to local mobility in the crystalline polyethylene lamellae, which mobilizes the chain translationally and affects the amorphous regions tied to the crystal. Perhaps it is possible that this motion is linked to an abnormal heat-capacity increase in polyethylene, not caused by fusion. The β-relaxation is linked to the glass transition, typically broadened in a semicrystalline polymer. Figure 15 shows the glass transition of polyethylene, as derived from heat capacity. The agreement in temperature (237 K) is quite reasonable.

Finally, the γ-relaxation at low temperature again occurs in the amorphous phase. It is a broad relaxation in the frequency or time domain. The molecular interpretation links this relaxation to a localized crankshaft-like motion of the backbone of the chain. The slow increase of heat capacity of amorphous polyethylene above about 100 K, as shown in Fig.

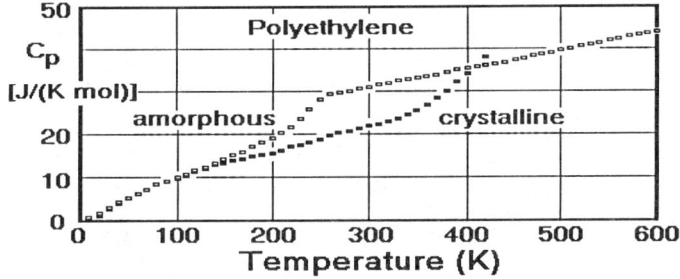

FIG. 15—*Heat capacity of amorphous and crystalline polyethylene. (Derived form over 100 data sets collected in the ATHAS data bank, by extrapolation to* $w_c = 0$ *and 1.0.)*

15, is an indication of this motion [33]. This example indicates that the melting and glass transitions are both visible in DMA. The lower-temperature, secondary relaxations may be linked to local motion that shows only a small effect on the thermodynamic properties as measured by heat capacity.

As with the dynamic mechanical relaxations, it is also possible to check the dielectric behavior of the sample. In order to see a dielectric effect, a dipole must be connected with the molecular motion. In this way dielectric relaxation may be more specific than DMA. A combination of DMA, dielectric measurements, and DSC is often needed for a detailed interpretation of the properties of the materials (see for example Ref 34).

Conclusions

The glass transition is well understood as the transition between a (noncrystalline) solid and the corresponding mesophase or liquid. Its time dependence permits the determination of the thermal history (sometimes also called thermal aging) of samples via their hysteresis behavior. A glass transition is always found if the solid state contains one or more of the three types of disorder: positional, orientational, and conformational. For well ordered crystals, the gain or loss of liquid-like motion is coupled with the melting/crystallization equilibrium. In the vicinity of the glass transition, frozen-in strain is lost, giving most often rise to exothermic or endothermic heat release and/or shrinkage. A change in glass transition temperature is not only caused by changes in the molecular structure, but also by changes in the phase structure. Nanophase and microphase separation can result in considerable broadening of the transition (symmetric and asymmetric) and changes in heat capacity or expansivity (rigid amorphous fractions). More details about the techniques of measurement and additional examples can be found in Refs 9 and 35.

Acknowledgments

This work was supported by the Division of Materials Research, National Science Foundation, Polymers Program, Grant DMR 90-00520 and the Division of Materials Sciences, Office of Basic Energy Sciences, U.S. Department of Energy, under Contract DE-AC05-84OR21400 with Martin Marietta Energy Systems, Inc.

References

[1] Bridgman, P. W., *The Logic of Modern Physics*, New York, 1927.
[2] *Webster's New Twentieth Century Dictionary*, Cleveland, 1970.
[3] Ubbelohde, A. R., *The Molten State of Matter. Melting and Crystal Structure*, Wiley, New York, 1978.
[4] Grey, G. W., *Molecular Structure and the Properties of Liquid Crystals*, Academic Press, New York, 1962.
[5] *The Plastically Crystalline State*, N. Sherwood, Ed., Wiley, Chichester, 1979.
[6] Wunderlich, B. and Grebowicz, J., "Thermotropic Mesophases and Mesophase Transitions of Linear, Flexible Macromolecules," *Advanced Polymer Science,* Vol. 60/61, 1984, p. 1.
[7] Wunderlich, B., Möller, M., Grebowicz, J., and Baur, H., *Conformational Motion and Disorder in Low and High Molecular Mass Crystals*, Springer Verlag, Berlin, 1988, (*Advanced Polymer Science,* Vol. 87).
[8] Wunderlich, B., *Journal of Physical Chemistry,* Vol. 64, 1960, p. 1052 for a larger list see the ATHAS data bank which can be requested from the author and is reprinted in Ref 9.
[9] Wunderlich, B., *Thermal Analysis*, Academic Press, New York, 1990.
[10] Simha, R. and Boyer, R. F., *Journal of Chemical Physics,* Vol. 37, 1962, p. 1003.
[11] Miller, A. A., *Journal of Chemical Physics,* Vol. 49, 1968, p. 1393.
[12] Tammann, G., *Der Glaszustand*, Leopold Voss, Leipzig, 1933.
[13] Peyser, P., *Glass Transition Temperatures of Polymers,"* Vol. VI, pp. 209-277, with over 1200

references, in *Polymer Handbook*, Third ed., J. Brandrup and E. H. Immergut, Eds., J. Wiley and Sons, New York, 1989.
[14] Richardson M. J. and Savill, N. J., *Polymer*, Vol. 16, 1975, p. 753; Tool, A. Q., *Journal of the American Ceramic Society*, Vol. 29, 1946, p. 240.
[15] Illers, K. H., *Kolloid Z. Z. Polymere*, Vol. 251, 1973, p. 397.
[16] Ferry, J. D., *Viscoelastic Properties in Polymers,*" Third ed., J. Wiley, New York, 1980.
[17] Wunderlich, B., Bodily, D. M., and Kaplan, M. H., *Journal of Applied Physics*, Vol. 35, 1964, p. 95.
[18] Schwartz, A., *Journal of Thermal Analysis*, Vol. 13, 1978, p. 489.
[19] Weitz, A. and Wunderlich, B., *Journal of Polymer Science, Part B, Polymer Physics Edition*, Vol. 12, 1974, p. 2473.
[20] Eisenberg, A. and Trepman, E., *Proceedings, 10th NATAS Conference*, 1980, p. 39.
[21] Gaur, U. and Wunderlich, B., *Macromolecules*, Vol. 13, 1980, p. 1618.
[22] Kovacs, A. J. and Hobbs, S. Y., *Journal of Applied Polymer Science*, Vol. 16, 1972, p. 301.
[23] Wood, L. A., *Journal of Polymer Science*, Vol. 38, 1958, p. 319; Couchman, P. R., *Macromolecules*, Vol. 16, 1983, p. 1924; Suzuki, H. and Mathot, V. B. F., *Macromolecules*, Vol. 22, 1989, p. 1380.
[24] Schneider, H. A. and DiMarzio, E. A., *Polymer*, Vol. 33, 1992, p. 3453.
[25] Ueberreiter, K. and Kanig, G., *Z. Naturforschung*, Vol. 6A, 1951, p. 551; Fox, T. G and Loshaek, S., *Journal of Polymer Science*, Vol. 15, 1958, p. 371.
[26] Lau, S. F., Pathak, J., and Wunderlich, B., *Macromolecules*, Vol. 15, 1978, p. 1278.
[27] Gaur, U. and Wunderlich, B., *Macromolecules*, Vol. 13, 1980, p. 1618.
[28] Cheng, S. Z. D., Cao, M. Y., and Wunderlich, B., *Macromolecules*, Vol. 19, 1986, p. 1868.
[29] Suzuki, H., Grebowicz, J., and Wunderlich, B., *British Polymer Journal*, Vol. 17, 1986, p. 1.
[30] Cheng, S. Z. D. and Wunderlich, B., *Macromolecules*, Vol. 20, 1987, p. 2801.
[31] Menczel, J. and Wunderlich, B., *Journal of Polymer Science, Polymer Letters Edition*, Vol. 19, 1981, p. 261.
[32] Boyd, R. H., *Polymer*, Vol. 26, 1985, pp. 323 and 1123.
[33] Wunderlich, B., *Journal of Chemical Physics*, Vol. 37, 1962, pp. 1203, 1207, and 2429.
[34] Boyd, R. H. and Karasz, F. E., *Dielectric Properties of Polymers,*" Plenum Publ. Co., New York, 1971; *Dielectric Properties of Polymers*, F. E. Karasz, Ed., Plenum Press, New York, 1972.
[35] Wunderlich, B., "The Basis of Thermal Analysis," in *Thermal Characterization of Polymeric Materials*," E. Turi, Ed., Academic Press, New York, 1981, second edition to be completed 1994.

Cornelius T. Moynihan[1]

Phenomenology of the Structural Relaxation Process and the Glass Transition

REFERENCE: Moynihan, C. T., "**Phenomenology of the Structural Relaxation Process and the Glass Transition,**" *Assignment of the Glass Transition, ASTM STP 1249,* R. J. Seyler, Ed., American Society for Testing and Materials, Philadelphia, 1994, pp. 32–49.

ABSTRACT: The glass transition is a kinetic phenomenon caused by the inability of the liquid structure to equilibrate on an experimental timescale at sufficiently low temperatures. This results in changes during heating or cooling in the temperature dependence of macroscopic properties such as volume, enthalpy, dielectric constant and loss, etc., over a narrow range in temperature generally referred to as the "glass transition region." During the past 25 years, fairly straightforward semi-empirical models have been developed for the behavior of liquids and glasses in the glass transition region. These models are able to describe both qualitatively and, if the system is not too far from equilibrium, quantitatively the time and temperature dependence of properties during cooling, heating, and annealing. In addition, analysis of the structural relaxation process using irreversible thermodynamics has shown that different properties, e.g., volume and enthalpy, are expected to exhibit different time dependencies in the glass transition region. The relevance of these models and theories to a meaningful assignment of a "glass transition temperature" is discussed.

KEYWORDS: dielectric relaxation, differential scanning calorimetry (DSC), fictive temperature, glass transition, stress relaxation, structural relaxation, viscosity

Structural relaxation, which is responsible for the glass transition, is the kinetically impeded rearrangement of the structure of a liquid in response to changes in external thermodynamic variables such as temperature, pressure, shear stress, or electric field [1–6]. It manifests itself, for example, as a time dependence of properties such as enthalpy H or volume V following a rapid temperature change. The timescale for structural relaxation increases rapidly with decreasing temperature. The glass transition region is the temperature region where the structural relaxation time falls in the range of a few seconds to tens of minutes, so that relaxational effects become perceptible on a human timescale. Since liquid-like degrees of freedom are involved in the structural relaxation process, the glass transition region may also be thought of as the temperature range above which an amorphous material behaves as a fluid or, for a polymer, in a rubbery fashion and below which it behaves as a solid.

The location of the glass transition region is traditionally specified in terms of the glass transition temperature, T_g. Since a kinetically impeded process underlies the glass transition, T_g cannot have the straightforward interpretation one generally associates with a first-order thermodynamic transition temperature such as the melting point of a crystal. Rather, whatever definition one chooses for T_g, its significance can only be assessed properly in the context of a thorough understanding of the structural relaxation process and how this process is revealed by various experimental probes.

[1]Materials Engineering Department, Rensselaer Polytechnic Institute, Troy, NY 12180-3590.

A good illustration of the types of problems inherent in defining T_g is found in the definition of the recrystallization temperature T_R of a cold-worked or strain-hardened metal [7]. Cold-working is plastic deformation of the metal at a low temperature. The plastic deformation introduces point defects and dislocations, which in turn cause an increase in strength and a decrease in ductility, electrical conductivity, and thermal conductivity. If the metal is annealed at a higher temperature, its properties will revert to their initial values via the process of recovery, in which point defects diffuse out to grain boundaries and disappear, followed by recrystallization, in which new dislocation free crystals nucleate and grow at the original grain boundaries. The recrystallization temperature T_R is defined as the temperature at which the mechanical properties (strength and ductility) of a cold-worked metal are restored in a time of 1 h. Since the recovery and recrystallization processes are kinetic in nature and require atomic diffusion, the question of time immediately enters into the definition of T_R. Restoration of the mechanical properties can be accomplished at temperatures less than T_R if the annealing time is greater than 1 h and at temperatures greater than T_R for annealing times less than 1 h. The reason for this is, of course, the rapid increase in the diffusional mobilities with increasing temperature. But the situation is more complicated. For a given annealing time, the temperature for restoration of mechanical properties will be lower for a heavily cold-worked or initially fine-grained metal than for a lightly cold-worked or initially coarse-grained metal. Moreover, the time and temperature for restoration of properties depend upon the property that is being monitored. The electrical and thermal conductivities, which are most sensitive to the point defects eliminated during the initial recovery process, will be restored at lower temperatures or shorter times than will the strength and ductility, which are determined primarily by the dislocations eliminated during the subsequent recrystallization process. Hence knowledge of the recrystallization temperature T_R is of minimal utility in the absence of a thorough understanding of the physical processes underlying the phenomenon.

In the present paper we intend to document, mostly on a macroscopic level, the physical behavior and corresponding phenomenology underlying the structural relaxation process associated with the glass transition.

Structural Relaxation in Response to Temperature Changes

Generic Behavior

In this section we discuss the kinetics and phenomenology of structural relaxation in response to temperature changes of ordinary thermodynamic properties such as enthalpy H and volume V. More detailed coverage of this material can be found in Refs *1* through *6* and *8* through *10* and papers cited therein.

Shown schematically in Fig. 1 is the response of the enthalpy or volume of a glassforming melt initially in equilibrium to a step change in temperature from T_1 to T_2 imposed at time $t = 0$. The melt initially exhibits a "fast" or glass-like change in H or V associated primarily with the vibrational degrees of freedom. This is followed by a "slow" or kinetically impeded further change in H or V associated with the structural relaxation process and the liquid-like degrees of freedom. This structural relaxation progresses until equilibrium is reached at the new temperature. At the top of Fig. 1 the molecular events attending the fast and slow processes are conceptualized for a network oxide liquid. In the fast or glass-like process there is a decrease in the amplitude of the atomic vibrations, which are presumed to be anharmonic. This leads to a net decrease in the distance of separation of the atoms, but no change in their relative positions. In the subsequent slow structural

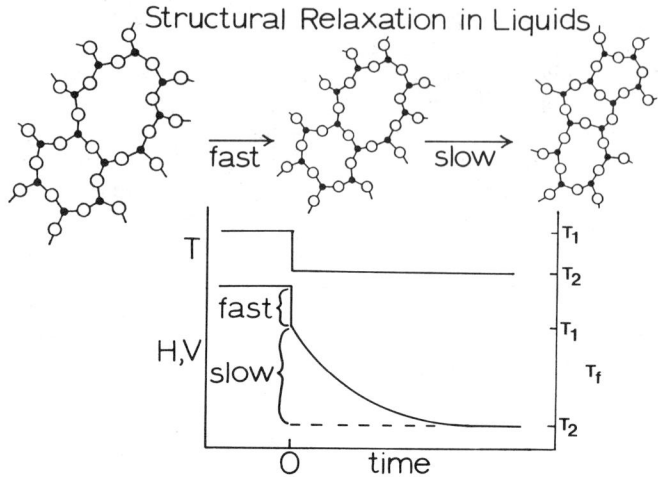

FIG. 1—*Schematic plot of enthalpy or volume versus time during isothermal structural relaxation in response to a step change in temperature.*

rearrangement there is a change in the liquid structure, suggested in Fig. 1 as a decrease in the average ring size.

Results of experiments of this sort are frequently described in terms of the fictive temperature, T_f. T_f is defined as the relaxational part of the property of interest expressed in temperature units. For an equilibrium liquid, $T_f = T$, and during heating or cooling $dT_f/dT = 1$. For a glass whose structure is frozen in well below T_g, T_f = constant, and during heating or cooling $dT_f/dT = 0$. As shown in Fig. 1, during the course of the slow structural relaxation following a step change in temperature from T_1 to T_2, the fictive temperature varies from T_1 to T_2 in parallel with the change in H or V. It should be emphasized that T_f is not a quantity of fundamental significance, but is merely an auxiliary variable defined for computational convenience. Moreover, the value of T_f is not necessarily a unique specification of the structural state frozen into a glass. For example, T_f values calculated from different properties (e.g., H and V) for a rate cooled glass will in general be different.

The rate of structural relaxation may be described by a structural relaxation time, τ. As a crude first approximation, for the isothermal relaxation of enthalpy H shown in Fig. 1, τ can be defined by the following expression for the relaxation function $\phi(t)$

$$\phi(t) \approx \exp(-t/\tau) \qquad (1)$$

where

$$\phi(t) \equiv (H - H_e)/(H_0 - H_e) = (T_f - T_2)/(T_1 - T_2)$$

H_0 is the enthalpy at time $t = 0$ immediately following the fast glass-like response of the material, and H_e is the equilibrium enthalpy approached at long times at temperature T_2. For small departures from equilibrium and over a short temperature range in the transition region the temperature dependence of τ can be approximated by an Arrhenius expression

$$\tau = \tau_0 \exp(\Delta H^*/RT) \qquad (2)$$

where τ_0 is a pre-exponential constant, ΔH^* an activation enthalpy, and R is the ideal gas constant. Hence τ increases rapidly with decreasing temperature.

Cooling or heating of a liquid or a glass at rate $q = dT/dt$ can be thought of as a series of small temperature steps ΔT followed by isothermal holds of duration $\Delta t = \Delta T/q$. In Fig. 2a is shown the behavior of the enthalpy H of a glassforming liquid during a stepwise cool through the glass transition region, followed by a stepwise reheat over the same range. The dashed line represents both the temperature T and the equilibrium enthalpy H_e. The solid line represents the actual enthalpy H. Following the first downward step in temperature, the relaxation time τ is sufficiently short compared to the time interval Δt ($\Delta t >> \tau$) that the system is able to equilibrate and exhibits a liquid-like response for the enthalpy H. Following the second downward step, however, the relaxation time is now longer at the lower temperature, and the system is unable to equilibrate completely ($\Delta t \sim \tau$). The system is now in the glass transition region. The extent of equilibration in time Δt becomes less and less after each subsequent downward step, so that after the last two such steps virtually no structural relaxation occurs during time interval Δt ($\Delta t << \tau$), the system exhibits only the fast glass-like change in H (compare Fig. 1), and the relaxational part of the enthalpy is "frozen" at a value much greater than the equilibrium value. At this point the system is behaving as a glass. Following the first upward step in temperature during reheating, the

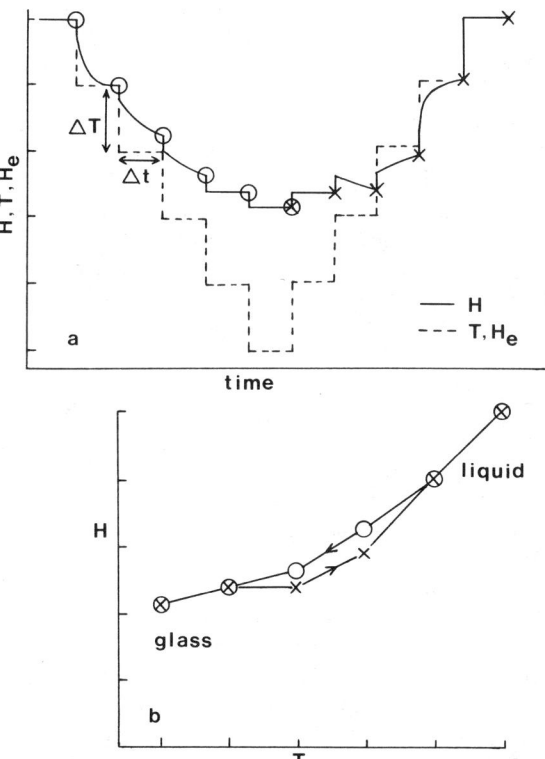

FIG. 2—*Schematic plot of* (a) *variation of temperature T, equilibrium enthalpy H_e and experimental enthalpy H with time, and* (b) *H versus T during stepwise cooling and reheating in the glass transition region.*

relaxation time τ is still too long to allow structural relaxation, so that the enthalpy exhibits only the fast glass-like change, but now in an upward direction. Following the second upward step, the system can exhibit partial relaxation. Since, however, H is above the equilibrium value at this point, it exhibits a downward relaxation, even though the system is now being heated. Following the third and subsequent upward temperature steps, H is below the equilibrium value H_e. Hence it relaxes upwards and is eventually able to equilibrate in time Δt at the highest temperatures, so that the system has returned to liquid-like behavior.

The enthalpy H at the beginning and end of each time/temperature step in Fig. 2a is plotted versus temperature in Fig. 2b. Note that in the transition region the H versus T heating curve is different from the H versus T cooling curve, so that there is hysteresis between the two curves when a liquid is cooled and subsequently reheated through the transition region. Similar behavior would be observed if other properties, e.g., V, were monitored. This hysteresis between cooling and heating curves is in no way unique to structural relaxation and the glass transition. Rather it is expected for any chemical or physical process (e.g., a gas phase chemical reaction or formation and disappearance of point defects in a crystal) whose equilibrium state and whose rate of approach to equilibrium are both temperature dependent.

As should be obvious from Fig. 2, a change in the heating or cooling rate q will change the timescale Δt over which the liquid is allowed to relax following each temperature step ΔT and hence change the temperature at which the liquid finds itself unable to equilibrate completely in time Δt. In particular, an increase in $|q|$ will shift the glass transition region upwards in temperature, and a decrease in $|q|$ will shift it downwards. This is shown schematically in the upper part of Fig. 3 for H versus T heating and cooling curves [8]. Also shown in the lower part of Fig. 3 is the heat capacity $C_p (= dH/dT)$ during cooling and subsequent reheating. On cooling the C_p versus T curve exhibits a sigmoidal shape in the transition region. On reheating the C_p versus T curve passes through a maximum at the upper end of the transition region. Since $C_p = dH/dT$, this maximum corresponds to the steepest part of the H versus T reheating curve and hence is an artifact of the kinetics of the relaxation process and of the expected hysteresis between the cooling and reheating curves.

The glass transition temperature T_g is usually taken as some characteristic point on a plot of a property (such as H) or of its temperature derivative (such as C_p) versus temperature during cooling or heating through the transition region. For example, as shown in Fig. 3, T_g might be taken as the extrapolated temperature of onset of rapid rise of H or of C_p during reheating. We can thus glean from Fig. 3 two important points relevant to the definition of T_g. The first is that T_g depends upon the cooling or heating rate. The second is that T_g measured using a property versus temperature curve (e.g., H versus T) will in general differ somewhat from T_g measured from a plot versus temperature of the derivative of that property (e.g., C_p versus T). Other things being equal (which they generally are not; see below), one would expect a similar disagreement between T_g's obtained in this way from two different properties. For example, T_g from the length or volume curve obtained during heating by dilatometry or thermomechanical analysis (TMA) should differ somewhat from T_g assessed from the C_p versus T heating curve obtained by differential scanning calorimetry (DSC).

In Fig. 4 are shown DSC traces for heating of As_2Se_3 glass through the transition region at three different rates [8]. The DSC output is proportional to the heat capacity C_p. Prior to measurement of each DSC heating curve the specimen was given a controlled and specified thermal history by cooling it through the transition region at a rate equal to the subsequent heating rate. Aside from some distortion of the DSC curves at low heating rates due to baseline curvature, the three DSC curves agree with the schematic curves shown in Fig. 3. As the heating rate is increased, the curves are shifted uniformly upwards in temperature

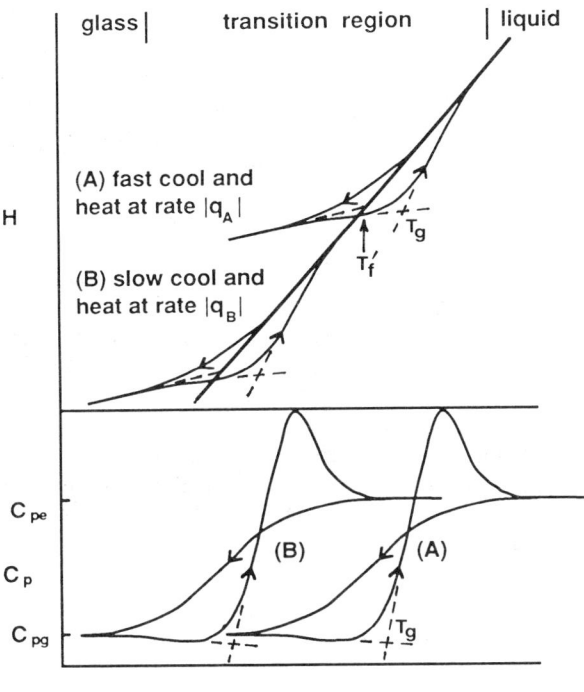

FIG. 3—*Schematic plot of enthalpy H and heat capacity C_p during cooling and reheating through the glass transition region at two different rates. From Ref 8.*

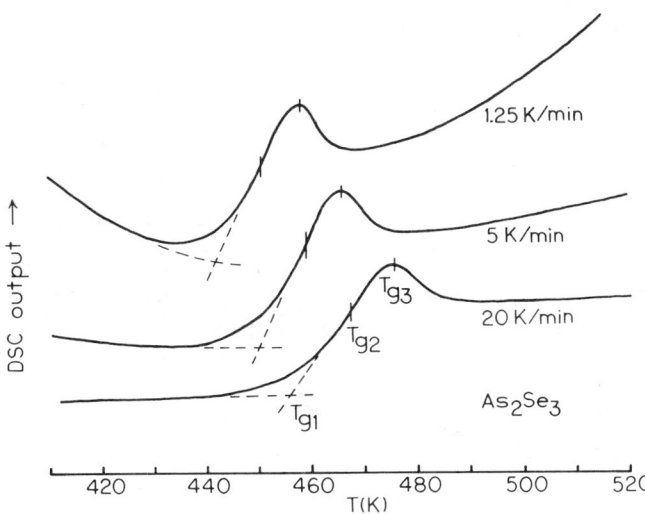

FIG. 4—*Differential scanning calorimeter traces for reheating of As_2Se_3 glass through the transition region at various rates after cooling through the transition region at the same respective rates. From Ref 8.*

without any change in shape. The glass transition temperatures, which might be taken from any characteristic point on the curves, such as a T_{g1}, T_{g2} or T_{g3}, likewise shift uniformly upwards in temperature with increasing heating rate. In fact, it can be shown [8,11] that experiments of the type in Fig. 3 (heating rate proportional to prior cooling rate) can be used to deduce the activation enthalpy ΔH^* in Eq 2, which specifies the temperature dependence of the structural relaxation time in the transition region

$$d\ln q_h/d(1/T_g) = -\Delta H^*/R \qquad (3)$$

where q_h is the heating rate. (Equation 3 remains valid even when the nonlinear and nonexponential features of structural relaxation discussed below are taken into account.)

Also shown in Fig. 3 is a temperature designated T'_f, the temperature of intersection of the extrapolated equilibrium liquid and glass H versus T curves. T'_f, as defined in this fashion, is the limiting fictive temperature (defined here using the property enthalpy, H) attained by the glass on cooling from well above to well below the glass transition region [9]. As should be evident from Fig. 3, the value of T'_f both depends on and is determined solely by the cooling rate. Consequently, it has occasionally been proposed that T'_f be used to define the glass transition temperature. It may be shown [9,11] that the dependence of T'_f on cooling rate q_c is given by an expression analogous to Eq 3

$$d \ln |q_c|/d(1/T'_f) = -\Delta H^*/R \qquad (4)$$

One useful feature of using T'_f as a definition of the glass transition is that, since the glass and equilibrium liquid H versus T curves are the same for cooling and subsequent reheating, T'_f may be determined from the reheating curves. Any reheating rate may be used for this determination and will be presumably chosen to optimize experimental accuracy and precision. If, as in a DSC experiment, the instrumental output is proportional to the temperature derivative of the property ($C_p = dH/dT$), determination of the extrapolated intersection of the glass and equilibrium liquid H versus T curves corresponds to a matching of areas under the measured C_p versus T reheating curve. Determination of T'_f in this fashion is illustrated

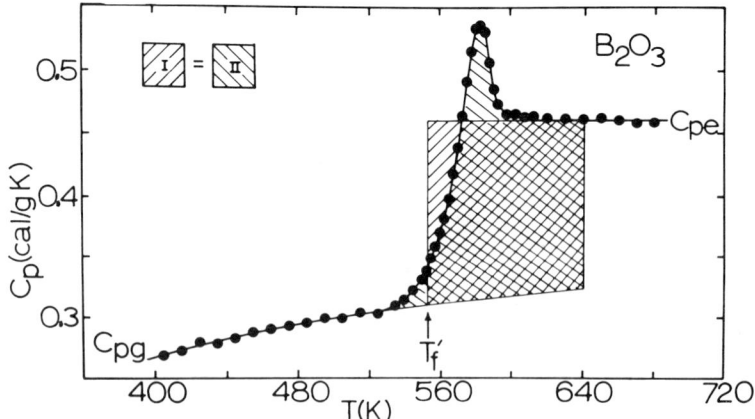

FIG. 5—*Heat capacity versus temperature for B_2O_3 glass during heating at 10 K/min following a rate cool at 10 K/min through the glass transition region. T'_f is the limiting fictive temperature attained by the glass on cooling at 10 K/min. From Ref 9.*

in Fig. 5 for B_2O_3 glass previously cooled through the glass transition region at 10 K/min [9]. The equation governing the area match is

$$\int_{T>>T_g}^{T'_f} (C_{pe} - C_{pg}) \, dT = \int_{T>>T_g}^{T<<T_g} (C_p - C_{pg}) \, dT \tag{5}$$

where C_{pg} is the glass heat capacity observed below the transition region and C_{pe} is the equilibrium liquid heat capacity observed above the transition region. Shaded area I in Fig. 5 corresponds to the left side of the expression and shaded area II to the right side. As should be clear from Fig. 5, this determination of T'_f by area matching can be done "by eyeball" to a sufficient degree of accuracy for many purposes and, if the prior cooling and reheating rates are approximately the same, T'_f corresponds very closely to T_g taken (as in Fig. 3) as the extrapolated onset temperature of rapid rise of the C_p versus T reheating curve.

Kinetics of Structural Relaxation in the Glass Transition Region

A detailed and quantitative description of structural relaxation in response to temperature changes requires that two additional features of the process be taken into account (again cf. Refs *1–6, 8–10*). The first of these is the *nonlinear* character of the relaxation process (in the sense that the rate of relaxation cannot be described by a linear differential equation). This is illustrated in Fig. 6 for the isothermal relaxation of the density of a soda lime silicate glass [*4,12*]. In the upper curve the glass was initially equilibrated at 500°C (so that its initial fictive temperature T_f = 500°C), then upquenched to 530°C and allowed to relax at that temperature. In the lower curve the glass was initially equilibrated at 565°C (initial $T_f \approx$ 565°C), then downquenched to and allowed to relax at 530°C. At long times both samples have come to equilibrium (final T_f = 530°C). (The densities in Fig. 6 were actually measured at room temperature after dropping the samples out of the furnace following

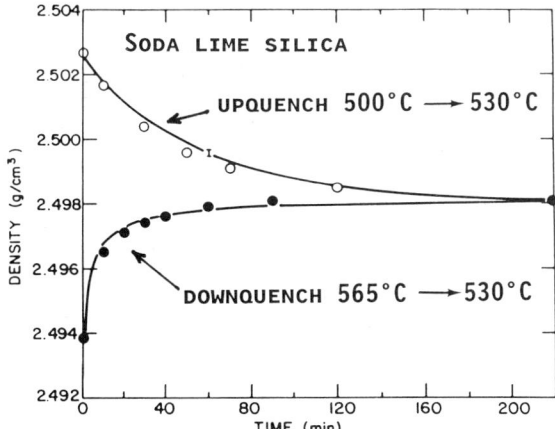

FIG. 6—*Isothermal relaxation at 530°C of the density of a soda lime silicate glass following an upward step change in temperature from 500°C (upper curve) and a downward step change in temperature from 565°C (lower curve). Solid lines are calculated from Eqs 6 and 9 using the parameters* τ_0 = 3.9 × 10^{-37}s, ΔH^* = 607 kJ/mol, x = 0.45, *and* β = 0.62. *From Ref* 12. *(Reprinted with permission from the American Ceramic Society.)*

annealing for various times at 530°C. Hence the fast, "glass-like" contribution to the density change (see Fig. 1) has effectively been subtracted from all the data points.) Even though the initial magnitude of the departure of the density from equilibrium is the same for both specimens and even though they are relaxing at the same temperature (530°C), the downquenched sample with the initially higher T_f ($\approx$ 565°C) clearly relaxes more quickly than the upquenched sample with the initially lower T_f (= 500°C). In terms of Eqs 1 and 2, this means that the structural relaxation time τ depends not only on temperature T, but also on the instantaneous structure or fictive temperature. One common way of modifying Eq 2 to account for this is via the so-called Tool-Narayanaswamy equation

$$\tau = \tau_0 \exp\left[x\Delta H^*/RT + (1-x)\Delta H^*/RT_f\right] \tag{6}$$

where x ($0 \leq x \leq 1$) is the nonlinearity parameter. Note that Eqs 2 and 6 become identical in the limit of small departures from equilibrium ($T_f \rightarrow T$). The nonlinear character of the relaxation process is sometimes referred to as "self-retarding" (for the downquench experiment in Fig. 6 T_f decreases and hence τ increases with increasing time) or as "autocatalytic" (for the upquench experiment in Fig. 6 T_f increases and hence τ decreases with increasing time).

The second of the aforementioned features of the structural relaxation process is its *nonexponential* character or the need for a *distribution of relaxation times*. To introduce this feature, the relaxation function $\phi(t)$ in Eq 1 may be modified to give

$$\phi(t) = \sum_i g_i \exp\left[-\int_0^t dt'/\tau_i\right] \tag{7}$$

where the g_i are temperature-independent weighting coefficients for a distribution of relaxation times τ_i. Each τ_i is given in turn by an expression of the form of Eq 6

$$\tau_i = \tau_{i0} \exp\left[x\Delta H^*/RT + (1-x)\Delta H^*/RT_f\right] \tag{8}$$

Hence the various τ_i differ only in their pre-exponential factors τ_{i0}. This, along with the presumed temperature independence of the g_i, leads to a condition known as *thermorheological simplicity*. The need for an integral over time in Eq 7 is due to the variation of T_f and hence of the τ_i with time during relaxation.

The number of adjustable parameters in Eqs 7 and 8 may be considerably reduced if one selects a continuous spectrum of relaxation times, the shape of which can be specified by a single parameter and the location on a logarithmic timescale of which can be specified by a reference relaxation time τ (e.g., the most probable relaxation time). One then need worry only about the T and T_f dependence of the reference relaxation time, which is presumed to be of the form of Eq 6. A common choice for the shape of the spectrum of relaxation times, which seems to give a good fit to most data, is that corresponding to the so-called Kohlrausch-Williams-Watts (KWW) relaxation function

$$\phi(t) = \exp\left[-\left(\int_0^t dt'/\tau\right)^\beta\right] \tag{9}$$

where β ($0 < \beta \leq 1$) is the nonexponentiality parameter. Hence, to describe structural relaxation in response to temperature changes for large departures from equilibrium one requires a minimum of four adjustable parameters. In Eqs 6 and 9 these are τ_0, ΔH^*, x, and β. (For most glasses a "large departure from equilibrium" means a difference between T_f and T greater than about 2 K.) The solid lines in Fig. 6 were calculated using Eqs 6 and 9 [12].

Relaxation in response to a complicated thermal history, e.g., a temperature change ΔT_1 at time t_1, ΔT_2 at t_2, ..., ΔT_m at t_m, may be dealt with by using the Boltzmann superposition principle in conjunction with Eqs 6 and 9. In this sort of situation the fictive temperature $T_f(t)$ at time $t(>t_m)$ is given by

$$T_f(t) = T_0 + \sum_{j=1}^{m} \Delta t_j [1 - \phi(t,t_j)] \tag{10}$$

where T_0 is an initial temperature at which the sample is at equilibrium.

An example of the application of Eqs 6, 9, and 10 to structural relaxation for a complex thermal history is given in Fig. 7 [5,10]. The solid lines are plots of the experimental heat capacities C_p of B_2O_3 glass obtained by DSC while reheating through the transition region at 10 K/min following cooling through the transition region at a variety of rates. The dashed lines were calculated using Eqs 6, 9, and 10. T_0 in Eq 10 was taken as a temperature well above the transition region ($T_0 = 630$ K) where the relaxation time τ is short enough that the liquid remains in equilibrium during the cooling and heating rates employed. Equation 10 was used to calculate T_f and dT_f/dT during the initial cool through the transition region starting at T_0, followed by the subsequent reheat. C_p was then obtained from the relation (compare Eq 5)

$$dT_f/dT = (C_p - C_{pg})/(C_{pe} - C_{pg}) \tag{11}$$

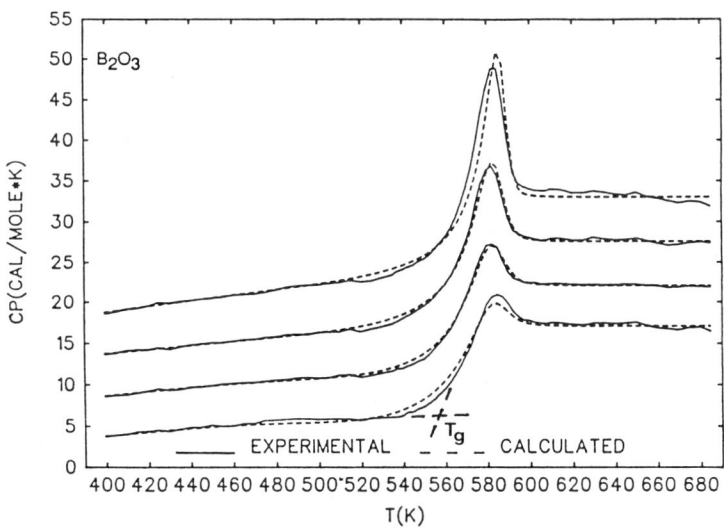

FIG. 7—*Heat capacities of B_2O_3 glass measured during heating at 10 K/min after cooling through the transition region at rates of 0.62, 2.5, 10, and 40 K/min (top curve to bottom curve). Heat capacity scale is correct for top curve; other curves have been displaced downward for clarity. Solid lines are experimental data. Dashed lines are calculated from Eqs 6, 9, and 10 using the parameters $\tau_0 = 1.3 \times 10^{-33}$s, $\Delta H^* = 377$ kJ/mol, $x = 0.39$, and $\beta = 0.62$. From Ref 5.*

As might be expected, changes in prior thermal history (cooling rate, in this case) affect the behavior of the system during subsequent reheating. In Fig. 7 this is manifested mainly as an increase in the maximum in C_p with decreasing prior cooling rate, which in turn reflects via Eq 4 the decrease in T'_f with decreasing cooling rate.

The calculated density and C_p curves in Figs. 6 and 7 agree with the measured curves within experimental error. This is generally found to be the case when the departure from equilibrium during actual relaxation is not extremely large. When this condition is not met, e.g., during annealing of a glass at temperatures very far below T_g, the quantitative agreement between the experimental and calculated relaxation data deteriorates somewhat, but all of the qualitative features of the experimental data are accurately predicted by Eqs 6, 9, and 10.

If we take T_g as the extrapolated onset of rapid rise in C_p during heating, it may be seen from Fig. 7 that T_g measured at a given heating rate is relatively insensitive to modest changes in prior thermal history (e.g., variations in the prior cooling to reheating rate ratio $|q_c/q_h|$ over the range $0.2 \leq |q_c/q_h| \leq 5$). This is not true for more drastic changes in prior thermal history. In Fig. 8 are shown DSC heat capacity scans for a ZBLA ($58ZrF_4$-$33BaF_2$-$5LaF_3$-$4AlF_3$) glass that was cooled through the transition region at 10 K/min, annealed isothermally somewhat below T_g for various amounts of time, and then reheated at 10 K/min [13]. Increases in the annealing time cause marked upward shifts in the apparent T_g defined as the onset temperature for the rise in C_p. In this case the upward shift in the onset temperature can be explained mostly on the basis of Eq 6. Isothermal annealing below T_g lowers T_f, increases τ, and hence increases the temperature at which the glass can start to relax on reheating. Sub-T_g isothermal annealing can cause even more bizarre effects. As shown in the DSC heating curves of Fig. 9 for poly(methylmethacrylate) (PMMA) [6], sub-T_g annealing can give rise to an additional endotherm in the DSC reheating curves, which in

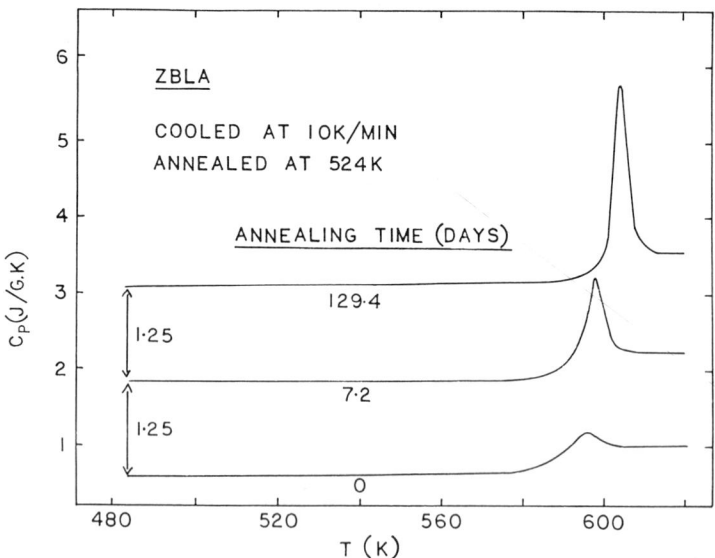

FIG. 8—*Heat capacities of $58ZrF_4$-$33BaF_2$-$5LaF_3$-$4AlF_3$ (ZBLA) glass measured during heating at 10 K/min after cooling through the transition region at 10 K/min and sub-T_g annealing at 524 K for various times. Heat capacity scale is correct for bottom curve. Other curves have been displaced sequentially upwards by 1.25 J/gK for clarity. From Ref 13.*

this case lowers the apparent T_g. The lines in Fig. 9 were calculated using Eqs 6, 9, and 10 or closely related expressions, so that in terms of these equations the additional endotherm in Fig. 9b is predictable. Whether sub-T_g annealing causes an apparent increase or decrease in T_g depends on the prior thermal history and the temperature and duration of the anneal [6,13].

DSC heating curves for glass samples (thin fibers or films) which have been quenched initially very rapidly through the transition region can also exhibit apparently aberrant behavior. Here the C_p heating curve can depart from the glasslike curve in a downward direction at temperatures well below T_g. This is illustrated in Fig. 10 for an alkali lime silicate glass (NBS 710), where the dashed lines are C_p heating curves for bulk samples and the solid lines are for fibers rapidly quenched during the drawing process [14]. In this case the fiber C_p data depart from the glass-like curve and observable relaxation commences at about 400 K, some 450 K below T_g measured using the bulk sample data. Qualitatively this

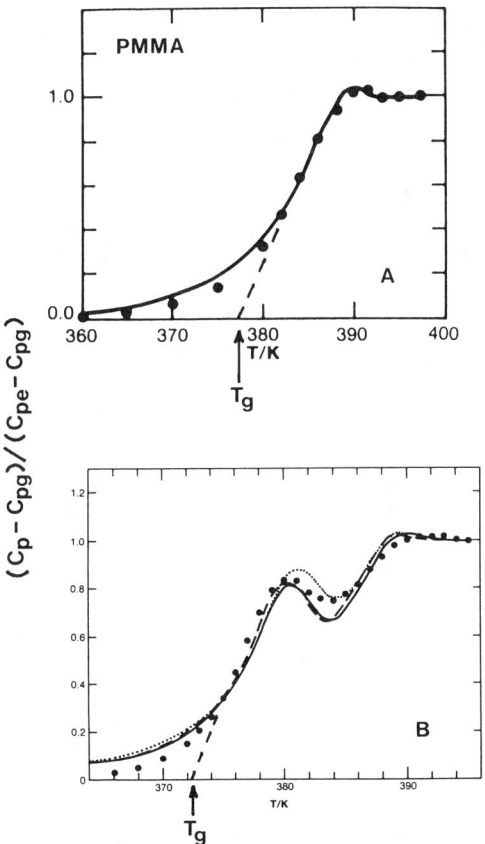

FIG. 9—*Normalized relaxational heat capacities of poly(methylmethacrylate) (PMMA) measured during heating at 10 K/min after* (A) *cooling through the transition region at 40 K/min and* (B) *cooling through the transition region at 40 K/min and sub-T_g annealing at 350 K for 17 h. Points are experimental data. Solid, dashed, and dotted lines are calculated using Eqs 6, 9, and 10 or closely related expressions. From Ref 6. (reprinted with permission from the American Chemical Society).*

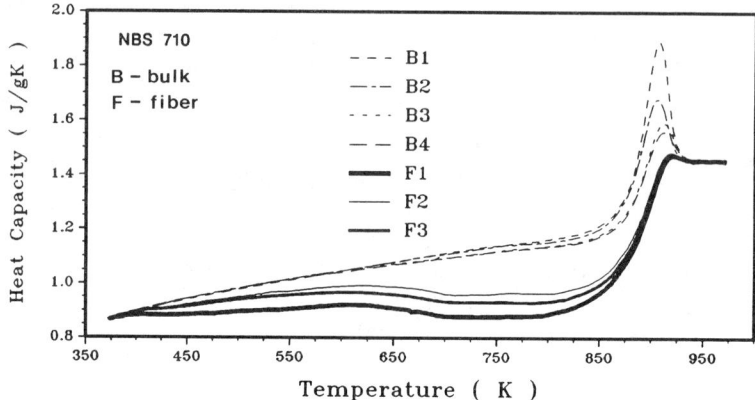

FIG. 10—*Heat capacities of NBS 710 alkali lime silicate glass measured during heating at 40 K/min. Dashed lines are for bulk samples previously cooled through the transition region at rates ranging from 2 to 40 K/min. Solid lines are for fiber samples previously quenched through the transition region during drawing at rates of the order of 5×10^6 K/min. From Ref 14.*

behavior can be explained in terms of Eq 6. The rapidly quenched fibers have much higher fictive temperatures T_f and hence much shorter relaxation times τ than do the more slowly cooled bulk samples. Hence the fibers may commence relaxation on reheating at much lower temperatures than the bulk samples, and this relaxation will lead initially to a large decrease in enthalpy H (compare Fig. 2) and to a measured C_p less than the glasslike value C_{pg}.

From the above discussion and from Figs. 7 through 10 it should be clear that T_g measured during heating can be sensitive to prior thermal history. Consequently, if T_g is to be defined in terms of a heating experiment, a standard prior thermal history (e.g., cooling through the transition region at a specified rate) must be part of the definition.

In Table 1 are listed T_g's for a variety of glasses ranging from network oxides to organics and polymers. These were all taken as the onset temperature (as in Fig. 7) on a DSC heating

TABLE 1—*Glass transition temperatures T_g taken as the onset temperature (see Fig. 7) on a DSC curve during reheating at 5 to 20 K/min following a cool through the transition region at a comparable rate. $\langle\tau_H\rangle$ is the mean equilibrium enthalpy relaxation time (i.e., for $T_f = T$) at T_g, η is the shear viscosity at T_g, G_∞ is the solid-like shear modulus, and $\langle\tau_s\rangle$ is the mean shear stress relaxation time at T_g. Data from Ref 15 except for ZBLA [13], 5-phenyl-4-ether [1] and poly(vinylacetate) [16].*

Glass	$T_g(°C)$	$\langle\tau_H\rangle(s)$	$\log\eta(Pa\cdot s)$	$G_\infty(GPa)$	$\langle\tau_s\rangle(s)$
Alkali Lime Silicate (NBS 710)	566	200	11.5	25	13
Lead Silicate (NBS 711)	449	190	11.6	23	17
$0.25Na_2O-0.75SiO_2$	483	250	11.3	25	8
B_2O_3	284	390	11.4	6.4	39
$58ZrF_4-33BaF_2-5LaF_3-4AlF_3$ (ZBLA)	313	690	11.6	22	18
As_2Se_3	181	242	10.8	7.0	9
5-phenyl-4-ether	−28	80	9.6	1.4	2.6
Poly(vinylacetate)	39	270	−	−	3.0

curve during reheating at 5 to 20 K/min following cooling through the transition region at a comparable rate. Also listed are the mean equilibrium (for $T = T_f$) enthalpy relaxation times $\langle \tau_H \rangle$ at T_g obtained via analysis of the C_p heating curves using Eqs 6, 9, and 10. (For the enthalpy relaxation function of Eq 7, $\langle \tau_H \rangle$ is given by $\langle \tau_H \rangle = \Sigma_i g_i \tau_i$, while for the relaxation function of Eq 9 it is given by $\langle \tau_H \rangle = (\tau/\beta)\Gamma(1/\beta)$.) Having standardized the property monitored (enthalpy, in this case), the method of measurement (onset of rapid rise in $C_p = dH/dT$ during heating), the timescale (set by the heating rate), and the prior thermal history, it may be seen that there is a definite correspondence between T_g and the structural relaxation time of the glass. For these experiments T_g is the temperature at which $\langle \tau_H \rangle$ is approximately 200 s.

Dependence of Transition Region Behavior on Property Monitored

In the previous subsection we have concentrated mostly on the relaxation of H in response to temperature changes. A relevant question is whether or not we would see identical behavior if we monitored a different temperature-dependent thermodynamic property, such as V. Without going into the details, we will simply state here that the irreversible thermodynamics relevant to this situation have been worked out [*17-19*] and compared with experimental data for a number of glasses. The predictions are that the rate of structural relaxation in response to temperature changes should not be the same for H and V. Hence if both properties were monitored simultaneously during heating, the V versus T curve would give a somewhat different value of T_g than would the H versus T curve.

Experimental relaxation results testing this prediction are rather sparse, but tend to support it. One such result is shown in Fig. 11 [*16*]. The mean equilibrium relaxation time $\langle \tau_V \rangle$ for relaxation of the volume in poly(vinylacetate) (PVAc) in response to temperature changes is about a factor of two shorter than the corresponding enthalpy relaxation time $\langle \tau_H \rangle$. While real, this difference is not extremely large when expressed in equivalent temperature units. If, as suggested by Table 1, T_g were defined as the temperature at which the equilibrium relaxation time attained a certain value, then T_g for PVAc obtained by monitoring the volume would be only about 2 K lower than T_g obtained from enthalpy measurements.

Relaxation in Response to Changes in Other Variables

We consider in this section the question of defining T_g in terms of the timescale for the response of the glass to perturbations in other thermodynamic driving forces such as mechanical stress and electric field. In the inorganic glass community T_g has often been taken roughly as the temperature at which the shear viscosity η is equal to 10^{12} Pa·s, a temperature very close to the conventional annealing point. In Table 1 are listed the values of $\log \eta$ at T_g, where again T_g was measured by DSC at heating rates of 5 to 20 K/min. For the high T_g inorganic glasses $\log \eta(T_g)$(Pa·s) is roughly constant; a more complete tabulation [*15*] gives $\log \eta(T_g)$(Pa·s) = 11.3 ± 0.4. For organic glasses and for some low T_g inorganics, $\log \eta(T_g)$(Pa·s) can be much lower than this, e.g., 9.6 for 5-phenyl-4-ether in Table 1.

The mean shear stress relaxation time $\langle \tau_s \rangle$ is related to the shear viscosity by the expression

$$\eta = G_\infty \langle \tau_s \rangle \qquad (12)$$

where G_∞ is the solid-like shear modulus. Values of G_∞ and $\langle \tau_s \rangle$ at T_g are also listed in Table 1 for the various glasses. As is evident, there is a definite timescale discrepancy between the shear stress and enthalpy relaxation rates. The enthalpy relaxation time $\langle \tau_H \rangle$ is longer than $\langle \tau_s \rangle$ at T_g by factors ranging roughly from 10 to 100. What this implies temperaturewise is again illustrated in Fig. 11 for PVAc, where $\langle \tau_s \rangle$ has also been plotted as a function of temperature. If T_g were defined as the temperature at which the equilibrium relaxation time attained a certain value (e.g., 200 s), T_g for PVAc measured in terms of shear stress relaxation would be about 7 K lower than T_g measured in terms of enthalpy relaxation.

A similar picture emerges if we consider dielectric relaxation, which for nonconducting materials measures the timescale for orientation of polar molecules or polar polymeric chain segments in response to an electric field. Most commonly dielectric relaxation measurements are carried out by measuring the real and imaginary parts, ϵ' and ϵ'', of the complex permittivity or dielectric constant as a function of frequency f at different temperatures. Typical results are shown for PVAc in Fig. 12 [*16*]. The mean dielectric relaxation time $\langle \tau_D \rangle$ may be obtained approximately from the frequency f_m of the maximum in the ϵ'' versus f curve at a given temperature

$$\langle \tau_D \rangle \approx 1/2\pi f_m \tag{13}$$

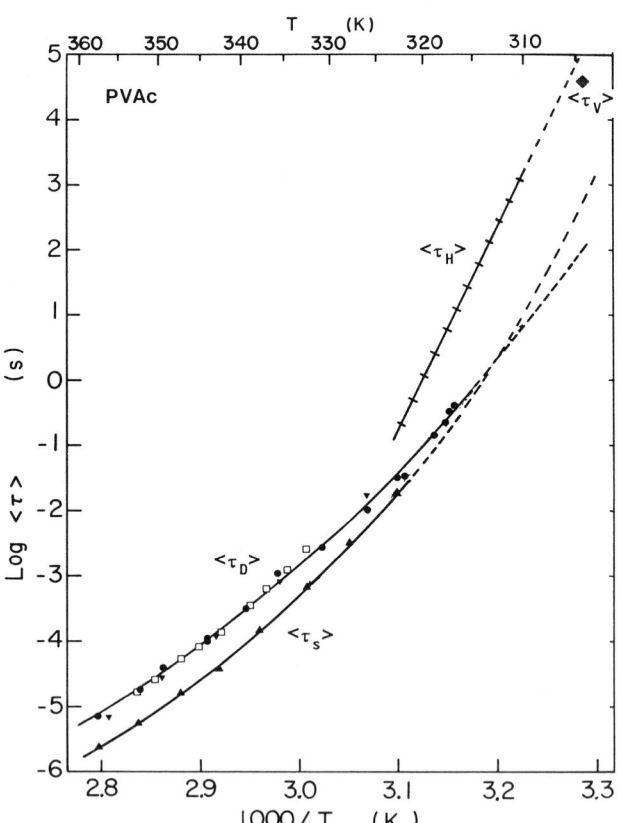

FIG. 11—*Arrhenius plots showing the temperature dependence of mean equilibrium relaxation times near and in the glass transition region of poly(vinylacetate) (PVAc).* $\langle \tau_H \rangle$ *is enthalpy relaxation time,* $\langle \tau_V \rangle$ *volume relaxation time,* $\langle \tau_s \rangle$ *shear stress relaxation time, and* $\langle \tau_D \rangle$ *dielectric relaxation time. From Ref 16.*

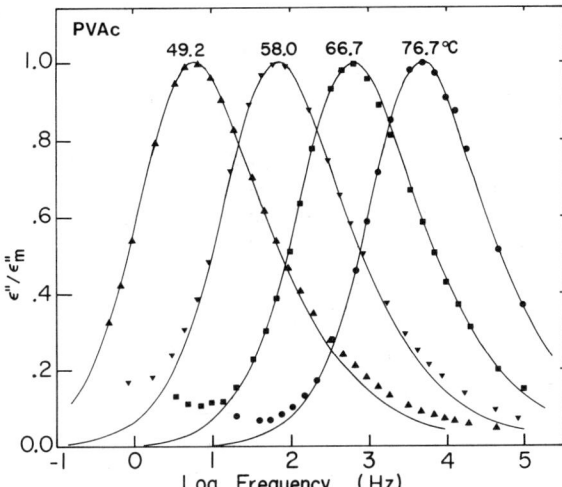

FIG. 12—*Normalized dielectric loss ϵ''/ϵ''_m as a function of frequency at various temperatures for PVAc. ϵ'' is the measured loss, and ϵ''_m is the measured loss at the maximum in the ϵ'' versus $\log f$ plot. From Ref 16.*

The temperature dependence of $\langle \tau_D \rangle$ for PVAc is also plotted in FIg. 11. If once again we defined T_g as the temperature at which the equilibrium relaxation time attains a certain value (e.g., 200 s), T_g for PVAc measured by dielectric relaxation would be about 10 K lower than T_g measured in terms of enthalpy relaxation.

Interpretation of the experimental results becomes exceedingly complex (if not totally intractable) if one decides to measure T_g (as suggested in several papers in this volume) by dynamic mechanical analysis (DMA) or by monitoring dielectric loss while heating through the transition region. In these experiments the real and imaginary parts of the complex tensile or shear moduli, $E^*(\omega)$ or $G^*(\omega)$, or of the complex permittivity $\epsilon^*(\omega)$ are monitored at fixed angular frequency ω ($= 2\pi f$) during heating. T_g then might be taken as the extrapolated temperature of onset of rapid decrease of the storage moduli, $E'(\omega)$ and $G'(\omega)$, or of rapid rise of the dielectric loss $\epsilon''(\omega)$. The first complication with these determinations is that they are highly sensitive to the probe frequency. For example, using Fig. 12 it is not difficult to deduce that changing the probe frequency by a factor of 10 would shift T_g of PVAc measured in terms of the onset of rapid rise of $\epsilon''(\omega)$ by about 9°C. The sensitivity of these determinations to probe frequency alone is, however, a complication not too different from that encountered with the dependence of T_g measured by DSC on heating rate.

A more subtle (and probably more serious) complication with DMA and $\epsilon''(\omega)$ heating experiments becomes evident if one realizes that during these measurements the sample is responding simultaneously to two externally imposed forces: the mechanical stress or the electric field plus the temperature changes. The complication of the response to the temperature changes comes about because the characteristic relaxation times for mechanical or electrical response, e.g., $\langle \tau_s \rangle$ and $\langle \tau_D \rangle$, which in turn determine the values of $G'(\omega)$ or of $\epsilon''(\omega)$, are also governed by Eq 6, where T_f is the fictive temperature associated with volume or enthalpy. As the sample is heated through the transition region and enthalpy and volume relaxation occur, T and T_f do not change proportionally, leading to a similarly nonmonotonic dependence of $\langle \tau_s \rangle$ and $\langle \tau_D \rangle$ on temperature. For the interested reader, a very lucid example of effects of this sort is to be found in Scherer's paper [20] on the equilibrium and isostructural viscosities η (which are proportional to $\langle\tau_s\rangle$) respectively above and below T_g for an alkali lime silicate glass. For the present purposes, however, we will conclude this section with the following caveats. Dielectric and mechanical relaxation data are best measured, reported, and interpreted in the form shown in Figs. 11 and 12. Values of T_g obtained from the temperature dependence during heating of the real or imaginary parts of $E^*(\omega)$, $G^*(\omega)$, or $\epsilon^*(\omega)$ measured at fixed frequency are not susceptible to the (comparatively) straightforward interpretation of T_g values obtained from H or V heating curves. Consequently, the former techniques should probably be avoided if one is looking for a relatively fundamental experimental specification of T_g.

Conclusions

In the context of this paper, the glass transition temperature T_g might be defined as the temperature at which some characteristic relaxation time attains a particular value. Alternatively and probably preferentially from an operational standpoint, it may be defined as some characteristic point on a cooling or heating curve along which the liquid falls out of or falls back into structural equilibrium. What is important to remember is that we are attempting here to characterize a complex, temperature-dependent kinetic process in terms of a single temperature, T_g. The value of T_g will be meaningful only if taken in the context of a broad appreciation of the structural relaxation process in liquids and how this is manifested experimentally. In particular, the following points should be kept in mind.

1. If T_g is measured during cooling, the cooling rate must be specified.
2. If T_g is measured during heating, the heating rate and prior thermal history in the glass transition region must be specified.
3. T_g obtained from a curve of property Q versus temperature will differ somewhat from T_g obtained from a plot of dQ/dT versus T.
4. Measurements of different properties will in general give somewhat different and sometimes sizeably different T_g's.
5. Experiments in which the sample is responding simultaneously to perturbations in more than one thermodynamic driving force are probably not best suited for T_g determinations.

References

[1] Moynihan, C. T., et al., "Structural Relaxation in Vitreous Materials," *Annals of the NY Academy of Science*, Vol. 279, 1976, pp. 15–35.
[2] Scherer, G. W., *Relaxation in Glass and Composites*, Wiley-Interscience, New York, 1986.
[3] Gupta, P. K., "Models of the Glass Transition," *Reviews of Solid State Science*, Vol. 3, No. 3–4, 1989, pp. 221–257.
[4] Scherer, G. W., "Theories of Relaxation," *Journal of Non-Crystalline Solids*, Vol. 123, 1990, pp. 85–89.
[5] Moynihan, C. T., Crichton, S. N., and Opalka, S. M., "Linear and Non-Linear Structural Relaxation," *Journal of Non-Crystalline Solids*, Vol. 131–133, 1991, pp. 420–434.
[6] Hodge, I. M., "Effects of Annealing and Prior History on Enthalpy Relaxation in Glassy Polymers. 6. Adam-Gibbs Formulation of Nonlinearity," *Macromolecules*, Vol. 20, 1987, pp. 2897–2908.
[7] Callister, W. D. Jr., *Materials Science and Engineering*, 2nd ed., Wiley, New York, 1991, pp. 176–183 (or any other introductory materials science and engineering text).
[8] Moynihan, C. T., Easteal, A. J., Wilder, J., and Tucker, J., "Dependence of the Glass Transition Temperature on Heating and Cooling Rate," *Journal of Physical Chemistry*, Vol. 78, 1974, pp. 2673–2677.
[9] Moynihan, C. T., Easteal, A. J., and DeBolt, M. A., "Dependence of the Fictive Temperature of Glass on Cooling Rate," *Journal of the American Ceramic Society*, Vol. 59, No. 1–2, 1976, pp. 12–16.
[10] DeBolt, M. A., Easteal, A. J., Macedo, P. B., and Moynihan, C. T., "Analysis of Structural

Relaxation in Glass Using Rate Heating Data," *Journal of the American Ceramic Society*, Vol. 59, No. 1–2, 1976, pp. 16–21.

[11] Narayanaswamy, O. S., "Thermorheological Simplicity in the Glass Transition," *Journal of the American Ceramic Society*, Vol. 71, No. 10, 1988, pp. 900–904.

[12] Scherer, G. W., "Volume Relaxation Far From Equilibrium," *Journal of the American Ceramic Society*, Vol. 69, No. 5, 1986, pp. 374–381.

[13] Moynihan, C. T., Bruce, A. J., Gavin, D. L., et al., "Physical Aging of Heavy Metal Fluoride Glasses," *Polymer Engineering and Science*, Vol. 24, No. 14, 1984, pp. 1117–1122.

[14] Gupta, P. K. and Huang, J., "Enthalpy Relaxation in Glass Fibers," *The Physics of Non-Crystalline Solids*, L. D. Pye, W. C. LaCourse, and H. J. Stevens, Eds., Taylor and Francis, Washington, DC, 1992, pp. 321–326.

[15] Moynihan, C. T., "Correlation Between the Width of the Glass Transition Region and the Temperature Dependence of the Viscosity of High-Tg Glasses," *Journal of the American Ceramic Society*, Vol. 76, No. 5, 1993, pp. 1081–1087.

[16] Moynihan, C. T. and Sasabe, H., "Structural Relaxation in Poly(vinyl Acetate)," *Journal of Polymer Science: Polymer Physics*, Vol. 16, 1978, pp. 1447–1457.

[17] Moynihan, C. T. and Gupta, P. K., "The Order Parameter Model for Structural Relaxation in Glass," *Journal of Non-Crystalline Solids*, Vol. 29, 1978, pp. 143–158.

[18] Lesikar, A. V. and Moynihan, C. T., "The Order Parameter Model of Liquids and Glasses with Applications to Dielectric Relaxation," *Journal of Chemical Physics*, Vol. 73, No. 4, 1980, pp. 1932–1939.

[19] Moynihan, C. T. and Lesikar, A. V., "Comparison and Analysis of Relaxation Processes at the Glass Transition Temperature," *Annals of the NY Academy of Science*, Vol 371, 1981, pp. 151–164.

[20] Scherer, G. W., "Use of the Adam-Gibbs Equation in the Analysis of Structural Relaxation," *Journal of the American Ceramic Society*, Vol. 67, No. 7, 1984, pp. 504–511.

Harvey E. Bair[1]

Glass Transition Measurements by DSC

REFERENCE: Bair, H. E., "Glass Transition Measurements by DSC," *Assignment of the Glass Transition, ASTM STP 1249*, R. J. Seyler, Ed., American Society for Testing and Materials, Philadelphia, 1994, pp. 50–74.

ABSTRACT: This is a review of glass transition temperature (T_g) measurements of materials by differential scanning calorimetry (DSC). T_g is noted in DSC derived heat capacity (C_p) plots against temperature as the jump in C_p or (ΔC_p) that occurs usually over a temperature interval of a few tens of degrees (°C). Under conditions which are normal for DSC experiments it was found that the temperature gradient across and from top to bottom of a 0.6-mm-thick molded disk of polycarbonate (17 mg) was no greater than 0.6°C when heated at 15°C/min. Under these conditions a glassy polymer can be scanned for its T_g as defined by its half-vitrification temperature (½ ΔC_p) as well as the beginning and end of the transition to within 1°C. DSC T_g measurements are reviewed to show how the location, shape, and magnitude of the transition can be used to characterize the composition and behavior of a wide variety of polymeric materials ranging from homopolymers, blocks, grafts, and blends to networks. Also the effect of molecular weight, plasticizers, aging, crystals, and stress on the T_g of several polymers is re-examined.

KEYWORDS: glass transition temperature, fictive temperature, differential scanning calorimetry (DSC), polymer blends, blocks, grafts, plasticizers, physical aging, glass transition broadening, rigid amorphous phase

The glass transition of polymers has been described in numerous ways, from the temperature region upon cooling at which micro-Brownian motion freezes, to the transition zone where polymer segments (conformers) undergo relaxation with the cooperation of their intermolecular neighbors [1,2].

Regardless of how the glass transition is modeled it is well known that there is a "step" in heat capacity, C_p, at this transition with no heat accompanying it. Hence, the phenomenon is similar to a thermodynamic second order transition, but it is not a truly thermodynamic transition for it depends on time. For example, the slower an amorphous specimen is cooled from the melt (liquid state), the lower the transition temperature. The value of the glass transition temperature, T_g, for polymers usually decreases by 2 to 3 degrees per decade decrease in cooling rate. Obviously there is a lower limit to T_g at which the materials' excess entropy becomes zero and this temperature is referred to as T_2 [3]. Although one cannot wait long enough to measure T_2 directly, it can be estimated from thermodynamic measurements and for polystyrene it was found to be ~80°C below the experimental T_g [4]. Therefore, a recorded T_g corresponds only to a glass formed via a specific cooling history. In addition, T_g is not only influenced by how rapidly the material is cooled, but also, if measured during a heating experiment, it is affected by that rate as well. Fortunately, by following a few simple rules differential scanning calorimeter (DSC) measurements of C_p will give T_g values that characterize the state of the glass and are relatively independent of instrumental parameters. The placement of T_g by C_p measurements is described below.

---

[1]Member of technical staff, Plastics Research and Engineering Department, AT&T Bell Laboratories, Murray Hill, NJ 07974-0636.

DSC is probably the most popular example of the growing number of techniques that fall under the broad heating of thermal analysis, a generalization for the measurement of some material response to a programmed change of temperature or frequency. The property that is monitored can range from simple weight or length to more complex phenomena such as dynamic mechanical or dielectric response. DSC measures C_p directly [5].

The term DSC was coined in 1963 at Perkin-Elmer to describe a new thermal analyzer they had developed [6]. This instrument operates on a power compensated null balance principle, in which energy absorbed or liberated by the specimen is exactly compensated by adding or subtracting an equivalent amount of electrical energy to a heater located in the sample holder. Platinum resistance heaters and thermometers are used in this DSC to carry out the temperature and energy measurements from $-170°$ to $725°C$ at rates which are variable from 0.1 to 500°C/min in 0.1°C/min increments. The constant, automatic adjustment of heater power necessary to keep the sample holder temperature equal to that of the reference holder provides a varying electrical signal that is tantamount to the changing thermal behavior of the specimen. This signal from the Perkin-Elmer DSC is proportional to the specimen's C_p [5]. Other types of commercial DSCs abound and their methods of operation have been reviewed elsewhere, but since the measurements presented here were made on Perkin-Elmer DSCs the key features of the other DSC instruments will not be discussed [7,8].

This review of T_g measurements by DSC will only focus briefly on a few essentials of instrument calibration, specimen preparation, and data analysis that are essential to the collection of reproducible and accurate T_g values. The remainder of this report will attempt to show by example how the glass transition can be characterized by DSC to not only gain quantitative insight into the composition of macromolecular materials, but also, in certain cases, to recognize various thermomechanical or chemical conditions that existed during vitrification or service in the field. Unfortunately, when multiple history effects are present it is usually impossible to separate the results. Nevertheless, a number of thermoanalytical techniques together with other complimentary methods hold great power for characterizing and predicting the long-term behavior of plastics.

Experimental Effects of Specimen Size and Heating Rate

The DSC must be calibrated for temperature and energy. Typically, two high-purity melting point, T_m, standards that bracket the temperature range of experimental interest are selected and T_m is taken as the temperature where the leading edge of each specimen's melting curve extrapolates into the baseline at the desired heating rate [9]. In addition, the heat of fusion, ΔH_f, of indium (integral of area under indium's melting curve) is determined. Typically these experimentally determined T_m's and ΔH_f's are placed into a T_m and ΔH_f software calibration routine and the specimens are remeasured. If necessary the values in the calibration menu changed until T_m and ΔH_f values agree with literature values within at least 0.2°C and 0.5% of ΔH_f.

With the DSC's temperature scale calibrated for a heating rate of 15°C/min the variation in temperature across the floor of the DSC's sample cell was checked near the center of the cell and then at the four major outer positions of the compass, namely, north (N), south (S), east (E), and west (W) with the fusion of a minute specimen of indium ($T_{m_0} = 156.61°C$) enclosed in an aluminum pan and covered with a lid. The extrapolated onset of melting, T_m, occurs at 156.65°C with the specimen at the cell's center. When indium is placed at the outer circumference of the cell's round bottom at points W, E, N, and S, T_m is found at 156.51, 156.56, 156.61, and 156.71°C, respectively, (Fig. 1). Hence, the maximum difference in

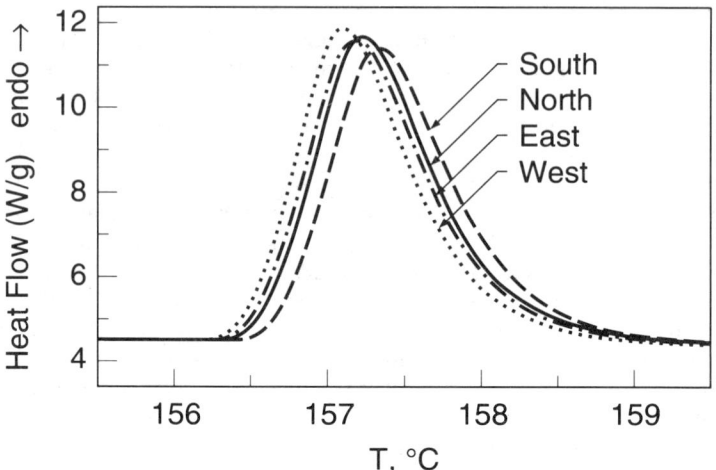

FIG. 1—*Fusion curves of indium placed at the four outer compass points: N, S, E, and W in the bottom of a DSC cell (heating rate = 15°C/min). (Data from H. E. Bair.)*

temperature from the center of the sample cell to its outer reaches ranges from 0.04 to 0.14°C.

The glass transition is more subtle than a first order melting process and in a polymer blend individual T_g's may be difficult, if not impossible, to detect unless several steps are taken to optimize the DSC's output. The effect of heating rate on the vertical displacement of the DSC signal during melting is important as is demonstrated in Fig. 2, where a thin film of indium is heated at 0.15, 1.5, 15, and 150°C/min. Note this DSC's temperature scale is calibrated for a heating rate of 20°C/min. Hence, the onset of melting ranges from 153 to

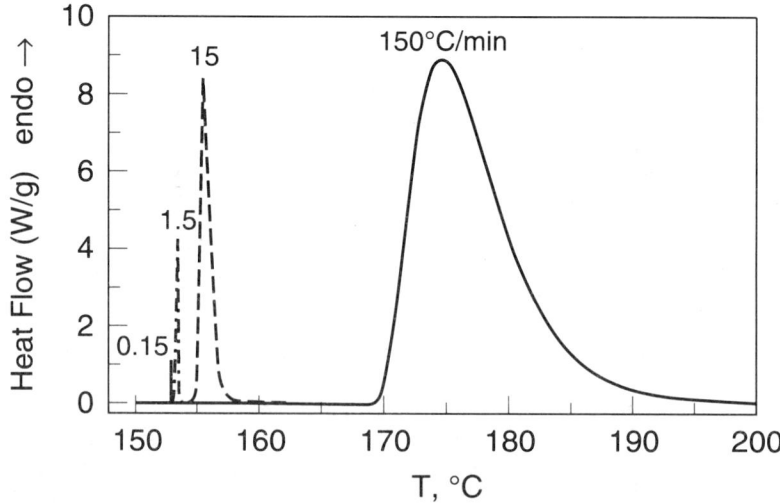

FIG. 2—*The effect of four heating rates (0.15, 1.5, 15, and 150°C/min) on the DSC's ordinate displacement and indium's melting curve. (Data from H. E. Bair.)*

170°C but most important the ordinate displacement or sensitivity is maximized near a heating rate of 15°C/min. At 150°C/min the ordinate output (~9 mW) is only slightly greater than that established at 15°C/min but the temperature lag, ΔT_{lag}, which is the temperature difference between the end of melting (apex in Fig. 2) and the DSC signal's return to baseline is about 25°C. ΔT_{lag} equalled 2.4, 0.2, and 0.07°C at heating rates of 15, 1.5, and 0.15°C/min. Since a heating rate of about 15°C/min optimizes the DSC's sensitivity and ΔT_{lag} is acceptable most DSC scans of T_g's are carried out at or near this rate.

It is also important to have a negligible temperature gradient within the specimen as it undergoes its glass transition in order to accurately assign not only the temperature where the transition commences but also where it terminates. In practice, samples of thin flat disks of the material, usually less than 0.6 mm thick, can be fashioned by milling injected molded bars if the polymer is solid at room temperature and then encapsulated inside an aluminum (thickness = 0.08 mm) pan with a lid crimped in place. As a final recommended measure to enhance the thermal contact between the encapsulated sample and the surface of the DSC's sample holder the encapsulated disk is placed on a smooth, flat metal surface and tapped gently with a metal cylinder whose end is of the same diameter as the sample. This latter step is intended to assure that the encapsulated package is relatively flat.

The temperature gradient within a specimen of polycarbonate (PC) is determined as it is scanned at 15°C/min through its glass transition. A thin disk (0.06-mm-thick) of indium is placed either below, in the middle, or on top of a 0.56-mm-thick PC disk. The PC disk is 3.1 mm in diameter and weighs 17.8 mg. The T_m is 156.64, 157.21, and 157.05°C when the indium is placed below, in the middle, or on top of the PC, respectively. Figure 3 shows the scan with the indium on top of the quenched PC disk. Note, above T_g of PC, the slope of C_p, as is customary, is less steep than below. The glass transition temperature, T_g, the temperature of half-vitrification (½ ΔC_p) is located at 146.6°C. This is the temperature at which C_p is halfway between the glass and liquid states ($\Delta C_p = C_{p_{\mathrm{liq}}} - C_{p_{\mathrm{glass}}}$ at T_g). In this case ΔC_p equals 0.24 J/g°C at T_g. In earlier work T_g and ΔC_p at T_g were determined for PC in an adiabatic calorimeter to be 142°C and 0.25 J/g°K at a heating rate of about 0.24°C/min [10].

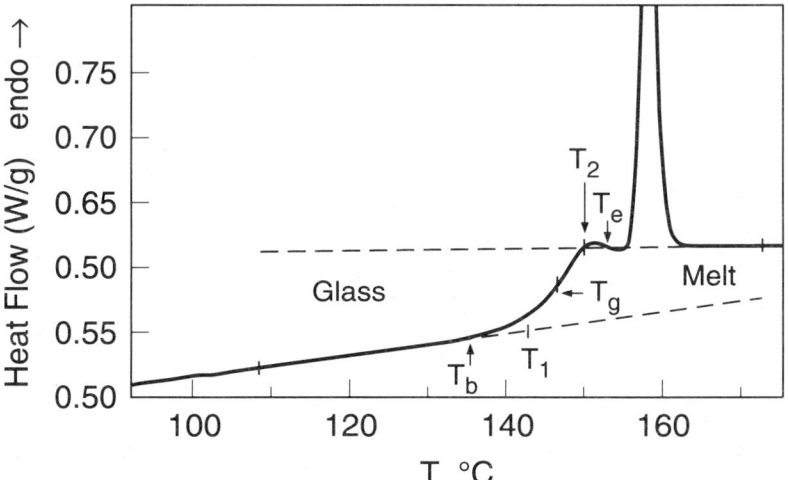

FIG. 3—*Melting of indium placed on top of a 0.6-mm polycarbonate disk and heated at 15°C/min. (Data from H. E. Bair.)*

In addition to specifying T_g and the specimen's prior thermal history the beginning of the transition, T_b, the end, T_e, the extrapolated beginning, T_1, and the extrapolated end, T_2, should be given as shown in Fig. 3. These values for the PC which was quenched at 200°C/min from the melt are 136, 142.8, 146.6, 149, 153°C for T_b, T_1, T_g, T_2, and T_e, respectively.

In the foregoing experiment the width of the PC glass transition interval, ΔT_{gt}, if taken as either the difference between T_2 and T_1 or T_e and T_b is 6.2 or 17°C, respectively. These values are typical for an amorphous thermoplastic homopolymer or random copolymer with a relatively narrow molecular weight polydispersity (i.e., M_w/M_n is 2 or less). The dispersity of this PC is 1.88 and its molecular weight (M_w) is about 30 000. Previous work shows that T_g of this condensation polymer decreases as M_w is reduced through hydrolysis [11]. This effect is shown in Fig. 4 where T_g fell from 148 to 127°C as M_w was reduced by hydrolysis from 28 000 to 8000. However, in contrast to the relatively narrow ΔT_{gt} behavior exhibited by the PC homopolymer it is well known that the breadth of the glass transition increases as polymers are mixed into miscible polyblends or covalently linked together into homogeneous copolymers. In this latter case a ΔT_{gt} equal to nearly 60°C has been measured in a homogeneous 1,4-polybutadiene-1,2polybutadiene diblock copolymer and this broad T_g is attributed to local segment density fluctuations [12].

A good example of how broad and complex T_g can become is illustrated in Figs. 5 and 6 where an uncured, low M_w, light sensitive acrylate adhesive's T_g is compared to that of the same material but after transformation via UV irradiation to a cross-linked network polymer.

C_p of the uncured adhesive increases linearly with temperature from $-105°C$ to $-75°C$ (solid line, Fig. 5). Beyond this temperature C_p rises abruptly in a sharp step that terminates near $-50°C$ and then proceeds to increase linearly again with increasing temperature to the end of the temperature scan at 80°C.

The C_p step between $-75°$ and $-50°C$ denotes the glass transition process of the uncured adhesive. It is characterized by an increase in C_p from the glassy to the liquid state, ΔC_p, of 0.53J°C^{-1}g and a glass transition temperature, T_g, of $-56°C$ at ½ ΔC_p at a heating rate of 15°C/min. T_1, T_2 and ΔT_{gt} equal -62, -50, and 12°C, respectively. In addition, another measure of the glass transition, which is termed the fictive temperature T_f, is presented in the

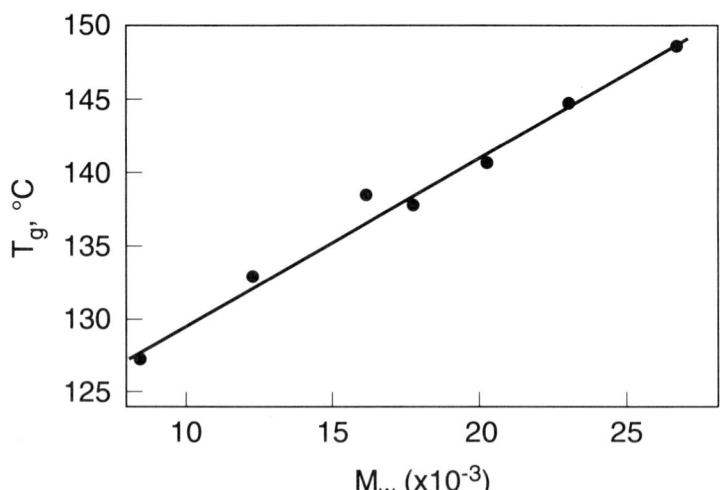

FIG. 4—T_g as a function of polycarbonate's molecular weight after Bair, et al. [11].

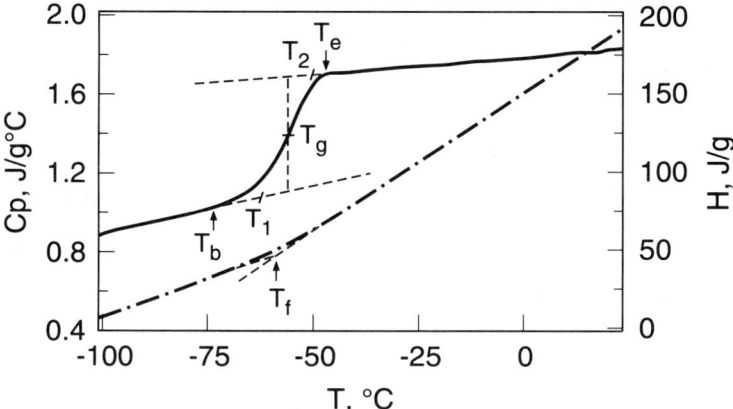

FIG. 5—C_p (solid line) and H (broken line) versus temperature for an uncured acrylate adhesive. (Data from H. E. Bair.)

figure as the extrapolated enthalpies of the liquid and the glass (Fig. 5, broken lines) [13,14]. In this case T_f equals $-58.7°C$. Prest has shown the usefulness of T_f measurements on annealed glasses, to establish the temperature where vitrification occurred during cooling [15].

A 15-mil-thick film of the adhesive cured at room temperature with a UV dose of 3 J/cm^2 is cooled to $-112°C$ and heated to 85°C at 15°C/min. In this case C_p increases linearly with temperature from $-100°$ to $-30°C$ and then begins to rise in a broad sweep until near 25°C (dashed line, Fig. 6). At this latter temperature, C_p starts to decrease anomalously and reaches a minimum near 50°C, then rises rapidly until it levels off above 76°C. The depression in C_p which initiates near room temperature is due to a small exotherm (5.5 J/g)

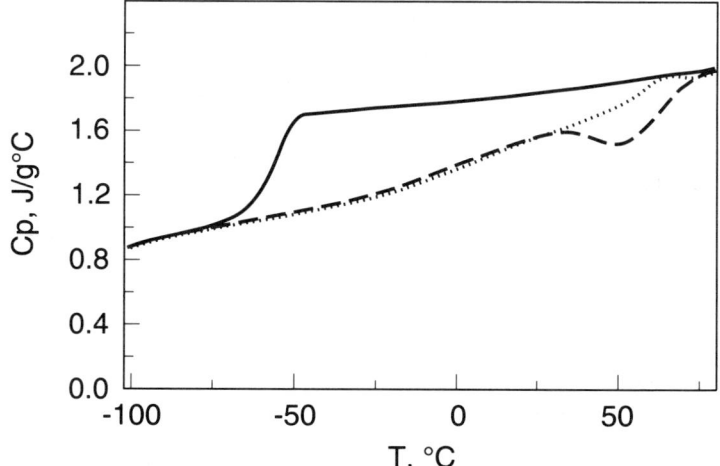

FIG. 6—Comparative C_p plots against temperature of three adhesive specimens. (Data from H. E. Bair.) ——— uncured; – – – first heating after irradiation at 23°C; ------ second heating.

associated with the liberation of heat that occurs when the adhesive is heated above T_g. Previously trapped unreacted species became free to react. From the residual heat of reaction we estimate about 3% of the reactive acrylate groups of the adhesive is involved in this process.

Upon cooling this specimen to $-112°C$ and reheating it to 80°C an extremely broad T_g interval is observed which starts near $-30°C$ and ends near 70°C (dots, Fig. 6). There is no sign of an exotherm on reheating. T_g for the cured adhesive is near 30°C at $\frac{1}{2} \Delta C_p$, where ΔC_p equals 0.37 $J°C^{-1}g^{-1}$ and ΔT_{gi} is now 97°C!

The large breadth of the T_g interval for the cured adhesive indicates the adhesive has a very heterogeneous structure with nearly half its elements possessing T_g's below room temperature. The reduction in ΔC_p with curing indicates the number of molecular configurational changes that can occur at T_g is reduced with crosslinking.

Although flat uniform solid disks are ideal samples to scan in a DSC, samples of liquids, powders, and fibers can be examined also. Usually liquids are sealed hermetically inside containers which will retain any volatiles that are liberated in an experiment. During a DSC heating scan partial volatilization of the sample is usually an undesirable endothermic process which can influence the material's transition behavior. Also, volatiles that escape and condense inside the DSC system can cause future shifts in the instrument's baseline or unwanted spurious peaks. Powers and fibers can have recovery of frozen-in mechanical stain history that will be released as heat during the first scan of the sample above T_g. These latter effects will be absent during subsequent heating scans. An example of this latter phenomenon is shown in Fig. 7 where excessively high stresses were molded into a fire retardant (FR) PC part during a two-shot injection molding process. During the first heating scan at 20°C/min a small endothermic peak occurred near the end of the glass transition followed by a larger exothermic process near 170°C. The net heat released by the FR PC during its first heating above T_g is approximately 4 J/g (solid line, Fig. 7). The second heating scan of this FR PC sample shows the normal "step" in C_p at T_g. Frozen-in stresses such as this can

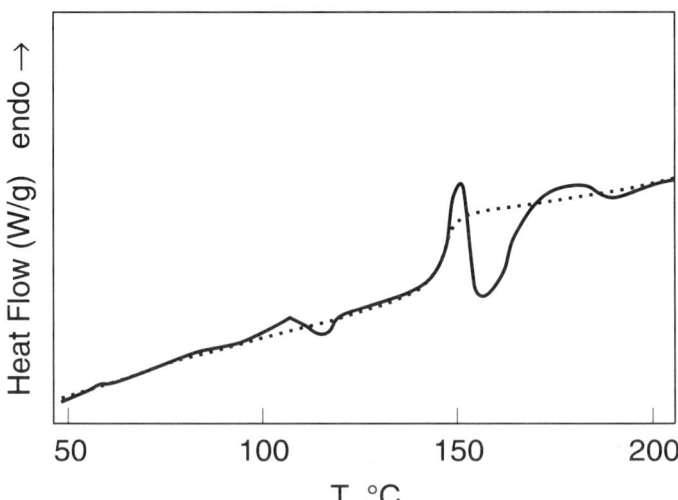

FIG. 7—*Two DSC curves of polycarbonate. (Data from H. E. Bair.)* —— *first heating of as-received specimen;* ------ *second scan.*

cause stress cracking of a molded part. Similar exothermic relaxations have been noted in polycarbonate fibers that were drawn near T_g [16].

T_g as a Function of Different Thermal Histories

As previously stated the rate at which a glass is cooled determines its T_g. This behavior is demonstrated schematically in Fig. 8 where enthalpy, H, is plotted versus temperature. Glass 1 is cooled rapidly (quenched) and its glass temperature, T_{g1}, is higher than that of Glass 2 which cooled more slowly and occurs at T_{g2} (note, Glass 2 can be created from Glass 1 by annealing below T_{g1}). However, if these glasses are heated at the same rate in a DSC an apparent paradox arises when considering the T_g's of the quenched and slow cooled or annealed samples—the anticipated transition order is reversed! This is because the longer relaxation time of Glass 2 causes the glassy state to persist to a higher temperature T_{g3} [17,18]. In order to reach the equilibrium rubbery state above T_{g3}, a delayed but sudden jump in enthalpy takes place, and an endothermic peak in C_p is observed. The difference in the initial values of enthalpy for Glasses 1 and 2 should be obtained by subtracting area A from area B (shaded areas in the upper half of Fig. 8). This annealing process occurs more rapidly the closer the annealing temperature is to T_g. Since most thermoplastic polymers, whether as pellets or injected molded parts, have been cooled rapidly (>100°C/min), the T_g

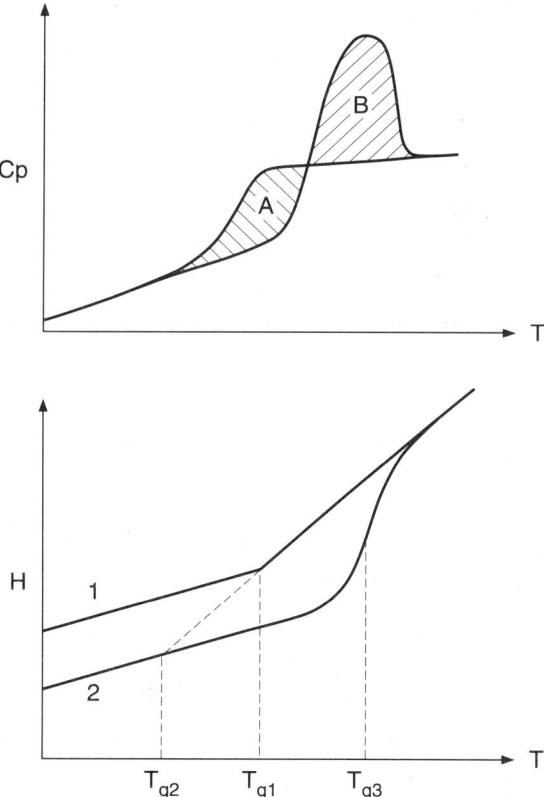

FIG. 8—*Schematic diagrams for heat capacity* (top) *and enthalpy* (bottom) *versus temperature for a quenched Glass 1 and an annealed Glass 2 (after Matsuoka and Bair* [18]).

determinations of these materials is normally carried out on samples with roughly comparable thermal histories. Large shifts in T_g can occur if storage, service, or processing conditions are such that isothermal annealing is possible. In general, without detailed knowledge of a polymer's prior thermal mechanical history it is desirable to heat the as-received material into the melt and then quench it below its T_g. In practice, one should record the first heating scan for it may yield insight into the samples past, revealing details of frozen-in stress, past thermal events, or even plasticization by moisture or other small molecules; however, the second heating after a rapid quench should be used to characterize the glass transition since it is made on a specimen with a known history.

PC whose T_g was characterized previously (Fig. 3) was vitrified at cooling rates of 0.10, 15, and 200°C/min and then reheated at 15°C/min (Fig. 9). Although the endotherm astride T_g grows in magnitude as the cooling rate decreases, T_g for the slow-cooled sample appears only about 2°C lower than that found for quenched PC. This apparent anomaly is probably associated with the small amount of enthalpy reduction that occurred by cooling at 0.1 versus 200°C/min. Most likely the glass transition of the sample cooled at 0.10°C/min should be shifted about 2°C below that of the PC cooled at 15°C/min as the latter sample is found ~2°C below the C_p step associated with the quench at 200°C/min. However the longer relaxation time connected with the slower cooled material causes the sample upon reheating to persist in the glassy state longer and thus the onset of its T_g step overlaps that of the sample formed by cooling at 15°C/min. Clearly, these results indicate that T_g determinations of polymers that are heated 15°C/min after quenching from the melt should give values that are consistent within about 1°C and agree from laboratory to laboratory so long as all instruments are properly calibrated and the idiosyncrasies of different thermal analyzers are taken into account.

The C_p behavior of annealed poly(methyl methacrylate) (PMMA) was compared against quenched PMMA. This polymer was polymerized in a manner described elsewhere [*19*]. In Fig. 10 (curve B) quenched PMMA is heated at 15°C/min and its glass transition begins near 80°C and terminates at 118°C with T_g at 111°C ($\Delta C_p = 0.34$ J/g°C) whereas when the same sample is annealed at 90°C for 45 days and then heated under the same conditions, its T_g is

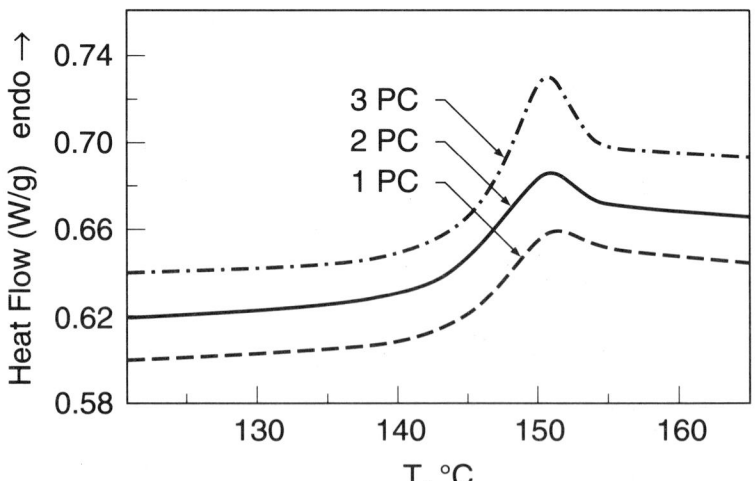

FIG. 9—*DSC heating scans (15°C/min) of PC specimens designated PC1, PC2, and PC3 cooled at 200, 15, and 0.10°C/min, respectively. (Data from H. E. Bair.)*

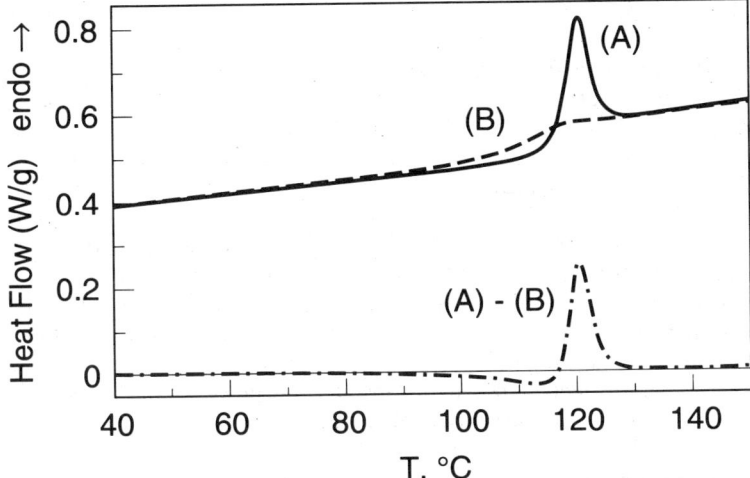

FIG. 10—*Upper DSC scans: PMMA aged at 90°C for 45 days* (A, solid line) *and quenched PMMA* (B, broken line). *(Data from H. E. Bair and Y. Koike.) Lower DSC curve: Difference curve = Curve A − Curve B.*

at 116°C—about 5°C higher then the quenched PMMA glass! The difference in the initial values of enthalpy, ΔH, for the annealed and quenched PMMA glasses is easily determined by subtracting the latter from the former (curve A − curve B, lower half of Fig. 10). From this difference curve −2.60 J/g are found below the zero line and 4.32 J/g above which indicates PMMA's excess enthalpy was lowered by 1.72 J/g by annealing the quenched glass for 45 days at 90°C.

The enthalpy difference curves for PMMA annealed at 90°C for various times are displayed in Fig. 11. ΔH values are 0.61, 0.77, 1.21, and 1.72 J/g for annealing times of 1,

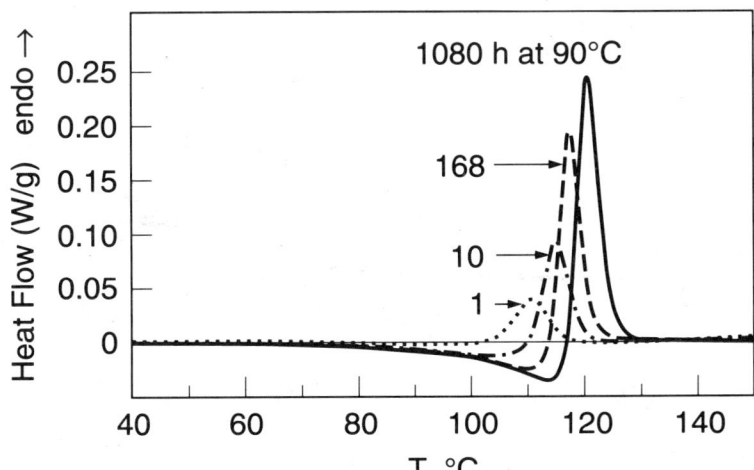

FIG. 11—*DSC difference curves for PMMA annealed at 90°C for 1, 10, 168, and 1080 h after subtracting a quenched scan from each. (Data from H. E. Bair and Y. Koike.)*

10, 68, and 1080 h, respectively. Note the peak temperature increases from 111 to 121°C as the annealed samples' relaxation time increases with longer annealing times. Hodge has developed a method for calculating C_p of annealed glasses such as PC and PMMA based on the Adam-Gibbs theory [20,21].

When a polymer's T_g is just above room temperature, storage near that temperature will cause T_g to increase rapidly with increased annealing time. In Fig. 12 quenched poly(vinyl acetate) (PVAc) which is held at 27°C for increasing periods of time and then scanned up in temperature at 20°C/min in a DSC exhibits T_g's of 44.0, 45.8, 48.0, and 50.5°C for annealing times of 1, 10, and 100 h, respectively. The glassy regions of partially crystalline polymers such as poly(butylene terephthalate) (PBT) or poly(ethylene terephthalate) (PET) have T_g's of ~50 and 80°C, respectively, and have been shown to be susceptible to the same physical aging process that occurs in noncrystalline polymers [22,23]. Annealing the amorphous areas of PBT and PET leads to embrittlement and a reduction in elongation [22,23]. Details of how physical aging affects a polymer's mechanical properties is beyond the intended scope of this work and is covered elsewhere [2,24].

Effect of Structure on ΔC_p

Wunderlich has shown that ΔC_p at T_g increases by about 11 J/mole of beads per°K [25]. The term "bead" refers to a model of a polymer chain as a string of beads with a bead being the smallest chain unit capable of rotation. The Gibbs-DiMarzio configuration entropy theory of glasses predicts the specific heat discontinuity at T_g is the sum of three terms [26]. The first two terms are configurational and arise from the changes of flexing and the number of holes with temperature. The third term is a vibrational contribution. Unfortunately, estimation of ΔC_p requires the polymer's structure be resolved into the number of flexible units and beads. PC can be visualized as having either two or four flexible units which leads to a predicted ΔC_p equal to 0.28 or 0.33 J/g°K [26]. The experimental value is 0.24 J/g°C (Fig. 3).

Hydrogenation of poly(1,3-dimethyl-1-butenylene) (PDMB) yields atactic poly(pro-

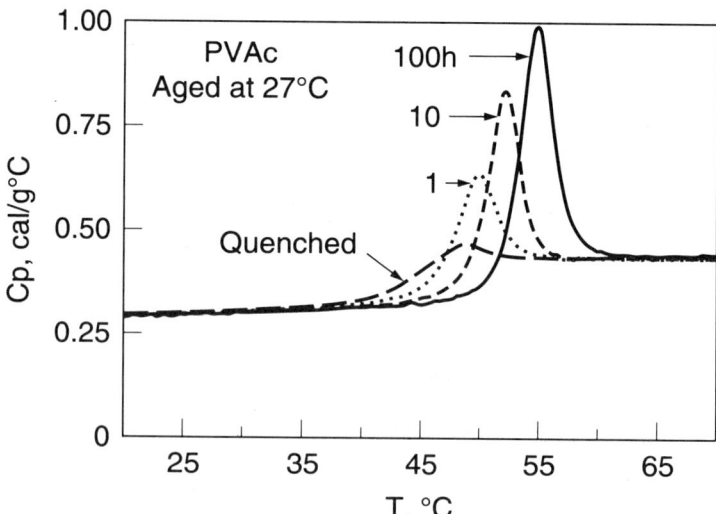

FIG. 12—*C_p curves for quenched PVAc and after aging at 27°C for 1, 10, and 100 h. (Data from H. E. Bair.)*

pylene) (PP) [this latter polymer's IUPAC name is poly(methylene) (PME) [27,28]. These two polymers

$$\begin{array}{c|c|c|c|c|c|c|c} | & | & | & | & | & | & | & | \\ C-C & =C-C & \longrightarrow & C-C-C-C \\ | & | & | & | & | & | & | & | \\ CH_3 & CH_3 & & CH_3 & & CH_3 \end{array}$$

are ideal to determine experimentally the effect of molecular structure on ΔC_p since hydrogenation and conversion of PDMB to atactic polypropylene results in an increase of flexible bonds in the polymer's backbone from three to four. C_p measurements at a heating rate of 10°C/min on PDMB shows T_1 and T_g at 2.6 and 4.7°C, respectively, with ΔC_p equal to 0.435 J/g°C. Conversion of the double bond to a single flexible bond yields PP with T_1 and T_g at −4.5 and −2.6, respectively, and ΔC_p increased by 28% to 0.531 J/g°C (Fig. 13).

O'Reilly has shown that the methyl groups in polypropylene can be neglected in calculating specific heat at T_g [29]. From theory ΔC_p for polypropylene is predicted to be 0.510 J/g^{-1} °C^{-1} [26]. If we reduce the configurational C_p terms in the DiMarzio-Dowell theory for PME by 25% which corresponds to the loss in flexible bonds in going from PME to PDMB, ΔC_p is calculated to decrease 0.122 to 0.406. Our experimental ΔC_p values of 0.536 and 0.418 J/g^{-1} °C^{-1} for PME and PDMB, respectively, are in good agreement with this predicted thermal behavior for the two polymers.

Thermosetting, polyfunctional molecules such as epoxy resins will react to yield infinite networks which restrict molecular motion and hence reduce ΔC_p. For example, the effect of high crosslink density at high conversions in the reaction of pentafunctional epoxy cresol novolac and cresol novolac hardener causes ΔC_p to fall nearly ten-fold from its initial of 0.427 J/g°C when the extent of reaction is 0.20 or less (Fig. 14) [30]. T_g in this reactive system is very sensitive to conversion particularly at high conversions (Fig. 15).

Hale and coworkers modeled T_g in this novolac system by assuming that the increase in

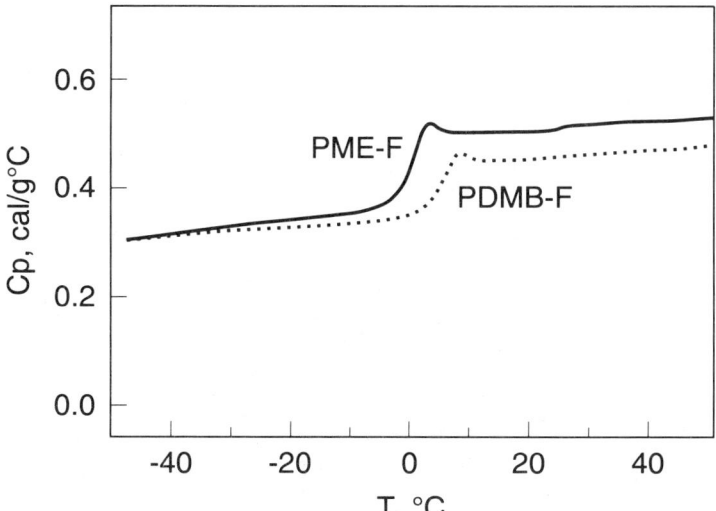

FIG. 13—*Heat capacity as a function of temperature for PDMB-F (dashed line) and PME-F (solid line) or atactic polypropylene (after Zhongde, et al. [27]).*

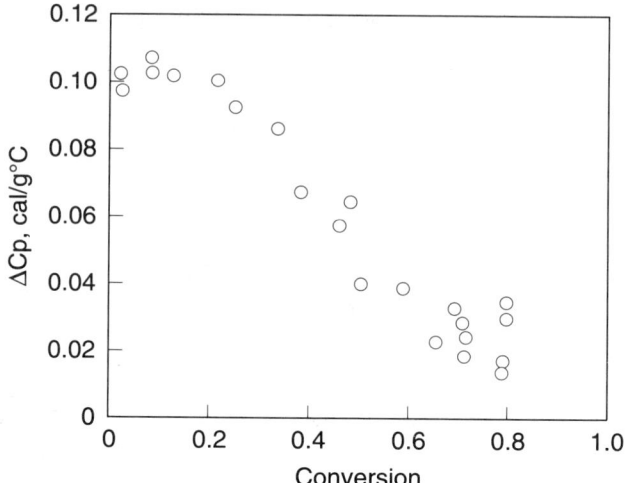

FIG. 14—ΔC_p *as a function of extent of reaction of an epoxy-novolac system (after Hale et al. [30]).*

T_g is caused first by a decrease in chain-end concentration; second, formation of effective crosslinks; and third, further decrease in the configurational entropy due to departure from Gaussian behavior at high crosslink densities [30]. The continuous line in Fig. 15 is calculated from this model of the glass temperature.

The T_g and ΔC_p at the glass temperature can be influenced by the crystals present in a semicrystalline polymer. Partially crystalline poly(phenylene sulfide) (PPS) exhibits this

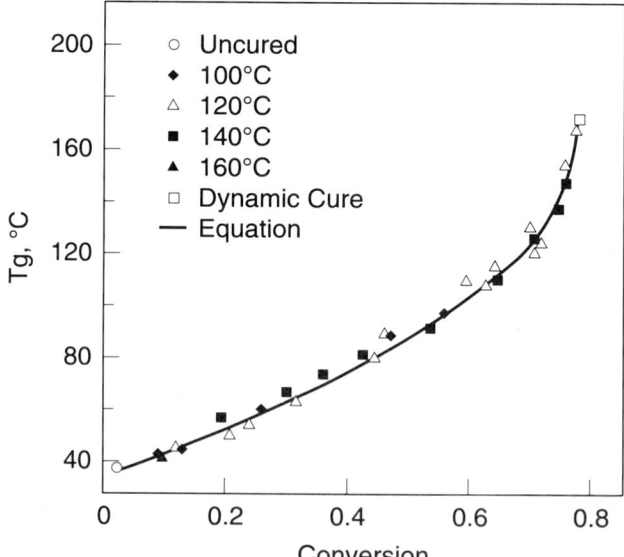

FIG. 15—T_g *as a function of extent of reaction for an epoxy-novolac material. The legend shows the temperatures at which the specimens were cured (after Hale et al. [30]).*

type of behavior. Glass-filled (GF, 40 wt.%) PPS was injection molded from the melt under two conditions: first into a "cold" mold ($T \simeq 60°C$), and second into a "hot" mold ($T \geq 100°C$). C_p measurements were made in a Perkin-Elmer DSC-2 on GF PPS samples weighing about 8 mg at a heating rate of 20°C/min [31].

The thermal behavior of the cold-molded GF PPS (filled circles) is compared with the hot-molded PPS (open circles) part as shown in Fig. 16. The onset of the glass transition starts near 83°C and T_g is at 92°C with ΔC_p equal to 0.12 J/g°C (filled circles, Fig. 16). Above T_g a large crystallization exotherm with a maximum at 134°C occurs. Approximately 14.6 J/g of heat is liberated as the sample crystallizes between 120° and 160°C. The bulk of the melting of the sample occurs between 250° and 280°C with an apparent heat of fusion, *ΔH_f, of 25.1 J/g. Note that only 42% (10.5/25.1) of the crystallinity is present prior to heating the cold-molded sample above 120°C. In contrast to this behavior C_p versus T for the sample injected into the hot mold is displayed in Fig. 16 as the open circles. Now, the jump in C_p at T_g is suppressed so that it cannot be detected on this plot. However, separate DSC runs at higher sensitivities indicates T_g occurs from 92° to 100°C. In addition the crystallization exotherm that was observed in the previous DSC run is absent. The main melting peak occurs again between 250° and 280°C with *ΔH_f equal to about 25.1 J/g.

The heat of fusion of PPS is about 71 J/g. Hence, the level of crystallinity in the samples from the hot and cold molds is 59 and 25%, respectively. This large difference in crystallinity for GF PPS when it is injected into either a hot or cold mold is apparently due to large

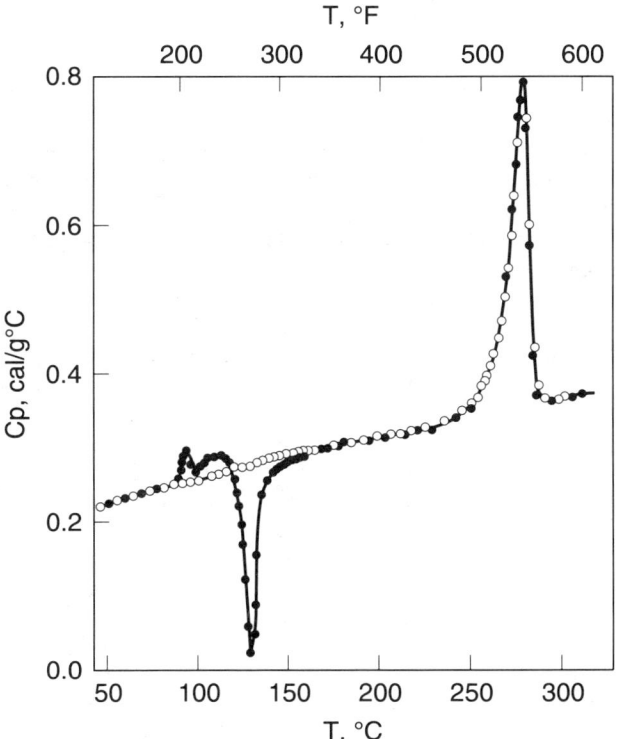

FIG. 16—C_p as a function of temperature for GF PPS specimens produced in a hot (open circles) and cold (filled circles) mold (after Bair et al. [31]).

differences in cooling rates during the molding cycle. If the crystal and glassy regions in PPS obeyed a simple two-phase model, the T_g of 59% crystalline GF PPS should be observable in Fig. 16. The results imply that some of the crystalline elements formed in the hot mold have suppressed a portion of the C_p rise at T_g. Wunderlich and coworkers have called this glassy material that does not contribute to ΔC_p at T_g the "rigid amorphous" polymer phase [*32*]. This failure of the two-phase model in some polymers has been recognized for a long time [*10,33*].

Compatibility of Polymer Blends

Binary polymer blends seldom form homogeneous, one-phase systems since the entropy of mixing, ΔS_{mix}, in the free energy of mixing expression

$$\Delta G_{mix} = \Delta H_{mix} - T\Delta S_{mix} \tag{1}$$

is usually too small to overcome the normally positive heat of mixing, ΔH_{mix}, that would produce a negative free energy of mixing ΔG_{mix}. The latter is a necessary condition for polymer compatibility. Experimentally the least ambiguous criterion for polymer compatibility is the detection of a single glass transition temperature, T_g, which is intermediate between those corresponding to the two component polymers. Phase separation is judged by the existence of two distinct glass transition temperatures.

Below T_g, phase separation is measured from the heats of solution of the polymers and their blends by Tian-Calvert microcalorimetry. Application of Hess's law to these values yields ΔH_{mix} [*34*]. Positive levels of ΔH_{mix} are a sufficient condition for phase demixing.

One of the best documented cases of a compatible polymer blend is that of polystyrene (PS) and poly(2,6-methyl-1,4-phenylene oxide) (PPO). A single T_g has been detected for all blend compositions by numerous thermal techniques. The effect of chemical modification of PPO and PS and their related copolymers on miscibility has been much investigated [*35*]. Karasz and MacKnight have reported the phase diagram for a blend of *p*-chlorostyrene/*o*-chlorostyrene copolymer with PPO as a function of PPO concentration [*36*]. Thus, if the phase boundaries are known, the processing conditions can be controlled to yield either a homogeneous polyblend at room temperature or a mixture containing two discrete phases. Additional tests must be carried out to determine the most desirable physical state.

Since identification of a single T_g or multiple T_g's in a mixture of polymers is the most widely applied criterion for determining phase behavior, the limiting experimental condition is how close can the pure components' T_g's be and still be recognized as separate transitions if the components are not miscible. Typically when T_g's are closer than 10°C a DSC cannot resolve phase differences clearly. Recently, the resolution of phase behavior in polymer blends with nearly coincident T_g's was accomplished by DSC enthalpy recovery measurements [*37,38*]. By this technique, blends of aromatic polyamides whose individual T_g's are closer than 3°C have been found to be immiscible. Representative thermograms of two blended nylons (50/50 w/w) with individual T_g's at 151 and 158°C are scanned at 20°C/min after various thermal treatments (Fig. 17). After 130 m of aging at 143°C two enthalpy recovery peaks are discernible indicating two phases are present.

Model diblock systems such as poly(ethylene-propylene)-poly(ethylethylene) PEP-PEE with molecular weights (M_w) ranging from 31 000 to 106 000 and with equal block lengths can be readily checked by DSC to discover if they are ordered (microphase separated) or disordered (homogeneous) by observation of their T_g behavior [*39*]. DSC traces for the quenched saturated hydrocarbon diblock copolymers and the corresponding homopolymers are shown in Fig. 18. Sample PEP-PEE-1 (M_w = 31 000) exhibits a single broad T_g

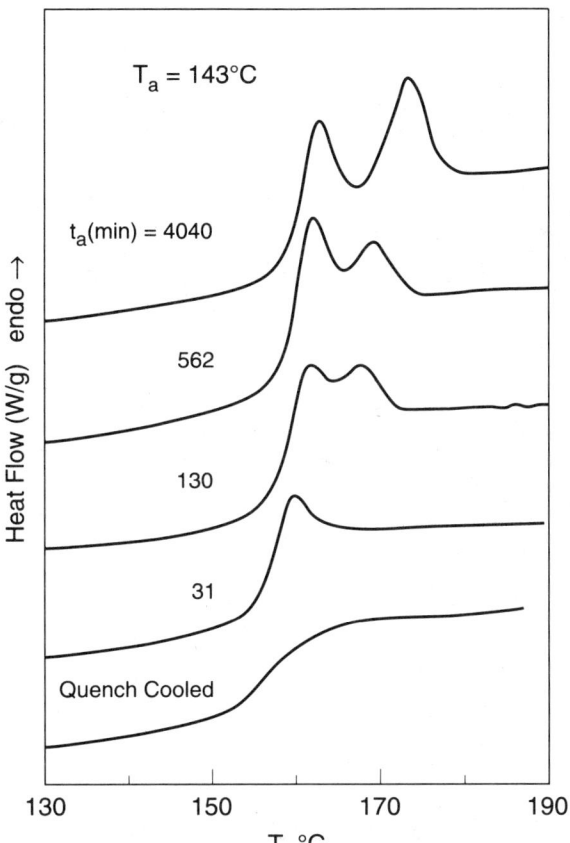

FIG. 17—*Comparative thermograms of blend of two aromatic nylons after quenching and then annealing at 143°C for 31, 130, 562, and 4040 min (after Ellis [38]).*

indicating the diblock is homogeneous. Note that ΔT_{gt} equals 23°C or more than three times the breadth of either corresponding homopolymer's glass transition interval. However, increasing M_w from 50 000 to 81 000 to finally 106 000 for samples PEP-PEE-2, -3 and -4, respectively, leads to the development of two discernible glass transitions near the T_g's for the homopolymers of PEP ($T_g \cong -56°C$) and PEE ($T_g \cong -20°C$). Small angle X-ray scattering traces obtained at 28°C on these same diblocks confirm the order-disorder state shown by the DSC measurements [39].

It should be noted that PEP-PEE-1 is calculated to undergo an order-disorder (microphase separation) transition (MST) at $-41°C$ and if cooled slowly to below T_g it will begin to phase separate [40]. MST is analogous to the upper critical solution temperature (UCST) behavior for homopolymer mixtures. However, quenching this diblock in the DSC at 250°C/min from room temperature to $-100°C$ will trap the copolymer in its homogeneous state when it is reheated at 15°C/min (upper curve, Fig. 19). At slower cooling rates the two phases are apparent on reheating. The MST for the higher molecular weight diblocks, PEP-PEE-2 and -3 is found to reside at 96 and 291°C, respectively [41].

Although at room temperature the PEP-PEE diblocks with molecular weights of 50 000

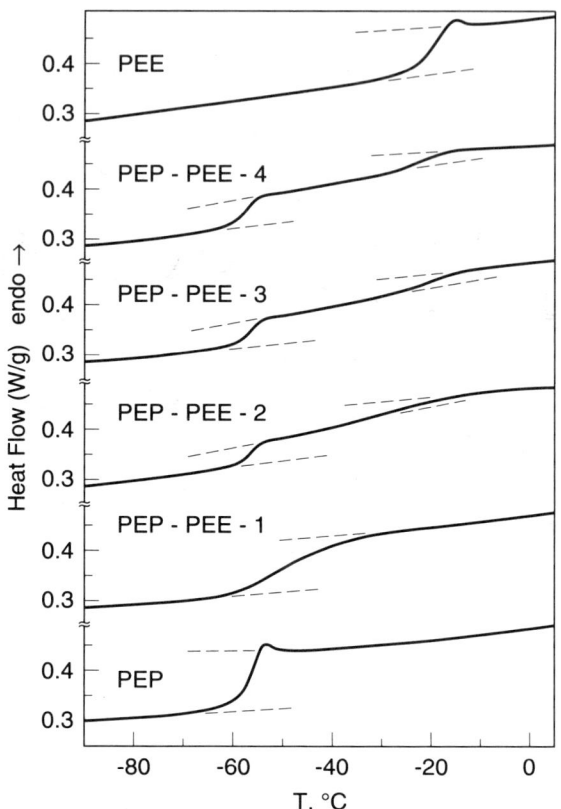

FIG. 18—*DSC traces for the saturated hydrocarbon diblock copolymers and the corresponding homopolymers. Homogeneous sample PEP-PEE-1 shows a single broad* T_g *whereas the other three microphase-separated diblocks are characterized by two* T_g*'s (after Bates, et al., [39]).*

and higher have two T_g's the upper T_g near $-24°C$ is much broader than the T_g at $-56°C$. This phenomenon suggests that some PEP is plasticizing a portion of the PEE block. Evidence for this phenomenon is shown in the C_p curves of the two homopolymers of PEP and PEE, the diblock PEP-PEE-4 and the calculated C_p for the diblock (weighted average of the two corresponding homopolymers' C_p's) (Fig. 20). The broadening of PEP-PEE-4's glass transition above the calculated curve and between $-50°$ and $-20°C$ is attributed to the partial mixing of PEP in PEE.

The structural recovery that occurs after annealing can be used to estimate the amount of interfacial material in diblock polymers whose separate block T_g's are sufficiently removed from each other [42]. By measurement of the enthalpy relaxation that results from annealing at a temperature, T_a, between the T_g's of the blocks, the content of interfacial material, F, can be calculated

$$F = \frac{2}{k_H} \frac{\Delta T_g}{(\Delta T)^2} \Delta H_e \qquad (2)$$

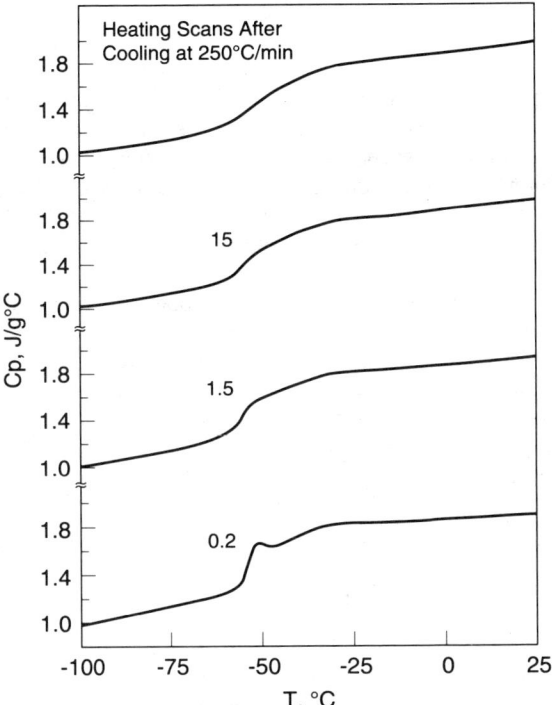

FIG. 19—*DSC heating (15°C/min) scans of 31 000 (M_w) PEP/PEE-1 diblock after cooling at 0.2, 1.5, 15, and 250°C/min. (Data from F. S. Bates and H. E. Bair.)*

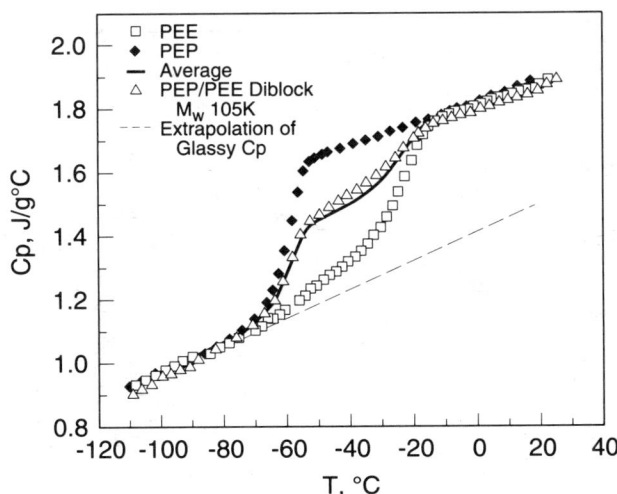

FIG. 20—C_p *versus* T *of PEP, PEE, PEP-PEE-4, and calculated* C_p *for the diblock. (Data from F. S. Bates and H. E. Bair.)*

where $\Delta T = T_{max} - T_a$, ΔT_g is the difference in T_g's of the microphases, k_H equals 0.073 cal/g°C, and ΔH_e is the excess enthalpy as determined in Fig. 21. In this figure the DSC curve of a styrene-isoprene-styrene (SIS) block blended with polyisoprene (PI) and annealed for 72 h at 23°C is plotted against temperature at a heating rate of 30°C/min. The blend composition is SIS/PI (90/10 w/w). The inset indicates the difference between traces of the same sample after annealing (dashed line, Fig. 21) and after quenching (solid line). T_g's of the styrene- and isoprene-rich microphases are indicated by T_gU (95°C) and T_gL (-33°C), respectively. ΔH_e is indicated by the cross-hatching in the inset. In this case the level of interfacial material is 0.11 which is in good agreement with measurements made by other techniques [42].

Besides DSC the compatibility of blends is often probed by other dynamic techniques and the resulting T_g behavior should be in agreement so long as all experiments are performed on the same time scale. Anderson and coworkers studied the compatibility of blends of poly(vinyl chloride) and a terpolymer of ethylene, vinyl acetate, and carbon monoxide by dynamic mechanical, dielectric, and DSC and found a single T_g by each method [43]. In addition, the transition temperature, as defined by the initial rise in E'' at 110 Hz, ϵ'' at 100 Hz and C_p at 20°C/min, agreed to with 5°C.

Quantitative Analysis

Multicomponent polymer blends that cannot be analyzed by conventional spectroscopic techniques due to overlapping absorption bands or the screening effects of additives can often be assayed by DSC without resorting to extraction procedures. If a polymer or a low-molecular-weight additive is incompatible with the base resin, it can be detected in a separate crystalline or glassy phase by either its melting point, T_m, or T_g and measured quantitatively from heat of fusion, ΔH_f, determinations at T_m or incremental change in heat capacity, ΔC_p, measurements at T_g. Conversely, in miscible amorphous blends, the composition can be estimated from T_g versus composition plots measured for that system. Likewise,

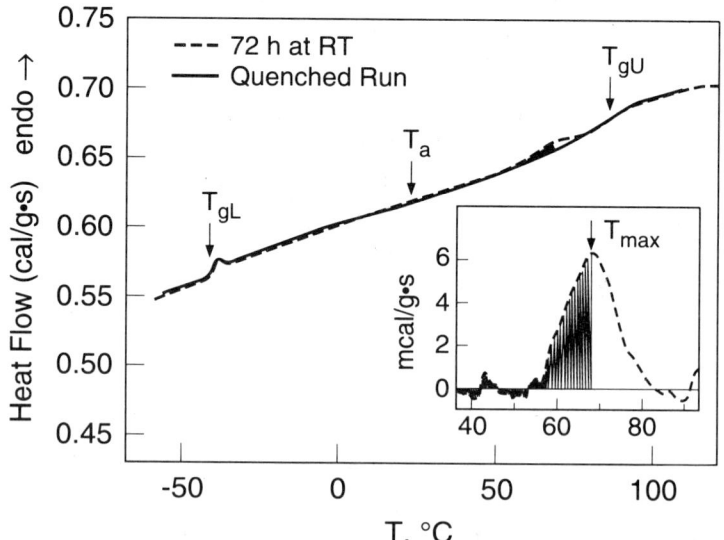

FIG. 21—*DSC trace of SIS/PI diblock and homopolymer blend (90/10 w/w) annealed for 72 h at 23°C and quenched (after Quan et al. [42]).*

when an additive is soluble in a polymer its concentration can be estimated from shifts in the T_m or T_g of the resin [44,45].

The typical C_p behavior of a commercial acrylonitrile-butadiene-styrene (ABS) resin is shown in Fig. 22. ABS is a rubber toughened plastic where a brittle glassy copolymer of styrene-acrylonitrile (SAN) is blended or chemically grafted to submicron spheres of polybutadiene (BD). In the low temperature insert, the discontinuity in C_p (ΔC_p) near $-90°C$ is identified with BD in the composite. The weight fraction of BD (0.13) is estimated from the ratio of the observed ΔC_p to the ΔC_p measured in a separate experiment on the unblended homopolymer. In the high temperature range, the T_g near 104°C is due to the SAN copolymer (76 wt.%), while the two first-order transitions near 70 and 150°C are due to the melting of a 0.24 wt.% fatty acid residue and 3.6 wt.% of a mold lubricant [46].

Further DSC studies have shown that the magnitude of ΔC_p at T_g for the BD phase of an ABS composite is reduced nonlinearly as the ratio of SAN grafted to BD is increased [47]. The attenuations of ΔC_p with increasing graft levels appear to be caused by the reduced number of configurational changes that can occur in the grafted BD molecules at T_g. Thus, at high graft levels calorimetric determination of the rubber content of an ABS resin does not give a measure of the total rubber but indicates the actual concentration of rubber which is effective in modifying the deformational behavior of the SAN matrix.

Plasticization of a Polymer and Its Effect on T_g

The water absorbed by PVAc at 23°C was found to exist in two states. The first, which can account for up to 4 wt.%, is bound to the polymer. The second is in a freezable or clustered form. The latter type of water had no effect on PVAc's T_g, whereas the former kind plasticized the glass transition [48].

In Fig. 23 the C_p of two PVAc samples is plotted versus temperature. One sample contained 1.5 wt.% water and the other had no detectable amount of water. T_g of the sample containing water was reduced by about 8°C below that of the dry PVAc sample at 43°C, although the shape of the C_p curves is identical except for their enthalpic relaxation peaks.

The plasticizing effect of the water absorbed at 23°C continues until a concentration of 4

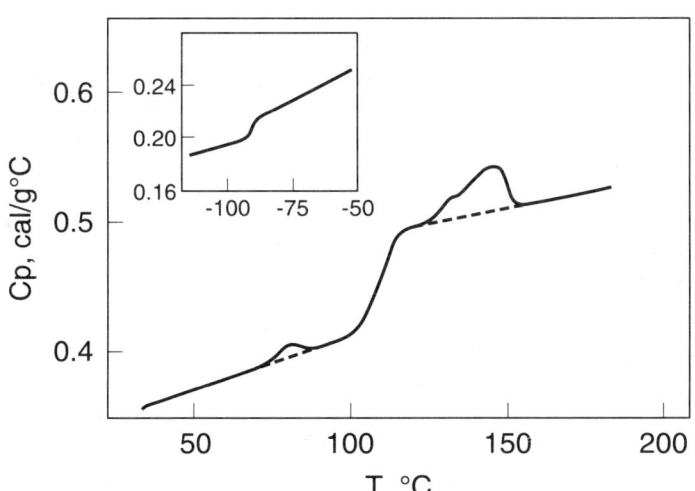

FIG. 22—C_p as a function of temperature of an ABS resin (after Bair et al. [46]).

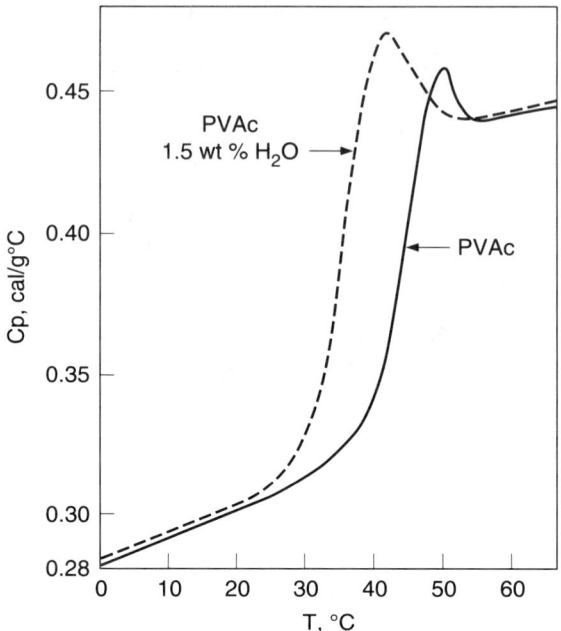

FIG. 23—C_p curve versus T showing the plasticizing effect of water on PVAc's T_g (after Bair et al. [48]).

wt.% water is reached as is shown in Fig. 24. This concentration of water where further plasticization ceases is equivalent to about 1 mole of H_2O to 5 moles of PVAc. All additional water which is absorbed by the PVAc is found to be freezable by DSC. The difference between the total water (W_T) measured coulometrically and the amount of freezable water (W_F) equals the concentration of bound water (W_B) in this system.

$$W_B = W_T - W_F \qquad (3)$$

This result is supported by the observation that PVAc's β transition as measured by dielectric analysis was found to increase in direct proportion to the amount of bound water. The freezable water gave rise to a separate loss peak. Moisture can be easily removed from most polymers by simply holding the material near or above its T_g in a dry atmosphere. Other plasticizers, such as small amounts of monomer or other contaminants, may be present in many commercial resins that may require working (shearing) the polymer in the melt under vacuum in order to remove the unwanted plasticizing compounds [49].

Specific heat measurements of blends of diisodecylphthalate (DIDP), tricresyl phosphate (TCP), di(2-ethyl-hexyl)azelate (DOZ), plasticizers, and commercial poly(vinyl chloride) (PVC) resins have revealed the presence of two distinct glassy phases [50]. The T_g of the lower-temperature phase decreases markedly with increasing plasticizer content in the usual manner, whereas the previously unreported T_g of the higher-temperature phase is less sensitive to plasticizer concentration. This phenomenon is attributed to the preferential solvation of PVC by the plasticizer. The higher T_g is associated with the noncrystalline syndiotactic PVC sequences which are not readily solvated, and the lower one is connected with the plasticized remaining amorphous chain segments. In addition, a small syndiotactic crystalline phase was identified from its crystalline vibrational spectrum.

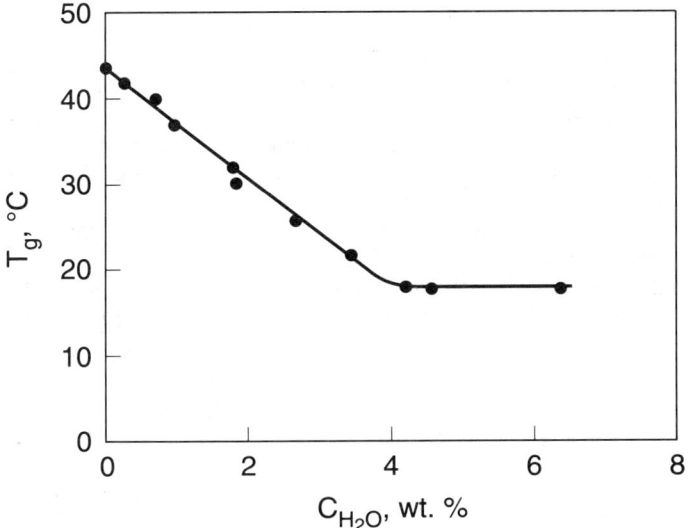

FIG. 24—*The influence of water on PVAc's T_g as determined by DSC at 200 C/min (after Bair et al. [48]).*

At low plasticizer levels, the high-temperature T_g occurs nearly 23°C above that of unplasticized PVC ($T_g = 89°C$), has a width double that of the low-T_g phase, and persists throughout the concentration range of 1 to 25 wt.% DIDP (Fig. 25). From ΔC_p values about 12% of the polymer is calculated to have participated in the higher T_g phase in the DIDP-PVC blend (1:99 w/w), whereas 33% was active in the DIDP-PVC mixture (25:75 w/w). Initial plasticization of the PVC does not increase the breadth of the lower PVC T_g until

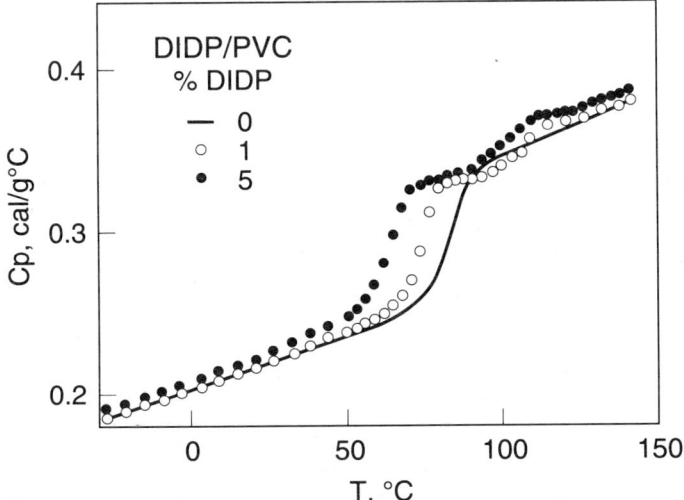

FIG. 25—*Comparative C_p curves versus temperature for PVC and two blends of DIDP and PVC (after Bair and Warren [50]).*

about 5% DIDP was added. Then the transition begins to broaden indicating the development of a heterogeneous glass structure. This latter phenomenon is accompanied by solvation of the high-temperature glassy phase.

The T_g of the neat PVC is found to be 89°C with ΔC_p equal to 0.326 J/g°C^{-1} at a heating rate of 20°C/min (Fig. 25). Note how the C_p curve for the two blends of 1 and 5 wt.% DIDP in PVC separates the glass transition into two with the high T_g phase near 110°C and the low T_g phase below 80°C. In Fig. 26 the effect of increasing the level of DIDP blended in PVC on the breadth of the glass transition can be readily seen.

NMR relaxation measurements tend to support this view of the heterogeneity that is present in these plasticized PVC samples. $T_{1\rho}$ data in particular indicate the presence of two types of material whose motions are drastically influenced by plasticizer [51]. In addition, T_2 values show the existence of a third component that remains rigid for temperatures below 120°C and is unaffected by plasticizer. The two former phases are equated with the two glassy phases that were detected by calorimetry, while the latter phase is attributed to the crystalline regions detected by FTIR measurements. Dimensional information derived from the nmr work suggests that the inhomogeneous regions are on the order of 50 Å [51].

Summary

The temperature gradient across and from top to bottom of a 0.6-mm-thick disk of molded polycarbonate was found to be no greater than 0.6°C when heated at 15°C/min in a DSC. Under these conditions of rate and sample size it was demonstrated that a glassy polymeric material could be analyzed for its glass transition temperature, T_g, as defined by the half-vitrification point (½ ΔC_p) as well as by the temperature where the transition begins and ends to within 1°C. DSC derived C_p measurements as a function of temperature were reviewed to show how the location, shape, and magnitude of the glass transition can be characterized to gain insight into the composition and behavior of a wide variety of polymeric materials ranging from homopolymers to blocks, grafts, blends, and network formers.

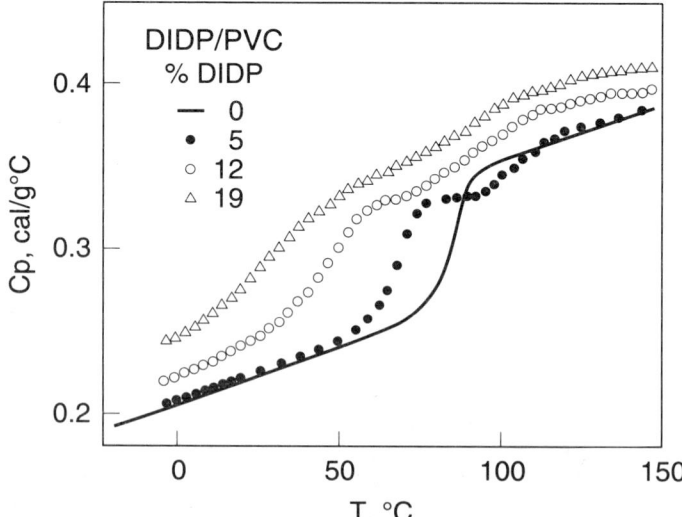

FIG. 26—C_p plotted against temperature for blends of DIDP and PVC containing up to 19 wt.% DIDP (after Bair and Warren [50]).

In particular, this review of T_g by DSC:

- shows the effect of a glassy polymer's prior thermomechanical history on its T_g behavior;
- describes how T_g is increased and ΔC_p at T_g is suppressed through chemical modification or crystal growth;
- demonstrates the way that C_p curves of polymer blends or blocks can be interpreted to indicate the presence or absence of miscibility;
- reveals how the breadth of the glass transition increases as polymers are mixed into miscible polyblends or covalently linked into homogeneous blocks;
- reviews the use of annealing experiments to estimate the amount of interfacial material in diblock polymers; and
- reveals the plasticizing effect of small molecules on the glass transition behavior of an important commercial polymer.

Acknowledgments

During my development as a thermal analyst, I have been fortunate to have many outstanding mentors including F. E. Karasz, J. M. O'Reilly, S. Matsuoka, and L. L. Blyler, Jr. In addition, I am grateful for the fruitful collaboration on various aspects of this review with a number of talented colleagues. Among those deserving of special mention are X. Quan, F. S. Bates, G. E. Johnson, and A. Hale.

References

[1] Matsuoka, S. and Quan, X., *Macromolecules,* Vol. 24, 1991, p. 2770.
[2] Matsuoka, S. in *Relaxation Phenomena in Polymers,* Hanser, Newiork, 1992.
[3] Gibbs, J. H. and DiMarzio, E. A., *Journal of Chemical Physics,* Vol. 28, 1958, p. 373.
[4] Karasz, F. E., Bair, H. E., and O'Reilly, J. M., *Journal of Physical Chemistry,* Vol. 69, 1965, p. 2657.
[5] Bair, H. E. in *Thermal Properties of Polymers, Encyclopedia of Materials Science and Engineering,* M. B. Mever, Ed., Pergamon Press, NY, 1986, p. 4945.
[6] Watson, E. S., O'Neill, M. J., Justin, J., and Brenner, N., *Analytical Chemistry,* Vol. 36, 1964, p. 1233, U.S. Patent 3,263,484.
[7] Gallagher, P. G. and Westlandt, W. in *Thermal Characterization of Polymeric Materials,* E. A. Turi, Ed., Academic Press, NY, 1981, p. 60.
[8] Wunderlich, B. in *Thermal Analysis,* Academic Press, Inc., NY, 1990, p. 132.
[9] McNaughton, J. L. and Mortimer, C. T., *DSC in IRS; Physical Chemistry Series 2,* Vol. 10, Butterworths, London, 1975 (this article can be ordered from Perkin-Elmer Corp., Norwalk, CT 06856, order number: L-604).
[10] Karasz, F. E., Bair, H. E., and O'Reilly, J. M., *Journal of Polymer Science,* Part C, Vol. 6, 1964, p. 109.
[11] Bair, H. E., Johnson, G. E., and Merriweather, R., *Journal of Applied Physics,* Vol. 49, 1978, p. 4976.
[12] Bates, F. S., Bair, H. E., and Hartney, M. A., *Macromolecules,* Vol. 17, 1984, p. 1987.
[13] Tool, A. Q. and Eichlin, C. G., *Journal of the American Ceramics Society,* Vol. 14, 1931, p. 276.
[14] Davies, R. O. and Jones, G. O., *Proceedings, Royal Society of London,* Ser. A 217, 1953, p. 26.
[15] Prest, W. M., Luca, D. J., and Roberts, F. J., Jr., in *Thermal Analysis in Polymer Characterization,* E. A. Turi, Ed., Heyden, Philadelphia, 1981, p. 24.
[16] Bair, H. E., Bull. Am. Phys. Soc., Vol. 26, No. 3, 1981, p. 431.
[17] Matsuoka, S., Aloisio, C. J., and Bair, H. E., *Journal of Applied Physics,* Vol. 44, 1973, p. 4265.
[18] Matsuoka, S. and Bair, H. E., *Journal of Applied Physics,* Vol. 48, 1977, p. 4058.
[19] Koike, Y., Matsuoka, S., and Bair, H. E., *Macromolecules,* Vol. 25, 1992, p. 4807.
[20] Hodge, I. M., *Macromolecules,* Vol. 20, 1987, p. 2897.
[21] Adams, G. and Gibbs, G. H., *Journal of Chemical Physics,* Vol. 43, 1965, p. 139.

[22] Bair, H. E., Bebbington, G. H., and Kelleher, P. G., *Journal of Polymer Science, Polymer Physics Ed. 1*, 1976, p. 2113.
[23] Mininni, R. M., Moore, R. S., Flick, J. R., and Petrie, S. E. B., *Journal of Macromolecular Science-Physics*, Vol. B8, 1973, p. 343.
[24] Struik, L. C. E., *Physical Aging in Amorphous Polymers and Other Materials*, Elsevier, NY, 1978.
[25] Wunderlich, B., *Journal of Physical Chemistry*, Vol. 64, 1960, p. 1052.
[26] DiMarzio, E. A. and Dowell, F., *Journal of Applied Physics*, Vol. 50, 1979, p. 6061.
[27] Zhongde, X., Mays, J., Xuexin, C., et al., *Macromolecules*, Vol. 18, 1985, p. 2560.
[28] Bair, H. E., Schilling, F. C., Pearson, D. S., and Fetters, L. J., *NATAS (15th) Conference Proceedings*, 1986, p. 126.
[29] O'Reilly, J. M., *Journal of Applied Physics*, Vol. 48, 1977, p. 4043.
[30] Hale, A., Macosko, C. W., and Bair, H. E., *Macromolecules*, Vol. 24, 1991, p. 2610.
[31] Bair, H. E., Kleinedler, G. E., and Machusak, D. A., *SPE Technical Papers 29*, 1983, p. 402.
[32] Cheng, S. Z. D., Wu, Z. Q., and Wunderlich, B., *Macromolecules*, Vol. 20, 1987, p. 2802.
[33] O'Reilly, J. M. and Karasz, F. E., *ACS Polymer Preprints*, Vol. 5, 1964, p. 351.
[34] Ryan, C. L., Karasz, F. E., and MacKnight, W. J., *Proceedings, 16th NATAS Conference 1:15*, 1980.
[35] MacKnight, W. J., Karasz, F. E., and Fried, J. R., in *Polymer Blends*, Vol. 1, D. R. Paul and S. Newman, Eds., Academic Press, New York, 1978, pp. 186–238.
[36] Karasz, F. E. and MacKnight, W. J., *Contemporary Topics in Polymer Science*, Vol. 2, 1977, p. 143.
[37] Grooten, R. and G. ten Brinke, *Macromolecules*, Vol. 22, 1989, p. 1761.
[38] Ellis, T. S., *ACS Polymer Preprints*, Vol. 31, No. 1, 1990, p. 285.
[39] Bates, F. S., Rosedale, J. H., Bair, H. E., and Russell, T. P., *Macromolecules*, Vol. 21, 1989, p. 2557.
[40] Almdal, K., Rosedale, J. H., and Bates, F. S., *Macromolecules*, Vol. 23, 1990, p. 4336.
[41] Rosedale, J. H. and Bates, F. S., *Macromolecules*, Vol. 23, 1990, p. 2329.
[42] Quan, X., Bair, H. E., and Johnson, G. E., *Macromolecules*, Vol. 22, 1989, p. 4631.
[43] Anderson, E. W., Bair, H. E., Johnson, G. E., et al., in *ACS Advances in Chemistry Series*, Vol. 176, S. L. Cooper and G. M. Estis, Eds., 1979, p. 413.
[44] Bair, H. E., in *Analytical Calorimetry*, Vol. 2, R. S. Porter and J. F. Johnson, Eds., Plenum Press, NY, 1970, p. 51.
[45] Bair, H. E., in *Thermal Characterization of Polymeric Materials*, E. Turi, Ed., Academic Press, NY, 1981, p. 845.
[46] Bair, H. E., *Polymer Engineering and Science*, Vol. 14, 1974, p. 202.
[47] Bair, H. E., Shepherd, L., and Boyle, D. J., in *Thermal Analysis in Polymer Characterization*, E. A. Turi, Ed., Heyden, Philadelphia, 1981, p. 114.
[48] Bair, H. E., Johnson, G. E., Anderson, E. W., and Matsuoka, S., *Polymer Engineering and Science*, Vol. 21, 1981, p. 930.
[49] Reed, T. F., Bair, H. E., and Vadimsky, R. G., in "Recent Advances in Polymer Blends, Grafts and Blocks," *Polymer Science and Technology*, Vol. 4, L. H. Sperling, Ed., Plenum Press, NY, 1974, p. 359.
[50] Bair, H. E. and Warren, P. C., *Journal of Macromolecular Science-Physics*, B20, 1981, p. 381.
[51] Douglass, D. C., in *ACS Symposium Series, No. 142, Polymerization Characterization by ESR and NMR*, A. E. Woodward and F. A. Bovey, Eds., 1980, p. 147.

Charles M. Earnest[1]

Assignment of Glass Transition Temperatures Using Thermomechanical Analysis

REFERENCE: Earnest, C. M., "**Assignment of Glass Transition Temperatures Using Thermomechanical Analysis,**" *Assignment of the Glass Transition, ASTM STP 1249*, R. J. Seyler, Ed., American Society for Testing and Materials, Philadelphia, 1994, pp. 75–87.

ABSTRACT: The use of thermomechanical analysis (TMA) as a tool for assignment of glass transition temperatures, T_g, is discussed in this paper. An attempt to address the minimum requirements of TMA instrumentation necessary for T_g assignments is included. The subject of temperature calibration of the TMA instrument, which is a prerequisite for any temperature of transition assignment, is discussed. Methodology is offered for using the TMA instrument in the expansion mode for T_g assignments. In addition, TMA methodology employing a flat-tipped penetration probe for the assignment of the softening temperature, T_s, is also given here. For most amorphous materials, T_s is very close to the T_g value assigned by expansion TMA or differential scanning calorimetry (DSC).

KEYWORDS: thermomechanical analysis (TMA), glass transition temperature (T_g), softening temperature (T_s), linear thermodilatometry, penetrometry, temperature calibration

The technique of thermomechanical analysis (TMA) is one of the major tools that is used in the assignment of glass transition temperatures, T_g. By definition, TMA is a technique in which the deformation of the sample under non-oscillatory stress is monitored against time or temperature while the temperature of the sample, in a specified atmosphere, is programmed [1–3]. The stress may be applied in compression, tension, flexure, or torsion. Thus, the technique includes several different instrumental configurations which impose different types of applied stress. The major change associated with the instrumental configuration is that of the choice of probe design (e.g., expansion, penetration, etc.).

When the TMA experiment is conducted with the sample under "negligible" load, using a flat-tipped (expansion) probe, the technique is often referred to as *linear thermodilatometry*. The prefix "linear" means that the expansion is measured in only one dimension.

The subject of this paper is the assignment of the temperature(s) associated with the second order thermal transition known as "the glass transition." When amorphous or semicrystalline materials are heated at constant rate through the glass transition region of temperature, the material exhibits a marked loss in rigidity that is associated with the physical change in the material. The material undergoes a change from vitreous solid to amorphous fluid in the transition region of temperature.

When using thermomechanical instrumentation, this transition associated with the loss of rigidity may be readily recognized using compression or penetration modes of the analyzer.

[1]Department of Chemistry, Berry College, Mount Berry, GA 30149.

A change in dimension of the sample material on direct contact with the loaded probe will be observed.

It is also well established that an amorphous or semicrystalline material will exhibit a dramatic increase in its coefficient of linear thermal expansion, α, as it proceeds through the temperature region associated with the glass transition. Thus, the use of linear thermodilatometric means of assignment of the transition temperature, T_g, has been adopted by many laboratories for a variety of materials (see ASTM E 831, Test Method for Linear Thermal Expansion of Solid Materials by Thermomechanical Analysis).

As will be discussed later in the paper, the assignment of the T_g in either case involves the extrapolation of the slope of the TMA probe displacement curve before and after the transition. The temperature that corresponds to the point of intersection of these two extrapolations is assigned as the glass transition temperature, T_g.

The glass transition temperature, T_g, is dependent on the thermal history of the material to be tested. For many amorphous and semicrystalline materials the T_g gives important information relative to processing conditions, stability, mechanical and electrical behavior, as well as product acceptability. Since many material formulations contain only minor amounts of the amorphous component, the greater sensitivity of the TMA methods (relative to that of the differential scanning calorimetric, DSC, method [see ASTM E 1356, Test Method for Glass Transition Temperatures by Differential Scanning Calorimetry or Differential Thermal Analysis]) has made them the method of choice in many cases.

In this paper, methods of assignment of the transition temperatures associated with the glass transition using both compression and expansion modes of the TMA instrument will be presented. Particular emphasis will be placed on methodology that is presently under development by Task Group TM-01-20 of ASTM Committee E–37 on Thermal Measurements.

Instrumentation

Thermomechanical analyzers have been commercially available for several decades. All TMA instruments share the same objective of monitoring a dimensional change in the test sample as it is subjected to a controlled temperature program. In all TMA measurements, the sample is under some degree of stress which is applied in a nonoscillatory fashion. The dimensional change observed in the specimen is monitored versus the temperature of the sample.

Thus, the major components of the instrumentation required for achieving such studies include a sample loading platform, an appropriate sample probe, a displacement transducer of adequate sensitivity, a furnace with temperature programmer, etc., a temperature-measuring device, a means of applying a static load to the test specimen, a pneumatic system for purging the sample compartment, and a data-handling device for final presentation of the TMA thermal curve.

Figures 1 and 2 show two different arrangements that are commonly found in commercial thermomechanical analyzers which operate in a vertical (direction of applied force as well as dimensional monitoring of test specimen) fashion. Figure 1 describes an instrumental arrangement that employs a weight tray and float suspension in an appropriate buoyancy fluid for application of the applied (static) load. A linear variable differential transformer (LVDT), located above the sample probe, platform, and furnace, monitors the dimensional change in the test specimen on heating or cooling. A thermostated heat sink, which, during operation, surrounds the test specimen and probe tip. The choice of coolant used in this reservoir is one means of extending the temperature range of the analyzer into the

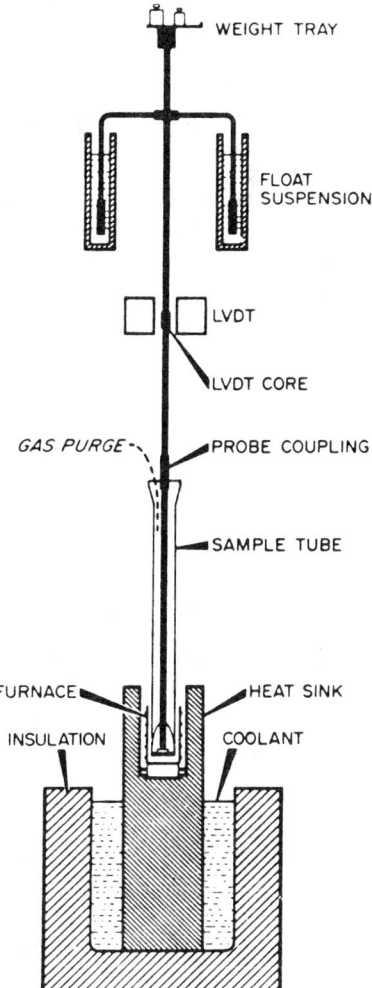

FIG. 1—*Diagram of thermomechanical analyzer employing weight tray for static loading of probe (from Ref 14).*

subambient. The sample temperature is monitored by a thermocouple (not shown) which extends down the sample tube and is positioned near the test specimen on the sample platform.

Figure 2 describes a somewhat different configuration that is similar to that presently employed by several instrument manufacturers [4,5]. In this vertical system, the weight tray is replaced by an electromechanical coil (force motor) for application of the static loading of the test specimen. Both the ordinate sensor (LVDT) and force motor are located beneath the sample, probe tip, furnace, and coolant reservoir in this case.

Figure 3 illustrates several probe geometries that are commonly used with thermomechanical analyzers. As can be seen in these diagrams, the probe type determines how

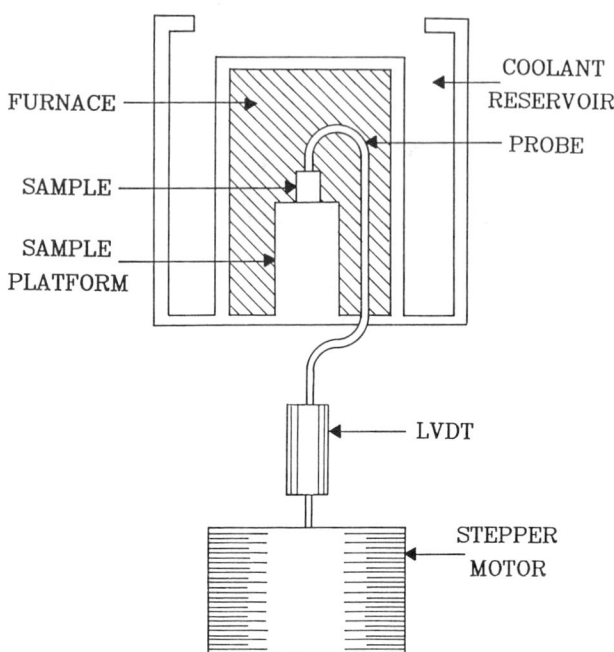

FIG. 2—*Diagram of thermomechanical analyzer employing an electromechanical coil for static loading of probe (from Ref 14).*

the stress is applied to the test specimen. The choice of probe type is often determined by the type of sample geometry. For example, the tension (or extension) mode is used primarily for fibers and thin film materials.

The present ASTM Task Group (TM-01-20) of Committee E-37 on Thermal Measurements has reviewed the general requirements of TMA instruments to achieve most T_g assignments. The working document presently reflects procedures and requirements for instruments that operate in a vertical fashion and have a typical operating range from -100° to 600° (depending upon choice of instrument).

In order to perform T_g assignments using the TMA methods described herewith, some requirements for the instrument are given below.

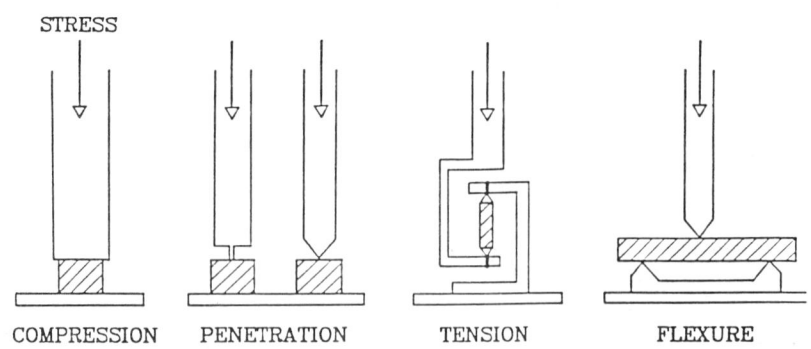

FIG. 3—*Examples of probe geometries or operating modes for TMA (from Ref 14).*

Specimen Holder or Platform

The instrument must provide a means for placement of the test sample in a manner that allows the measurement of either compression or expansion. The material from which the platform is constructed should be relatively inert and should exhibit low expansion itself.

Availability of Appropriate Probe Type

The methodology under development by the ASTM Task Group uses a rigid flat-tipped probe which makes good contact with the test sample under applied force. Probe diameters of 2 to 6 mm, depending upon the mode of the TMA employed, are recommended. The probe material itself should be made of an inert low expansivity material. A material commonly used in probe manufacturing is quartz.

Methods for expansion and penetration modes of the TMA are given later in this paper. Other modes, such as flexure and extension, have also been employed for observation of the glass transition by TMA.

Displacement Transducer (Sensing Element)

Most commercial instruments use a LVDT to monitor the vertical position of the probe resulting from changes in the thickness or hardness of the test specimen as it is heated through the temperature range of the study. The minimum displacement sensitivity the apparatus should possess is one micrometre (μm).

Force or Loading Device

Many of the older TMA instruments use weight trays for static loading of the probe. More recent commercial instruments use an electromechanical coil (force motor) for selecting the force applied to the test material. For performing the test methodology set forth in this paper, the instrument should be capable of delivering an applied force from 0 to 50 mN.

Furnace, Controller, and Associated Electronics

The furnace, controller, and associated power board, etc., must provide a means of uniformly heating the test specimen at a linear heating rate. Modern microcomputer-based controllers found in commercial TMA instruments are adequate for the TMA methods discussed here. The test methods under development by Committee E-37 require a heating rate of 5.0 ± 0.2°C/min. The temperature range of the analyzer must include the temperature region of the type of material to be studied. This will depend on the type of furnace employed by the analyzer and, in many cases, subambient capabilities of the instrument.

Temperature Measuring Device

Most TMA instruments employ a thermocouple for making the temperature measurement. Many instruments employ Type K (Chromel-Alumel) thermocouples which are covered with a stainless steel sheath. The thin stainless steel cover gives the thermocouple rigidity for placement purposes. The sheath also protects the thermocouple junction from any corrosive gases that might be encountered during the thermomechanical testing.

The Type K thermocouples may be used at temperatures ranging from -270° to ca 1370°C. This wide temperature range of applicability as well as its good thermoelectric power (40μV/°C) make the Type K thermocouple a popular choice of many commercial vendors.

For many high-temperature thermomechanical analyzers, such as those commonly used in the ceramic industry, the Type K thermocouple will not suffice. In these analyzers, either a Type S (Pt/Pt, 10% Rh) or Type R (Pt/Pt, 13% Rh) is employed as the temperature measuring device. Both of these offer an upper temperature limit of ca 1770°C.

The TMA methods of T_g assignment presented here are primarily for use in analyzers which operate in a vertical manner at temperatures up to 1000°C. The primary requirements of the temperature sensor are that it should be reproducibly positioned near the test sample and exhibit a sensitivity of 0.1°C or better.

Pneumatic System

The use of a dry inert purge gas in the sample chamber is common practice for most TMA measurements. The dynamic purge leads to a better coupling of the sample temperatures with that of the furnace: It facilitates heat transfer and minimizes gradients. The dynamic purge also removes any gaseous substances that might evolve from the test sample during heating.

Nitrogen, argon, or helium are often used when air oxidation of the sample is a concern. A good flow controller should be included with the purge system for setting the dynamic purge rate. Provision should be made for the removal of any water vapor that may be present in the purge gas. Commercially available molecular sieve traps are adequate for this purpose.

Data-Handling Device

Since the output of the temperature measuring device (thermocouple) is an emf and the ordinate sensing transducer (LVDT) also creates a voltage signal, potentiometric records of both the X-Y and Y-T varieties were commonplace data-handling components of the TMA System hardware until ca. 1980. These satisfied the minimum requirement of displaying changes in the analyzer probe position as a function of temperature. With the introduction of microcomputer based data-handling devices in 1979-1980, the potentiometric recorders have disappeared from the thermal analysis laboratory.

The microcomputer-based data-handling devices, associated software, and digital plotters have brought many time-saving conveniences to the thermal analyst. Probably the most time-saving advantage has been the ability to rescale the ordinate scale after the original thermal curve has been stored. Prior to the memory-based capability, the operator often had to replace the test specimen with a new one and repeat the TMA experiment using a more suitable ordinate scale sensitivity. The ability to store the TMA data and recall at a later date, for comparative purposes, is also a major convenience to the quality control laboratory. Through the use of microcomputer data-handling systems and custom software packages for TMA measurements, the graphical extrapolations leading to the assignment of the T_g temperatures are rapidly and reproducibly executed.

Temperature Calibration of Thermomechanical Analyzers

The assignment of glass transition temperatures is one of a large number of applications in which the major task of the TMA instrument is to assign the temperature of transition. In these experiments, the abscissa (temperature axis) of the TMA thermal curve requires a greater degree of calibration (accuracy) than does the ordinate (dimensional) scale. It is true that most TMA thermal curves generated from modern thermomechanical analyzers have temperature axes which arise from a single thermocouple positioned somewhere in close proximity to the test specimen in the instrument. This practice, although extensively employed, does not give the true sample temperature in most situations.

Earnest and Seyler [6] have recently described a method based on the use of melting point standards of accurately known transition temperatures. This method, ASTM E 1363 (Test Method for Temperature Calibration of Thermomechanical Analyzers), was developed by ASTM Committee E-37 on Thermal Measurements. The calibration method employs the TMA instrument in a penetration or compression mode. Experimentally, TMA thermal curves are obtained for two melting point standards that fall within the chosen temperature range of use for the TMA instrument. The penetration profiles of these melting standards are obtained using a recommended 50 mN of force on the TMA probe. Once the TMA thermal curves are generated, the observed extrapolated onset temperatures, T_o, are assigned for each of the melting point standards. Figure 4 describes the method for assignment of the extrapolated onset temperature, T_o, from the penetration TMA thermal curve.

An equation is then developed describing the linear correlation of the experimentally observed temperatures of transition and the actual melting temperatures as published by Rossini [7]. Thus, for this two-point calibration procedure, it is assumed that the relationship between the experimentally observed transition temperature, T_e, and the actual specimen temperature, T_t, is a linear one governed by the equation

$$T_t = ST_e + I$$

where S and I are the slope and intercept of a straight line, respectively.

In practice, the metal standards should be chosen as close to the upper and lower temperature limits of the actual analysis runs as is practical. Once the values of S and I are calculated, Eq 1 may be used to calculate the actual specimen transition T_t from any experimentally observed transition temperature, T_e, using the particular TMA instrument.

An alternate, one-point method of temperature calibration was also given by Earnest and Seyler [6]. This one-point method is intended for use over very narrow temperature ranges. A round-robin study using such a one-point approach has been conducted in Japan [8]. The two-point method is the preferred method in all cases, however.

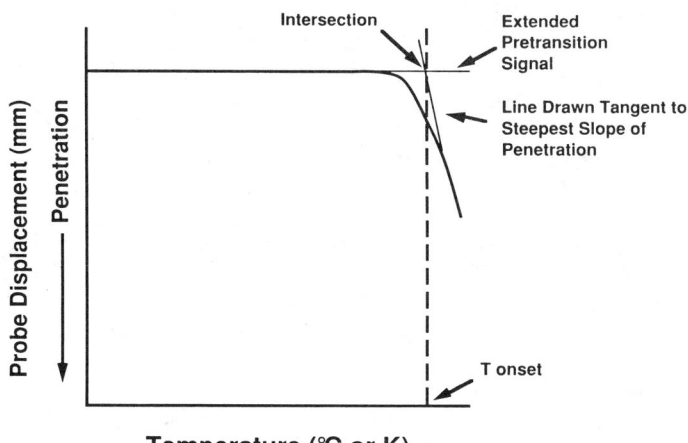

FIG. 4—*Assignment of extrapolated onset temperature, T_o, from a penetration TMA thermal curve.*

It is also possible to develop more elaborate calibration methodology using multiple (more than two) fusion standards and quadratic regression analysis. Since most modern instruments are capable of essentially linear heating rates in the region of use, the two-point method should be sufficient.

An interlaboratory round-robin test study was conducted using the two-point method. Ten industrial laboratories participated in the study using commercial thermomechanical analyzers from TA Instruments, Perkin-Elmer, Mettler, and Omnitherm. These results have been published [9] in the literature. According to this study, the repeatability which may be expected by a single analyst on different days was estimated to be 0.51°C. The between laboratory reproducibility standard deviation of results was estimated to be 1.47°C. The average deviation of results from the true value was 0.01°C with a within-laboratory confidence interval (95% CI) of ±1.16°C and a interlaboratory confidence interval of ±3.32°C.

Methodology for the Assignment of Glass Transition Temperatures by TMA

The literature reflects many variations in detection and graphical assignment of the T_g value by TMA. Dimensional changes associated with the T_g-region can be monitored by bending, extension, expansion, or penetration [10]. The method of choice varies from one laboratory to another and depends on the type of material and sample geometry. Moreover, even for a particular specimen, geometry, and TMA measuring device, there is considerable variation in experimental and analytical procedures [11].

In an effort to improve interlaboratory agreement and to put some degree of standardization into the assignment of glass transition temperatures by TMA, a task group (TM-01-20) was established by Committee E-37 to address this need. Although the working document for this test method has not been completely finalized or approved by balloting, some of the major components of the method are given below.

First, the procedure provides for methods utilizing both expansion and penetration modes using a flat-tipped probe of appropriate diameter in each case. Thus, the TMA experiment may be carried out under constant load from 0 to 50 mN of force depending on the mode and requirement of the sample specimen.

The analyzer must first be calibrated using ASTM E 1363. Metal standards used in the calibration procedure should be chosen to include the entire temperature range of the experiment. Once the analyzer is calibrated for temperature assignments, a preweighed specimen of 0.5 to 3 mm thickness is placed on the stage of the analyzer at room temperature. The analyzer probe may then be placed in contact with the test specimen at the temperature only if the T_g of the specimen is above room temperature. If this is not the case, the specimen is cooled, using the coolant reservoir, etc., to some temperature below the T_g prior to probe placement.

The specimen is then subjected to a controlled temperature program of 5°C/min using the probe of choice (expansion or penetration). Experiments conducted in the expansion mode generally employ flat-tipped probes having diameters of 4 to 6 mm. Typical static loading of the expansion probe will vary from 0 to 5 mN of force. On the other hand, experiments conducted in the penetration mode should use a flat-tipped probe of 2 to 4 mm in diameter. When using the penetration mode, a static loading of 20 to 50 mN of force is usually applied to the test specimen.

Figure 5 shows some typical expansion TMA thermal curves obtained in the region of the glass transition of an unspecified material. The lower curve (Run 1) shows a phenomenon that occurs in the region of the T_g for some test specimens. This unexpected transient, which is observed between the pretransition slope of the TMA curve and the post-transition slope

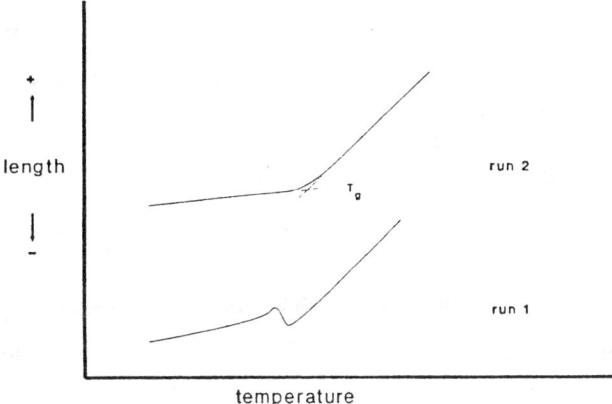

FIG. 5—*TMA thermal curves for glass transitions region of an unspecified material using an expansion probe.*

of the thermal curve, may be due to settling, stress relief within the material, or alteration of the specimen morphology. If this phenomenon is observed during the testing experiment, the heating program should be terminated ca. 20°C above the observed deflection. The probe should then be removed from the test specimen and the TMA furnace cooled to the original starting temperature. The probe is then placed in contact with the specimen and a second heating run is conducted on this test specimen.

The TMA expansion curve, labeled Run 2 in Fig. 5, represents the second heating of the test sample. Graphical extrapolation of the pre-transition and post-transition expansion curve lead to the T_g assignment. As is shown in Fig. 5, the glass transition temperature, T_g, corresponds to the temperature at which the extrapolated expansion curves intersect.

Figure 6 describes the TMA thermal curve that is observed when the glass transition region is studied using a penetration probe under a static load. Run 1 displays the transient

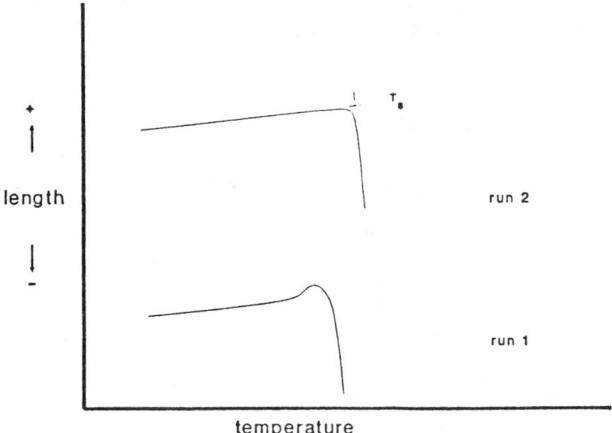

FIG. 6—*TMA thermal curves for glass transition region of an unspecified material using a penetration probe.*

which was discussed in the above paragraphs for the expansion methodology and its elimination from the TMA curve is handled in exactly the same manner as in the expansion studies.

As will be observed in Run 2 (upper thermal curve) of Fig. 6, the test specimen softens and the loaded penetration probe shows a downward deflection as the specimen material undergoes the glass transition. The extrapolated onset temperature is graphically assigned from the TMA thermal curve. In this case it is called the "softening temperature," T_s, rather than the glass transition temperature.

Factors that Affect the Assignment of Glass Transition Temperatures

There is a number of factors that affect the assignment of glass transition temperatures. Some of these have been listed in Table 1 of this paper. One will note that the author has attempted to divide these into two different categories. Those listed under "Physical and Experimental" include the experimental choices of sample loading and heating rate. Sample geometry can also be that chosen by the analyst but oftentimes it is not.

Effect of Heating Rate

Garn and Menis [12] have thoroughly examined the variation of T_g with heating rate for cases where positioning of the measuring transducer is some point removed from the test specimen. With the thermocouple placed near the sample specimen in the TMA instrument, the measuring point (thermocouple junction) will be heated more or less independently of the sample, yielding a poorly resolved relationship between the sample temperature and the measured temperature. In steady-state heating, the two temperatures may be very close. However, during extra absorption of heat by the sample, or even during a discontinuous increase of heat capacity during a glass transition, this relationship will change. The magnitude of the change can be expected to be heating rate dependent.

The best practice that the thermal analyst may follow in order to minimize these effects is to calibrate the TMA instrument for temperature assignment using exactly the same heating rate as that used with the test specimen under study. Once the TMA instrument is calibrated, the thermocouple position must not be changed from run to run. Consistency when placing the sample on the platform is also another recommendation that will help in this case.

TABLE 1—*Factors affecting T_g assignments.*

Physical or Experimental
• Heat or Thermal History of Specimen
• Sample Geometry
• Presence of Moisture
• Presence of Additives, Adhesives, or Skin Material
• Effect of Loading
• Heating Rate

Data Handling
• Extrapolation Methodology (Algorithm)
    One-point Tangent
    Two-point Extrapolation
    Abnormal Expansion Curve

Sample Geometry

Sample geometry is a variable that can also affect the TMA dimensional response to the glass transition. The use of a sample specimen which is too thick can lead to a thermal gradient in the test specimen itself. This will tend to broaden the TMA curve response over a wider temperature range. Thus, a thermal lag is built into the specimen itself. This is why the ASTM Task Group TM-01-20 has recommended that sample specimens not exceed 3 mm in thickness if at all possible. Specimens with smooth parallel surfaces are preferred so that maximum contact with the sample probe is achieved.

In some cases, the sample geometry is dictated by the end-use product for which the material is used. Thin film and fiber materials are excellent examples of this. Another good example was given in this symposium by Moltzan and Baker [*13*]. In their work with epoxy mold compound specimens used in electronic integrated circuits, the shape of the sample was found to affect the reproducibility of the assigned T_g and the coefficient of linear thermal expansion, α. Interesting enough, the thicker sample in this study gave the best results. The authors suspect that stress in the thinner samples (flame sticks) is greater than that in the thicker specimens (pellets) and that this may be a significant factor in these observations.

Effect of Loading

The level or amount of static loading used in the TMA experiment can influence the shape of the thermal curve that is obtained for the glass transition region of a test specimen. Obviously, studies carried out using the expansion mode of the TMA instrument use little or no loading. Experiments performed using the penetration mode are made with a range of forced loading or applied stress.

Figure 7 describes the thermal behavior of a silicone gum rubber specimen when heated at 10°C per min using a penetration probe having a flat tip with a 0.46 mm radius. One will observe that the upper curve was obtained with no net loading on the TMA probe. The

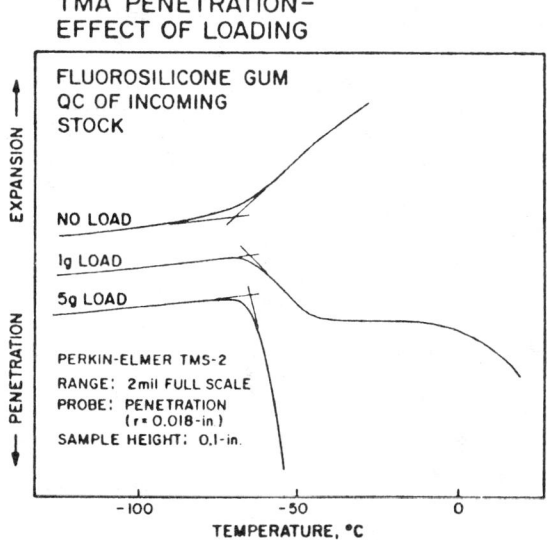

FIG. 7—*Effect of loading on TMA penetration thermal curve (from Ref 10).*

resulting TMA thermal curve, in this case, is the expansion profile for the temperature range of the glass transition. T_g is thus assigned, as was previously described, by the extrapolation of the pre-transition and post-transition expansion curves. The two penetration thermal curves given in Fig. 7 correspond to 1 g and 5 g (10 and 50 mN) loading of the probe, respectively. One will note the obvious increase in the rate of penetration of the TMA probe as the static load is increased. These data would imply that greater loading would be preferred due to sharper penetration curves. While this is true to a certain extent, the use of too much loading can lead to lower values for the T_g assignment. Thus, pre-transition softening can be forced to occur as the applied stress is increased.

Effects of Moisture

The presence of moisture can lead to some adverse effects on the T_g assignment. This is particularly true if the material understudy is to be tested at subambient temperatures. It has already been mentioned in this paper that proper steps must be taken in the pneumatic system to remove any water vapor that may be present in the purge gas. Other sources of moisture include that of adsorbed moisture on the test sample itself.

During subambient operation of thermomechanical analyzers, once the temperature of the sample environment reaches a value below the dew point, any moisture in the sample environment will start to condense on the specimen and its immediate surroundings. On further lowering of the temperature, such condensed moisture will eventually solidify (freeze). This solidified water not only affects the heat capacity of the total sample material but also will appear in the rising temperature TMA thermal curve as it undergoes the first order (fusion) transition on heating.

For these reasons, every effort should be made to ensure the absence of moisture from the test material and its immediate environment when performing TMA.

Conclusion

Thermomechanical analysis (TMA) offers a sensitive and convenient means of assignment of the glass transition (T_g) for a variety of sample types. Although only the penetration and expansion modes of the thermomechanical measuring device are described in the methodology presented here, other modes of operation, such as tension and flexure which are described in Fig. 3, have also been employed for T_g assignments. When the technique of TMA is employed as a tool for assignment of glass transition temperatures, calibration of the temperature axis of the TMA thermal curve is the major calibration requirement. The TMA instrument operator should take steps necessary to ensure the absence of moisture from the sample and its immediate environment in the measuring system. The thermal analyst should become familiar with those factors, given in Table 1, which affect the T_g assignment prior to actual laboratory experimentation.

References

[1] MacKenzie, R. C., *Journal of Thermal Analysis,* Vol. 8, 1975, p. 197.
[2] MacKenzie, R. C., *Thermochemica Acta,* Vol. 28, 1979, pp. 1-6.
[3] Hill, J. O., *For Better Thermal Analysis,* 2nd. ed., International Confederation for Thermal Analysis, 1990.
[4] Sauerbrunn, S. R. and Gill, P. S., "Thermomechanical Analysis: Advances in the State of the Art," *Material Characterization by Thermomechanical Analysis, ASTM STP 1136,* A. T. Riga and C. M. Neag, Eds., American Society for Testing and Materials, Philadelphia, 1991, pp. 120-125.
[5] Wiedemann, H. G., Reisen, R., and Boller, A., "Elasticity Characterization of Materials During Thermal Treatment by Thermal Mechanical Analysis," *Materials Characterization by Ther-*

momechanical Analysis, ASTM STP 1136, A. T. Riga and C. M. Neag, Eds., American Society for Testing and Materials, Philadelphia, 1991, pp. 84-99.

[6] Earnest, C. M. and Seyler, R. J., "Temperature Calibration of Thermomechanical Analyzers: Part I—The Development of a Standard Method," *Journal of Testing and Evaluation,* Vol. 20, No. 6, November 1992, pp. 430-433.

[7] Rossini, F. D., *Pure Applied Chemistry,* Vol. 22, 1970, p. 557.

[8] Takahashi, et al., *Thermochimica Acta,* Vol. 147, 1989, p. 387.

[9] Seyler, R. J. and Earnest, C. M., "Temperature Calibration of Thermomechanical Analyzers: Part II—An Interlaboratory Test of the Calibration Procedure," *Journal of Testing and Evaluation,* Vol. 20, No. 6, November 1992, pp. 434-439.

[10] Cassell, R. B., "Polymer Testing by TMA," *Thermal Analysis Application Study No. 20,* Perkin-Elmer Corporation, Norwalk, CT, 1977.

[11] Cassell, R. B. and Twombly, B., "Glass Transition Determination by Thermomechanical Analysis, A Dynamic Mechanical Analyzer, and a Differential Scanning Calorimeter," *Materials Characterization of Thermomechanical Analysis, ASTM STP 1136,* A. T. Riga and C. M. Neag, Eds., American Society for Testing and Materials, Philadelphia, 1991, pp. 108-119.

[12] Garn, P. D. and Menis, O., "Instrumental Effects of Glass Transition Temperatures," *Journal of Macromol Sci-Phys,* Vol. B13, No. 4, 1977, pp. 611-629.

[13] Moltzan, H. and Baker, C. T., "T_g: Does It Adequately Describe the Glass Transition Region and Its Affect on Integrated Circuit Reliability," Symposium on Assignment of the Glass Transition, 4-5 March 1993, Atlanta, GA, ASTM.

[14] Neag, C. M., "Thermomechanical Analysis in Material Science," *Material Characterization by Thermomechanical Analysis, ASTM STP 1136,* A. T. Riga and C. M. Neag, Eds., American Society for Testing and Materials, Philadelphia, 1991, pp. 3-21.

Richard P. Chartoff,[1] Peter T. Weissman,[2] and Anil Sircar[1]

The Application of Dynamic Mechanical Methods to T_g Determination in Polymers: An Overview

REFERENCE: Chartoff, R. P., Weissman, P. T., and Sircar, A., "The Application of Dynamic Mechanical Methods to T_g Determination in Polymers: An Overview," *Assignment of the Glass Transition, ASTM STP 1249*, R. J. Seyler, Ed., American Society for Testing and Materials, Philadelphia, 1994, pp. 88–107

ABSTRACT: Many factors are involved with establishing T_g values from Dynamic Mechanical Methods. These can be grouped as:

(a) instrumental factors
(b) test frequency
(c) material characteristics
(d) choice of T_g criterion

Instrumental factors may include temperature calibration, thermal gradients, sample size, clamping effects, and sample geometry. Increasing test frequency causes a shift of T_g to higher temperatures. Material characteristics include such factors as the degree of crystallinity in crystalline polymers, the degree of crosslinking in thermosets, the specific thermal and mechanical history of the material, and possible moisture effects. The T_g criterion selected relates to the subjective choice of viscoelastic function used to establish T_g. For example, the peak value of E'' will give a different result than the peak tanδ.

In this paper we will review these various aspects of determining T_g, citing examples in each category which illustrate problems in assigning appropriate T_g values.

KEYWORDS: glass transition, dynamic mechanical analysis (DMA), T_g criterion, instrumental factors, thermal/mechanical history, thermoplastics, crystallinity, thermosets, crosslinking, moisture

Linear viscoelastic test methods are widely used by thermal analysts for a number of purposes. Of the different methods for viscoelastic characterization, dynamic test methods are the most popular, since they are readily adapted for studies of polymeric solids and liquids. These methods are often referred to as dynamic mechanical analysis (DMA). In the dynamic mechanical experiment a known sinusoidal strain (or stress) is applied to the test specimen and both the magnitude and phase shift of the resulting stress (or strain) are measured. In most commercial DMA instruments strain is the controlled input, while the resulting stress is measured.

Complex notation for the generalized sinusoidal strain function is

$$\epsilon(t) = \epsilon_0 \, e^{i\omega t} \tag{1}$$

---

[1]Professor of materials engineering and research polymer scientist, respectively, Center for Basic and Applied Polymer Research, University of Dayton, Dayton, OH 45469-0130.
[2]UCB Radcure, Inc., Smyrna, GA 30080.

where ϵ_0 is the strain amplitude, ω is the radian angular frequency of the sine wave (rad/s), and t is the elapsed time. The steady state ratio of stress to strain is a complex quantity having in-phase and out-of-phase components

$$E^* = E' + iE'' \quad (2)$$

E^* is a complex dynamic mechanical tensile (or flexural) modulus where E' is the ratio of in-phase stress to the applied strain and E'' is the ratio of out-of-phase stress to strain. The out-of-phase stress leads the strain by 90°. Further, it can be shown that E' is related to the mechanical energy stored per cycle and E'' is related to the energy dissipated or converted to heat through viscous dissipation. As a result E' is referred to as the storage modulus and E'' is called the loss modulus. The material loss factor or loss tangent is the ratio of energy dissipated to energy stored per cycle of deformation

$$\tan\delta = E''/E' \quad (3)$$

One of the reasons for the popularity of dynamic mechanical measurements for viscoelastic characterization is because the data obtained are the two complementary functions representing both elastic and viscous material response. The viscous response, manifested in $E''(\omega)$ or $\tan\delta$, indicates through a series of maxima (or loss dispersion peaks) associated with each transition, a mechanical "fingerprint" or spectrum for the material being tested. Thus, historically the method has been referred to as "dynamic mechanical spectroscopy."

The glass transition generally is easily identified from dynamic data because of the sharp drop in storage modulus, E' (shear storage modulus G'), and the corresponding loss dispersion in E'' (shear loss G'') or $\tan\delta$ as shown in Fig. 1 [1]. It is evident that there is latitude in how the exact value of T_g is chosen from a set of dynamic data and this often leads to

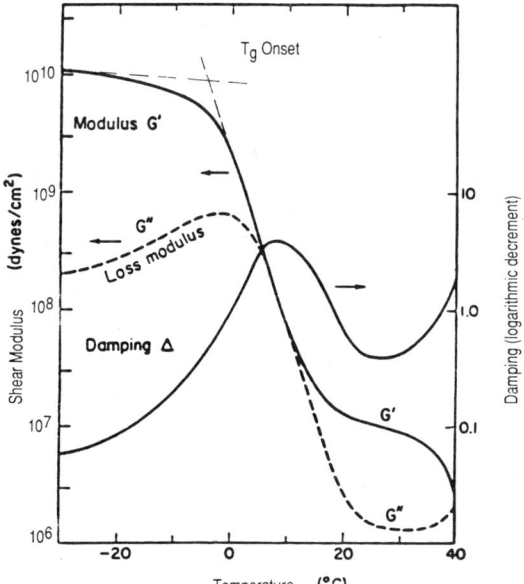

FIG. 1—*Typical dynamic mechanical behavior for viscoelastic functions,* G′, G″, *and* tanδ *(isochronal plot);* log *decrement* = π tanδ; *data taken at* ∼ 1 Hz [1].

confusion in the literature. The criterion for selection of T_g from DMA data is usually either the peak loss modulus, E'', or peak tanδ. The peak tanδ is the most prevalent criterion appearing in the literature. The peak tanδ value is several degrees higher than peak E''. It corresponds more closely to the transition midpoint while the peak loss modulus more closely denotes the initial drop from the glassy state into the transition. In this respect the peak E'' value of T_g is similar to the value defined by intersection of two tangents to the storage modulus curve, one originating from the glassy region and the other from the transition region. This is often referred to as the so-called "onset" temperature as shown in Fig. 1. The fact that tanδ peaks at a higher temperature than E'' results from Eq. 3. Mathematically the peak arises as a compromise between the largest value of E'' and the smallest (relaxed) value of E'.

ASTM D 4065, Practice for Determining and Reporting Dynamic Mechanical Properties of Plastics, asserts that the temperature of the maximum loss modulus is the appropriate standard value. This is reasonable from a practical point of view because the upper use temperature of many amorphous polymers is the "softening" point. It is clear that by the transition midpoint (peak tanδ) the softening point has been exceeded. It should be noted that even in the ASTM standards on DMA measurements some ambiguity persists. ASTM D 4092, Terminology Relating to Dynamic Mechanical Measurements on Plastics, defines T_g as the approximate midpoint of the temperature range over which the glass transition takes place.

For most linear amorphous polymers the transition region is rather narrow, covering around 15°C. In these cases the distinction between Peak E'' or tanδ is not substantial. As discussed later in this paper, there are cases such as highly crosslinked polymers where the T_g region is particularly broad and neither the peak in E'' or tanδ seem appropriate for assigning T_g.

Several table-top instruments are available commercially that measure the storage and loss modulus of solid polymers in tension, compression, shear, and flexure (Fig. 2) and liquid polymers in shear. In principle the data obtained in any of these geometries are equivalent. We do not intend in this paper to discuss individual instruments in detail but will focus on factors that influence identification of T_g that are common either to the dynamic method itself or to specific measurement geometries.

Viscoelastic properties are both time (or frequency) dependent and temperature dependent. From the thermal analyst's view a most useful approach to viscoelastic property characterization is the isochronal method where data are taken at constant time or frequency with temperature varying over a wide range. This is most often done by sweeping the temperature range at a constant rate of temperature increase (ramp). It is a particularly convenient mode for studying relaxation transitions such as the glass transition. However, using this method results in a certain ambiguity in the value assigned to T_g. Thus T_g values

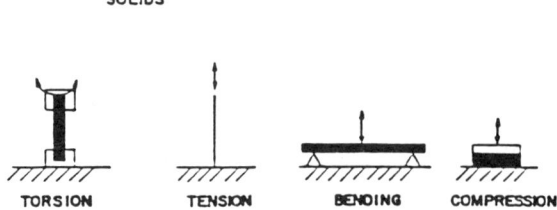

FIG. 2—*Measurement geometries used in instruments for DMA of polymeric solids. (Courtesy of Rheometrics, Inc.)*

determined at high frequencies will be higher than those determined at low frequencies because the rate of segmental molecular motion (or molecular relaxation) will not catch up with the higher test frequencies until the polymer sample reaches a higher temperature. Thus, for practical purposes the glass transition is correctly thought of as a kinetic transition (often referred to as a relaxation transition) because of its dependence on the imposed test rate (frequency). This aspect of T_g measurement by the dynamic method also will be discussed.

Instrumental Factors

There are several instrumental factors that affect the assignment of the glass transition temperature from DMA data: error in the thermocouple due to calibration or placement, temperature gradients within the sample chamber (or oven), temperature gradients within the sample, and exceeding the compliance limitations of the machine. These factors are specific to each individual instrument. Some instruments allow the user to input software corrections to correct errors in thermocouples but fail to allow their proper electronic calibration by the user. Different instruments have different size and shape ovens, employ different heating and cooling mechanisms, and have a wide variety of gas flow rates. Finally, the instruments are constructed with various rigidities and offer a variety of sample clamping configurations. The sample's modulus, the sample's dimensions, and the fixture all combine to determine if the compliance limits of the machine have been exceeded.

Making Accurate Sample Temperature Measurements

A typical modern automated DMA instrument uses two thermocouples in the oven compartment. One of these thermocouples is used in a feed-back loop to the computer and controls the oven's temperature. The second thermocouple is positioned by the user very near the sample and is recorded as the sample's temperature. It is critical that this thermocouple accurately represents the sample's temperature. However there are several issues that combine to give an overall error.

Errors in the Thermocouple Calibration—The thermocouple is interfaced with the computer via an analog to digital conversion board. Since a thermocouple's output is a millivolt signal, the interface board will linearize and set the limits of the signal. Many instruments also allow the operator to use a standard thermocouple calibrator to ensure the sample thermocouple is within specifications.

Oven Temperature Variations— This is highly specific to the oven and heating mechanism used in the instrument. Variations of up to 15°C have been reported for isothermal conditions in some commercial instruments [2]. For certain recently introduced instruments the manufacturers claim to have reduced the temperature variations within the oven to less than 1°C at heating rates of up to 20°C per min. An oven that has wide variations in temperature causes large temperature gradients within the sample as well as causing the sample temperature to depend on thermocouple placement within the oven. The user is advised to check this by methods such as those discussed by Mora and Macosko [2] in order to determine whether significant deviations exist.

Actual Sample Temperature—Since the sample thermocouple is not in contact with the sample, it is actually measuring the oven's atmospheric temperature. The actual sample temperature is a function of its thermal conductivity, the mass and composition of the fixture in contact with the sample, the uniformity of the oven environment, and the heating rate. Because all these variables are present to varying degrees in every instrument, a sample will have a thermal gradient within it. Even if the oven temperature were completely uniform, the sample temperature can lag behind the oven temperature due to heat transfer limitations.

Thermal gradients within the sample are greatest for thick samples (lower surface area to volume ratio), large high mass fixtures (such as parallel plates or cone and plate fixtures), and high heating rates. For large, thick samples these thermal gradients can have a significant effect on the measured T_g. Mora and Macosko [2] reported a 5.2°C increase in the peak tanδ temperature for a polyurethane sample as the thickness was increased from 0.5 mm to 4.0 mm.

The thermal gradients within the sample can be dramatically reduced by performing the experiment in discrete temperature steps instead of dynamic heating. If the oven's temperature profile is good, and equilibration times are sufficient, the only significant instrumental factor is due to electronic calibration of the thermocouple. Suitable equilibration times can be determined as a function of sample size and clamping fixture by embedding thermocouples into the sample and measuring the sample's temperature gradient as a function of time and temperature.

For samples that require dynamic heating (such as those containing volatile components, or polymers that might chemically react during the test) temperature calibration curves can be made by testing polymers of similar thermal conductivity and dimension but with well-established T_g values. Other techniques that have been used are tubes filled with sharp melting materials, fiberglass cloth impregnated with a calibration material, and cast bars of metals. However, in these methods the sample used to perform the calibration generally may not resemble the actual sample of interest in either dimensions or thermal conductivity. It is recommended that dynamic heating experiments be restricted to heating rates less than 2°C per min to minimize the factors noted above. This point is considered further in the following paragraphs.

The Double T_g Artifact

Under certain conditions a double T_g has been observed in measuring fiber reinforced composite specimens using a DMA instrument. This has been discussed in detail by Thomason [3] and is illustrated in Figs. 3 and 4, which are DMA scans taken for a unidirectional carbon fiber epoxy matrix composite at a scan rate of 10°C/min. The two samples differ in that one was thoroughly dried (Fig. 3) and the second humidity aged to an equilibrium moisture sorption (Fig. 4). In each case a second peak appears in the loss modulus and tanδ at a higher temperature than the T_g peak of the polymer matrix. The second peak is an artifact that appears primarily because the metal arms to which the sample is clamped extend out of the sample oven into an unheated area and act as a heat sink for the sample. The clamped portions of the sample exist at a lower temperature than the central portion and because of the thermal lag do not go through their T_g until the oven chamber is heated above the actual T_g of the polymer. The position of the second loss dispersion is dependent on the temperature gradient along the sample and the heating rate.

In general, the double T_g artifact is not observed unless heating rates exceed 2°C per min. Additional data for the same sample as Fig. 3 taken at 2°C per min (not shown here) verify this. It may not be observed for unreinforced polymer samples because the transition of the clamped portion can be detected only if the measured storage modulus of the entire sample is still within the detection capability of the instrument. However, for crosslinked thermosets with a high plateau modulus above T_g it is likely that the artifact will be observed. It is important to note that a double transition for composite specimens is not unique to one particular DMA instrument but may appear in any instrument where such thermal gradients exist. The guideline of 2°C per min as the maximum rate of temperature scanning in all DMA instruments is a useful one. This is the rate specified in the various ASTM standards

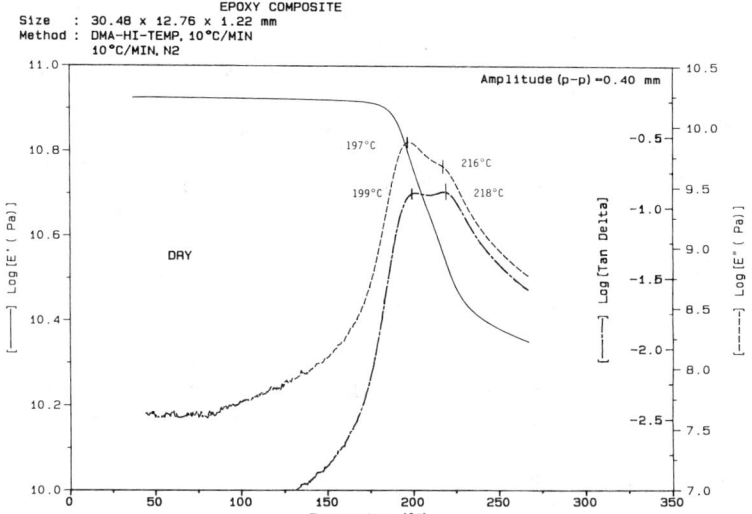

FIG. 3—*DMA flexural data for dry specimen carbon/epoxy unidirectional composite showing double T_g artifact; frequency 1 Hz (University of Dayton).*

for DMA. The authors recommend once again discrete temperature stepping instead of dynamic (ramped) heating.

Subambient Operation

In order to measure T_g for elastomers it is necessary to operate in a subambient temperature regime. This presents some unique problems. It is desirable to calibrate the instrument

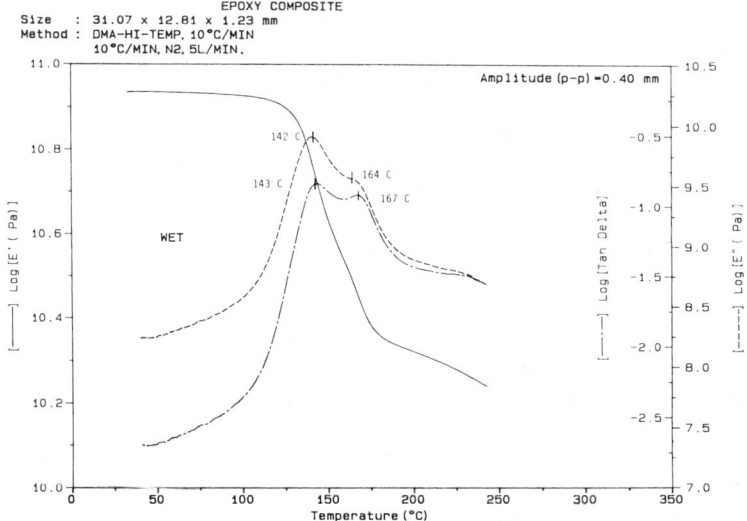

FIG. 4—*DMA flexural data wet specimen carbon/epoxy unidirectional composite of Fig. 3 showing double T_g artifact and lower T_g due to moisture; frequency 1 Hz (University of Dayton).*

in the same temperature range as the T_g of the elastomer which might range from slightly below room temperature to about -120°C for silicone rubber. Calibrants that may be used are: hexane (-95°C), cyclohexane (-83°C), octane (-56.8°C), decane (-29.7°C), dodecane (-9.6°C), water (0°C), and tetradecane (5.9°C). The calibrants should be as pure as possible, since the melting point is strongly dependent on percent purity. A calibrant is formed into a DMA sample by impregnating it into an inert substrate such as a rectangular strip of fiberglass cloth. The fiberglass/calibrant composite is then run as a normal DMA specimen, with the melting event providing a discontinuity in the DMA modulus signal at the transition.

Another similar procedure uses a small hollow tube of polymer filled with frozen calibrant. Here the tube material is chosen so that it is "inert" in the transition range of the calibrant and it is impervious to the calibrant. In the authors' laboratory a PEEK tube 6 to 8 cm in length, 4.25 mm diameter and 0.2 mm in wall thickness is used. The calibrant is pipetted into the tube, the ends are capped with an inert sealant and the "sample" is then cooled to freeze the calibrant. The calibration sample is then run as a normal DMA specimen with the sample thermocouple(s) placed as close as possible to the surface (0.1 to 0.2 mm) in the same place as for a normal DMA run. Several temperature calibration points are taken with various calibrants.

Clamping of samples for subambient operation should be done with the sample precooled below T_g after cooling the chamber. Our experience has been that a modest amount of frost buildup does not interfere with the measurement. Moisture condensation is a common problem in subambient operation of DMA instruments. Opening the instrument at subambient temperatures cannot be avoided for DMA experiments where the original length of the sample must be taken in the glassy state. In some DMA instruments the thermocouple shield further aggravates the problem.

Machine Compliance Issues

DMA instruments have a measurable stiffness associated with them. When a strain is imposed on the sample, both the sample and the instrument are deformed. The actual sample strain is typically determined by calibrating the machine's compliance (a constant, m/kg) with a very stiff material, such as steel. The deformation of the machine is then subtracted out of the total deformation yielding the deformation of the sample. However, if the stiffness of the sample and the test geometry combine to allow deformations in the sample that are close to or less than that of the machine, significant errors can result.

The most obvious error is an artificially low storage modulus and loss modulus. The storage modulus, for example, often can be suppressed to one half or more of its actual magnitude. It is noteworthy, however, that the data appear quite uniform and consistent. Unfortunately, not only will this give artificially low modulus data but it can have a dramatic effect on the position and shape of the loss modulus and loss tangent peaks. Figures 5 and 6 are data for a highly crosslinked acrylic polymer. The data in Fig. 5 were taken using a tensile fixture. The cross-section of the sample was considerably larger than machine compliance limits allowed. The sample polymer was tested again (Fig. 6) using the three-point bend configuration that ensured the test was performed within the machine's compliance limits. For the sample tested in tension, the loss modulus and loss factor peaks are relatively sharp and rounded compared to the broad transition peaks for the sample tested in bending. Additionally, the loss modulus peak is 30°C higher while the loss tangent peak temperature is 20°C lower than the data taken with the three-point bend fixture.

This can be explained by considering the effect that the moduli of the sample have on the experiment. For the sample tested in the tensile fixture, both E' and E'' are artificially

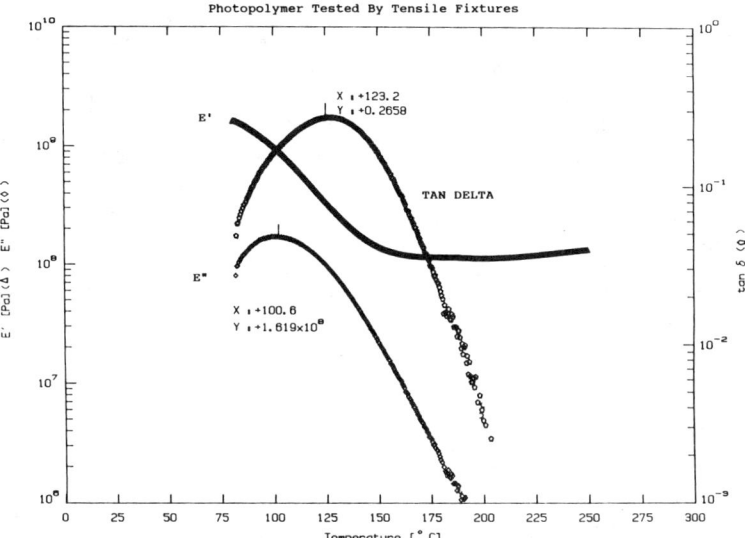

FIG. 5—*DMA tensile data for a highly crosslinked acrylic polymer showing abnormally sharp transition loss dispersion due to machine compliance limitation; frequency 1 Hz (University of Dayton).*

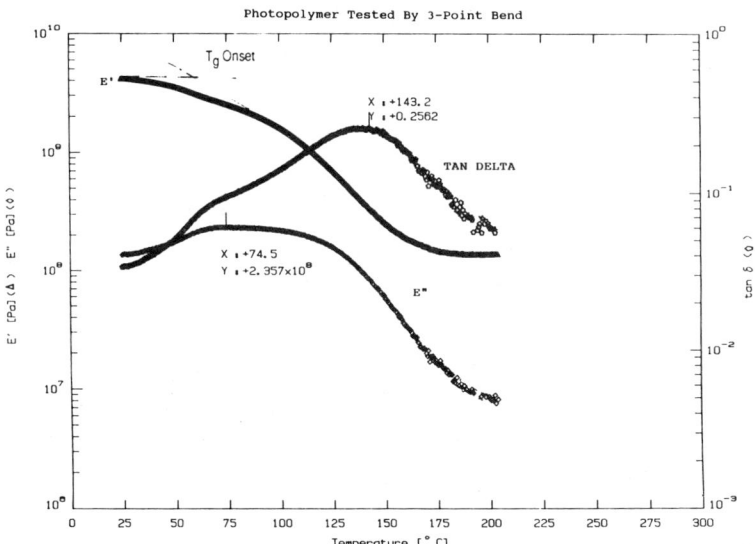

FIG. 6—*DMA flexural (3-point) data for the same crosslinked acrylic polymer of Fig. 3 showing accurate data taken within the machine compliance limits; frequency 1 Hz (University of Dayton).*

suppressed. Therefore, as the sample is being tested, no E' is measured until the actual E' of the sample has dropped to a value equal to the erroneous suppressed, measured value. By this time the actual E' of the sample in the transition region is dropping rapidly. At this point E'' (which was abnormally low) rises rapidly to reach its actual value. The result (Fig. 5) is a narrow, rounded E'' peak shifted to a higher temperature. Since the tanδ is the ratio E''/E', it too becomes distorted and, in this case, is actually shifted to a lower temperature than the true tanδ curve in Fig. 6.

Test Frequency

The glass transition is a kinetic transition, strongly influenced by the rate or frequency of testing. Molecular relaxation involving significant cooperative segmental motion occurs at T_g. The rate of segmental motion depends on temperature so that as test frequency increases the relaxations associated with T_g can occur only at a higher temperature. Thus T_g increases with frequency as illustrated in Fig. 7 [4].

Characteristic features of the transition illustrated by these data with increasing frequency are a general decrease in the intensity of tanδ, a broadening of the peak, and a decrease of the slope of the storage modulus curve in the transition region. These characteristics reflect a broadening of the relaxation spectrum in the T_g or α relaxation vicinity, coupled in many cases with the merging of relaxations associated with the β ($T<T_g$) transition. The temperature position of the E'' peak at various frequencies for the α and β transitions in poly(methyl acrylate) is shown in Fig. 8 [5]. Note that the α transition curve is nonlinear and does not follow the Arrhenius equation. Thus the variation in T_g increases with frequency. For practical purposes, however, in the limited frequency range covered by the commercial DMA instruments (0.01 through 200 Hz) the frequency variation of T_g may be assumed to follow an Arrhenius dependency.

Because of the frequency dependence of T_g the convention adopted for assignment of the

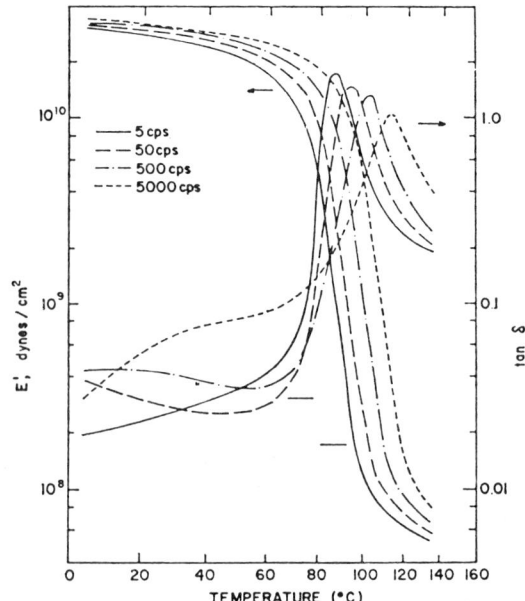

FIG. 7—*Variation in dynamic properties of PVC with frequency showing shift in T_g* [4].

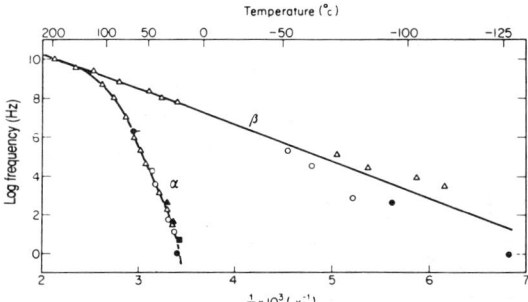

FIG. 8—*Diagram showing the frequency-temperature location for loss maxima for the α and β relaxations in poly(methylacrylate); data are composite curves derived from both DMA and dielectric measurements [5].*

glass transition temperature is an important consideration. Traditionally a frequency of 1 Hz has customarily been used as a standard value. Historically the torsion pendulum was the most widely used DMA technique in the early days of viscoelastic property measurements. The torsion pendulum is a free vibration technique with a natural frequency of approximately 1 Hz. The 1 Hz value of T_g also is reasonably close (within 10°C) to the T_g values determined by other widely used methods such as DSC and dilatometry. The relation between DMA and DSC T_g values will be discussed further later in this paper. Because of the ambiguity inherent in the kinetic nature of T_g it is most important that the test frequency be reported along with any T_g value determined by a DMA technique.

Material Characteristics

Crystallinity

In semicrystalline polymers the fact that the amorphous and crystalline phases are intimately related to each other has a significant effect on relaxation at the glass transition, resulting in broadening of the transition region. In polymers that develop a high degree of crystallinity the glass transition is suppressed so that the T_g loss dispersion appears as a relatively minor event and in some cases is difficult to detect at all. The subject of relaxation processes in crystalline polymers is treated in an excellent review paper by Boyd [6].

An example of the broadening of T_g in a moderately crystalline polymer (PET) is illustrated by the data shown in Fig. 9 [7]. The low crystallinity sample exhibits the glass transition relaxation at 80°C, while the higher crystallinity samples show the relaxation around 105°C. As the relaxation broadens, the loss peak shifts to a higher temperature and assignment of the T_g value becomes more difficult. The broadening of the transition is associated with the presence of a rigid amorphous phase at the interface between the crystalline phase and the mobile amorphous phase. This is discussed in more detail elsewhere in this volume by Wunderlich.

Figures 10 [8] and 11 [9] illustrate the suppression of relaxation at T_g in highly crystalline polymers. Although the identification of T_g in polyethylene is still a controversial subject, substantive arguments point to the β(-30°C) relaxation as the glass transition [8,10,11]. In Fig. 11 we note that for a series of polyethylenes with different amounts of short-chain branching, as crystallinity increases, the intensity of the β-loss peak diminishes until it is

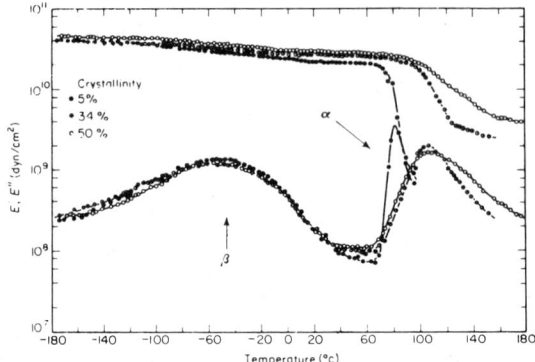

FIG. 9—E' and E" as a function of temperature at 138 Hz for PET samples of differing degrees of crystallinity [7].

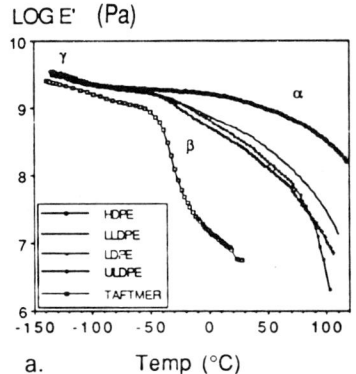

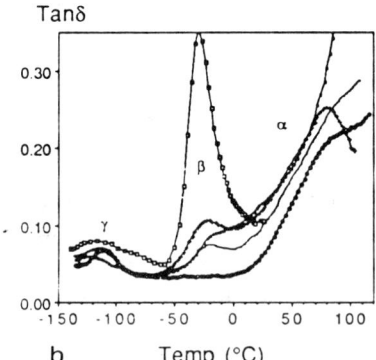

FIG. 10—(a) *Storage modulus* E' (b) *loss tangent versus temperature for various polyethylenes of different densities and frequency 3* Hz [8].

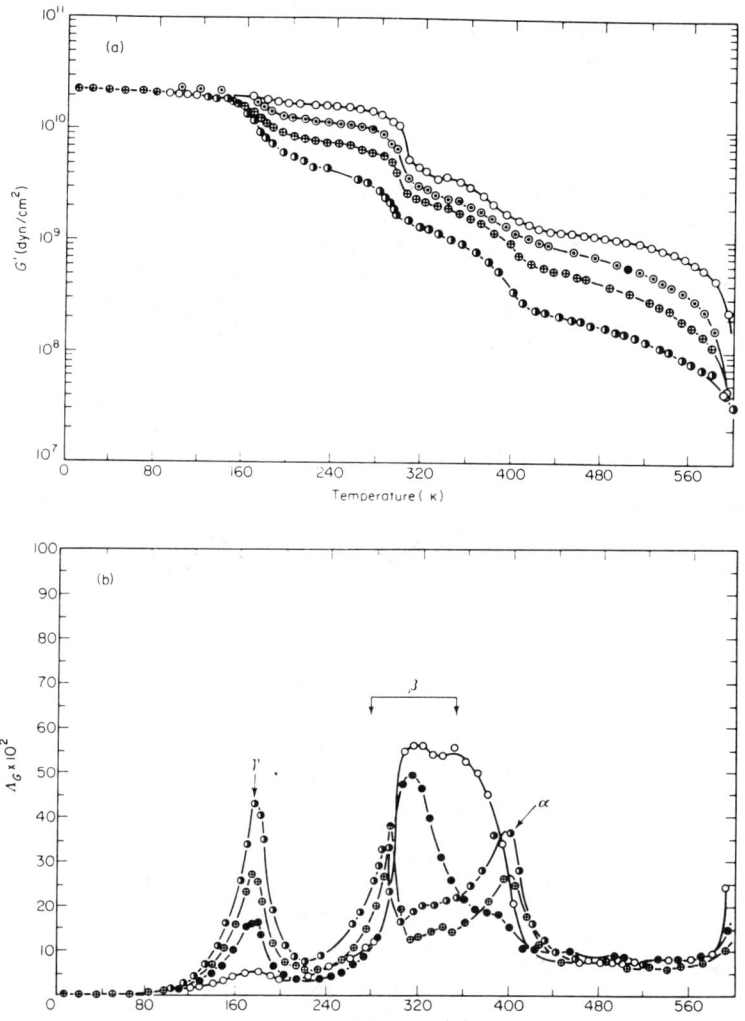

FIG. 11—*Temperature dependence of the shear storage modulus and damping decrement at ~ 1 Hz for PTFE samples of 90% (○), 76% (●); 64% (⊕); and 48% (◐) crystallinity. Log decrement = π tanδ. [9].*

hardly detected. Similarly in PTFE (Fig. 11) the glass transition, noted as β at around 140°C, disappears as the degree of crystallinity increases.

The difficulty in assigning T_g for highly crystalline polymers is that the T_g relaxation is a relatively minor process and there are often several $T < T_m$ relaxation processes which may be influenced by crystallinity. Thus the assignment of the particular type of molecular motion associated with the relaxation and relating it solely to the amorphous fraction is not a straightforward matter.

From a practical view, the physical and mechanical properties of highly crystalline polymers are dominated by crystallinity and their upper use temperature limit is T_m, not T_g.

Thus T_g has less significance in this class of materials than it does in amorphous polymers where T_g often represents the maximum use temperature.

Crosslinking

Highly crosslinked polymers are an important class of materials used as adhesives, coatings, and matrices for composites, to name just a few applications. Many of this class are thermosets and all are amorphous. The glass transition is a most significant parameter for this group and the difficulty in specifying T_g arises from the effect of crosslinking on relaxation in the transition region. The general effect on the glass transition by increasing crosslink density is shown in Fig. 12. The transition loss dispersion decreases in intensity, broadens and shifts to higher temperatures. Also, the transition slope of the storage modulus decreases. The broadening of the relaxation spectrum is further enhanced by network heterogeneity introduced in systems formed from mixtures of reactive monomers having different functionalities. The data of Fig. 6 are representative of such a system.

The highly crosslinked acrylic represented in Fig. 6 has a very broad transition region that well illustrates the problem involved in assigning T_g in such cases. The question to consider is: what criterion do we use for specifying T_g? The loss dispersion in Fig. 6 covers a span of approximately 100°C and the E'' peak is 70°C lower than the peak tanδ. Furthermore, the E'' peak may occur in the transition region above the temperature where E' begins to decrease gradually from the glassy state value. Our recommendation in such cases is based on practical considerations. Since network polymers are frequently used in structural applications as adhesives and matrix materials, the point where E' begins to drop from its glassy value is the most appropriate T_g assignment based primarily on the criterion of retention of structural integrity. This is most easily specified by the onset point, defined by intersection of the tangents to the storage modulus glassy state curve and transition region curve as shown schematically in Figs. 1 and 6.

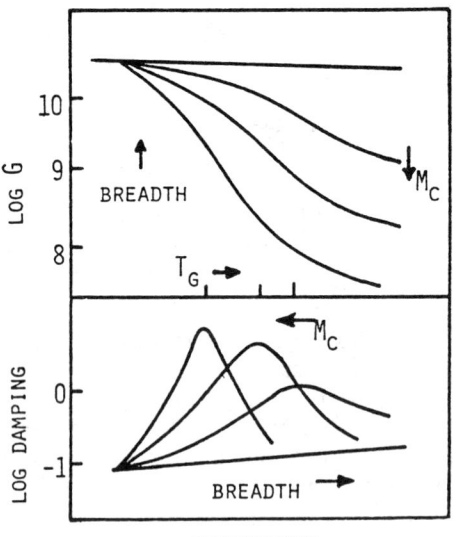

FIG. 12—*Schematic diagram showing effects of crosslinking on modulus-temperature and damping (loss tangent) behavior at T_g.*

Moisture

T_g values of polar polymers are seriously affected by the presence of even small amounts of absorbed moisture. The moisture acts as a diluent or plasticizer and lowers the glass transition. This is shown rather dramatically in Fig. 13 [6] for poly(vinyl alcohol), a crystalline thermoplastic and in Fig. 4 for an epoxy thermoset matrix resin. Comparing the data of Fig. 4 to Fig. 3 for the dry sample of the same polymer we see a reduction of 55°C in T_g. For epoxies it is well known that T_g may decrease by as much as 12 to 15°C for each percent moisture absorbed [12]. Normally equilibrium moisture sorption amounts for epoxies are in the 5 to 7% range. This is a serious problem for structural applications of epoxies in adhesives and composites. Thus it has resulted in substantial research directed toward synthesis of more moisture resistant modifications.

Another example of the effect of moisture on T_g is shown in Fig. 14 for Torlon 4203 an aromatic poly(amideimide) [13]. This is a thermostable, high T_g polymer. However, the data show that sorption of a few percent moisture lowers the apparent T_g by 100°C. The authors also show that incidental exposure to moisture by exposing a dry sample to ambient humidity will cause a reduction in T_g of around 25°C.

Polymers Containing Particulate Fillers

Particulate fillers can cause noteworthy changes in the linear viscoelastic properties of amorphous polymers in the vicinity of T_g. It is well established that fillers increase the storage modulus E'; usually the increase in E' is greatest in the transition and plateau regions of the viscoelastic spectrum [14,15]. At the same time the damping peak, tanδ, broadens and the peak position shifts to a higher temperature. All of these features are shown in Fig. 15. The amount of broadening increases with the volume fraction of filler or decrease in particle size for a given volume fraction, when the particles are spherical.

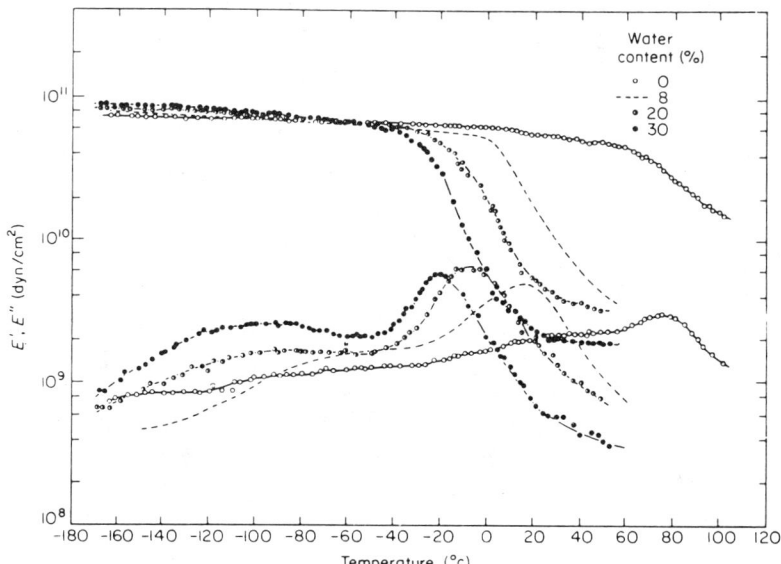

FIG. 13—*E' and E'' versus temperature at 138 Hz for PVOH with water contents indicated $T_g \sim 70°C$ for dry sample* [7].

102 ASSIGNMENT OF THE GLASS TRANSITION

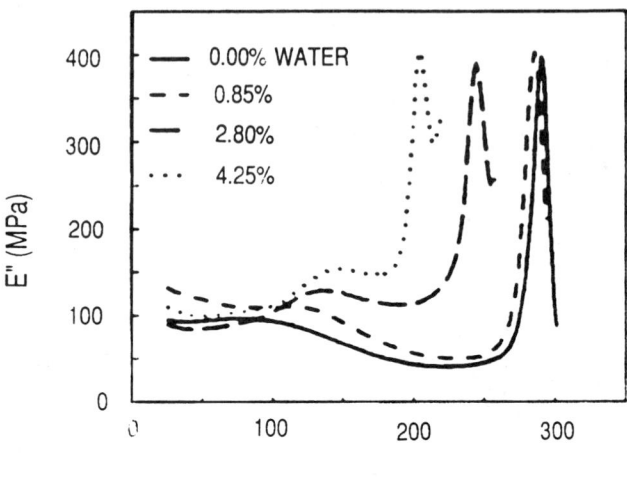

FIG. 14—*E″ versus temperature for samples of Torlon poly(amideimide) of various moisture contents* [13].

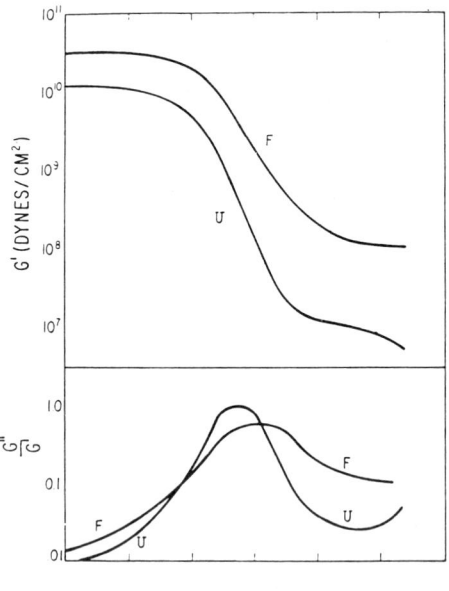

FIG. 15—*Typical dynamic mechanical behavior of a filled polymer (F) to the same unfilled polymer (U)* [14].

Depending on the type of filler and the interfacial bonding between the filler and the matrix, however, other effects have been noted [15,16].

Platelet fillers such as mica and graphite are more efficient than spherical particles in broadening the damping peak with no reduction in damping peak intensity at filler volume fractions as low as 5 to 10%. For platelet shaped particles the effects on viscoelastic properties are more pronounced as filler particle size increases. Chartoff [15,16] cites evidence that two new damping mechanisms affecting the relaxation spectrum are introduced in the case of platelet particles. These are particle-matrix slippage and polymer shear between filler particles. The latter is operative particularly when the platelets are oriented and parallel to each other. Chartoff notes that normally there is poor adhesion between graphite and the matrix polymer. When the graphite is treated with a silane coupling agent to create strong adhesion, the T_g damping peak narrows and the peak temperature decreases as shown in Fig. 16.

Thermal, Mechanical, and Orientation Effects

Thermal History—Harrell [17] studied thermal effects on the α and β transitions in PVC (Fig. 17) and noted that the width of the α relaxation is increased for samples quickly quenched from the melt to well below T_g compared with samples slowly cooled from T_g to ambient. While slowly cooling, the material is able to undergo considerable physical aging. The tanδ T_g peak position does not appear to be altered even though the relaxation associated with the transition begins at a greatly reduced temperature. Wetton [18] made similar observations for polystyrene and also found that the intensity of the tanδ peak for the aged sample is reduced compared to the quenched material. Harrell also showed that the β relaxation intensity of quenched PVC is reduced compared to an aged sample, a result of some of the β relaxation modes shifting upward into the α region because of the greater local free volume existing within the quenched material.

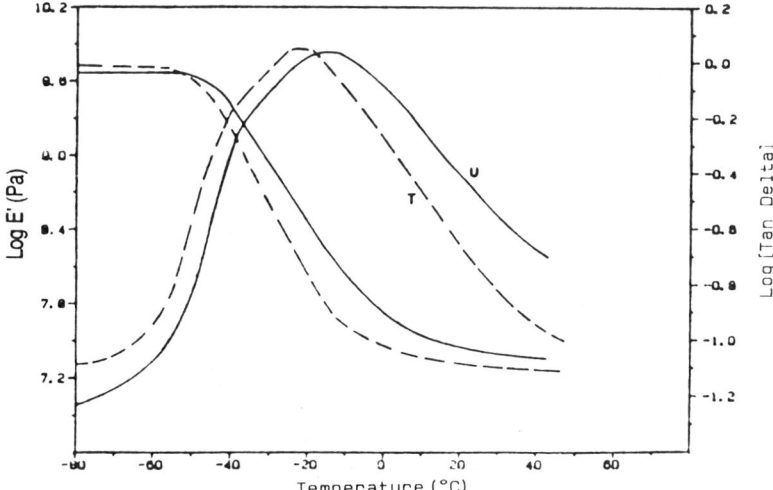

FIG. 16—*DMA data comparing E' and* tanδ *of butyl 1066 filled with graphite at 30% volume of both silane treated (T) and untreated (U) filler (data taken at fixed frequency of 1 Hz)* [16].

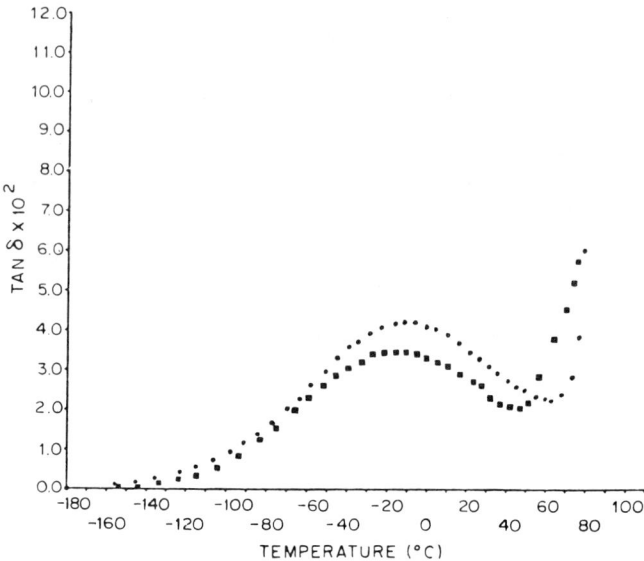

FIG. 17—*Variations in loss tangent between* (●) *slowly cooled and* (■) *quenched PVC at* 100 Hz [17].

Mechanical Fatigue—Amorphous, unoriented (bisphenol A) polycarbonate was cyclically fatigued in tension at several stress levels below the yield stress and then studied by dynamic mechanical testing [19]. The data (Fig. 18) reveal shifts in both the α (145°C) and β (-78°C) relaxations toward lower temperatures as the number of fatigue cycles increased. The density of the material also decreased with consecutive fatigue cycles. It was noted that as fatigue progressed, well dispersed microvoids were formed leading to modest decreases in density (0.08%). This suggests an increase in local free volume which could give rise to increased molecular mobility, thus lowering T_g.

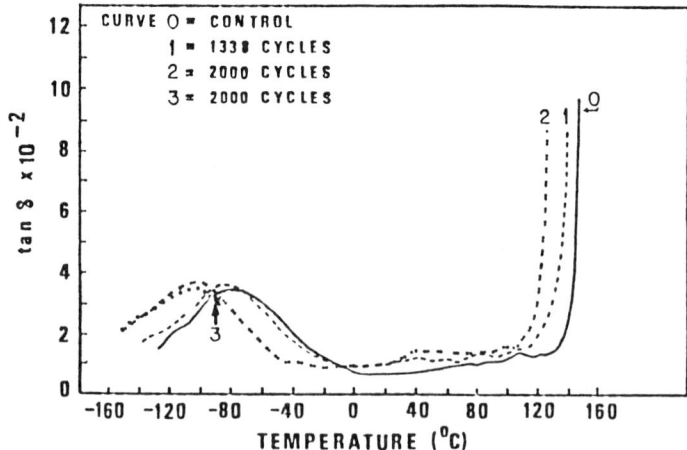

FIG. 18—*Tanδ versus temperature plot for unfatigued and fatigued PC. Frequency of test* = 35 Hz; *curve* O: *control sample* [19].

Orientation—Orientation affects the measurement of T_g by DMA methods. An example for a thermoplastic is the data in Fig. 19 for unoriented and biaxially oriented samples of the same unplasticized PVC. The peak loss modulus temperature is shifted 11°C higher for the oriented sample. Also the transition in the oriented specimen covers a broader temperature range with a shoulder appearing in the vicinity of the loss peak for the unoriented sample. The increase in T_g with orientation is related to the fact that the relaxation times of the longest range cooperative molecular motions contributing to the transition are increased due to the constraints imposed on them by orientation [20].

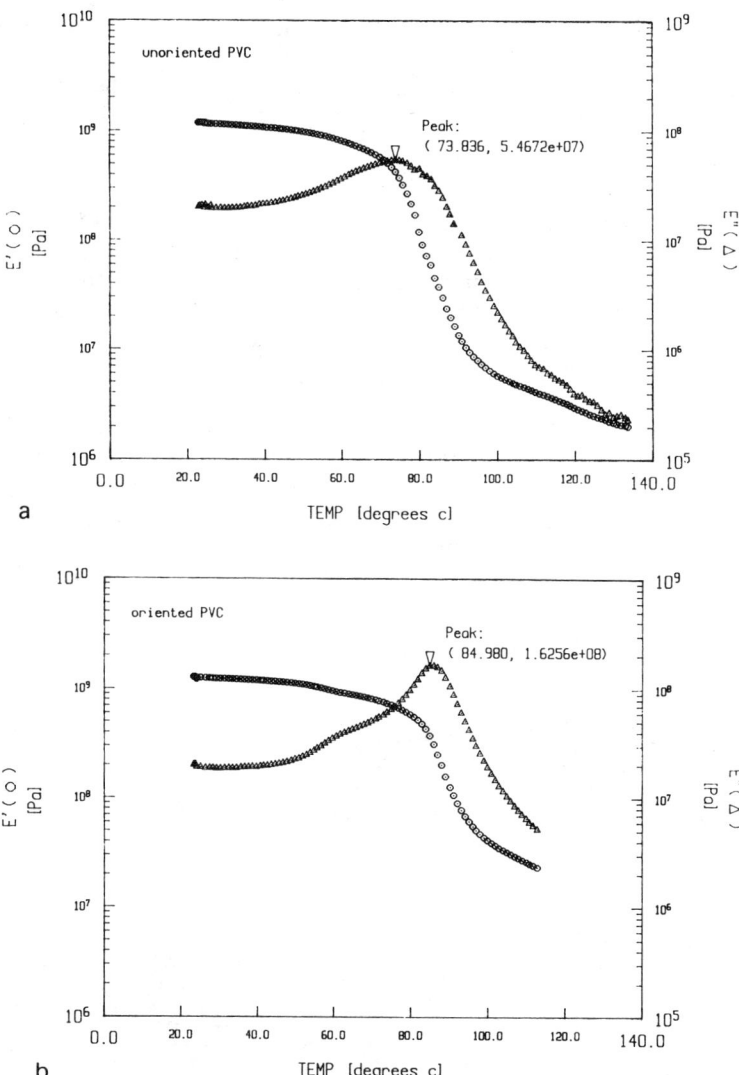

FIG. 19—*DMA data for a commercial rigid PVC sample* (a) *unoriented and* (b) *biaxially oriented at a draw ratio of 3:1; the glass transition increases in temperature with orientation (University of Dayton).*

The effects of orientation on mechanical relaxation at the glass transition were noted many years ago. For example, we cite various DMA studies on PET and polyamides [21].

In composite systems involving continuous fiber reinforcements the T_g relaxation in both thermoset and thermoplastic composites is affected by fiber orientation. For example, storage modulus data for a unidirectional glass fiber/epoxy laminate are shown in Fig. 20 [3]. As the fiber orientation for the measured DMA sample changes from 0 to 90°, T_g decreases and the transition intensity increases. The loss dispersion is expected to show a complementary sharpening and a shift to lower temperatures. The effect is similar to the case of particulate fillers discussed previously where the effect of the filler diminishes with decreasing volume fraction. As we progress from 0 to 90° in the composite of Fig. 20 we go from a fiber dominated to matrix dominated viscoelastic relaxation process. The fact that the 45° and 90° glassy state storage modulus values coincide in Fig. 20 is puzzling, since the 90° glassy modulus should be lower than the 45° value. Nevertheless, the data illustrate the correct trend for T_g. Similar results are presented by Jankowski et al., later in this volume.

Comparison of DMA and DSC T_g Measurements

DMA and DSC methods are the most popular among thermal analysts for identifying the glass transition. Both techniques have strengths and weaknesses. In this paper we have discussed the use of DMA methods to identify T_g. However, DSC methods are more widely used. Thus it is of interest to compare the results obtained by DMA and DSC. Such comparisons are considered in detail in two recent papers [22,23].

DSC uses very small samples (5 to 10 mg) and yields quantitative thermodynamic data for ΔC_p at T_g but it has relatively poor sensitivity. In order to improve resolution for measuring T_g one frequently increases sample size and heating rate (above 10°C/min) in order to sharpen the T_g baseline shift. Identifying T_g's using DSC is particularly a problem when T_g is a relatively minor event such as in crystalline polymers, crosslinked systems, and multicomponent systems where more than one T_g exists. In the later case, T_gs of minor components are very difficult to detect because ΔC_p is small.

DMA signal strength is approximately 1000 times greater for detecting T_g [22,24] and the

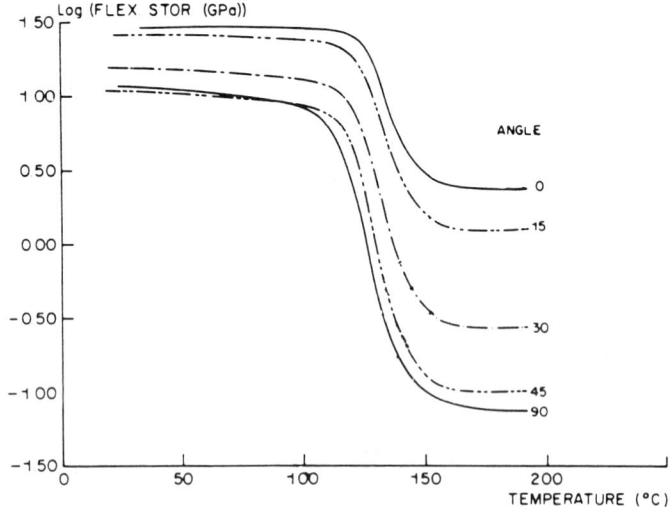

FIG. 20—*Dependence of composite storage modulus (glass/epoxy composite) on fiber orientation angle; frequency ~ 5 Hz [3].*

technique also provides useful mechanical property data. But it is limited to larger, usually solid samples. Because of the thermal gradient problems discussed previously, heating rates in DMA should be limited to no more than 2°C/min. However, the data obtained for T_g at a fixed frequency are independent of scanning rate in the absence of system thermal gradients.

In general, the DSC T_g value will be lower than the DMA value. The ΔT_g's for the two methods are influenced by the criterion used for selecting T_g. Most frequently DSC T_g values are taken as the endotherm onset (ASTM D 3418, Test Method for Transition Temperatures of Polymers by Thermal Analysis) and the DMA T_g is identified from the peak tanδ [22,23]. However, if the DMA T_g is taken as the peak loss modulus, ΔT_g is slightly less. The zero heating rate DSC T_g agrees with DMA loss peaks in the range of 10^{-4} to 10^{-3} Hz. The 1 Hz DMA data may correspond to a DSC heating rate in the 20 to 40°C range. Sircar [23] discusses methods for calculating DMA ΔT_g values from DSC data. This requires a knowledge of the Arrhenius activation energy for the temperature shift of the DMA loss peak with frequency.

References

[1] Nielsen, L. E., *Mechanical Properties of Polymers,* Reinhold, New York, 1962.
[2] Mora, E. and Macosko, C. W., in *Proceedings of the 20th NATAS Conference,* 1991, p. 506.
[3] Thomason, J. L., *Polymer Composites,* Vol. 11, No. 2, 1990, p. 105.
[4] Becker, G. W., *Kolloid Z.,* Vol. 140, 1955, p. 1.
[5] McCrum, N. G., Read, B. E., and Williams, G., *Anelastic and Dielectric Effects in Polymeric Solids,* Wiley, New York, 1967, p. 289.
[6] Boyd, R. H., *Polymer,* Vol. 26, 1985, p. 323.
[7] Takayanagi, M., *Proceedings of the Fourth International Congress on Rheology,* Part 1, Interscience, New York, 1965, p. 161; see also Illers, K. H. and Bruer, H. J., *Journal of Colloid Science,* Vol. 18, 1963, p. 1 for additional results and discussion on PET.
[8] Sha, H., Harrison, I. R., and Zhang, X., *Proceedings of the 19th NATAS Conference,* 1990, p. 179.
[9] McCrum, N. G., *Journal of Polymer Science,* Vol. 34, 1959, p. 355.
[10] David, G. T. and Eby, R. K., *Journal of Applied Physics,* Vol. 34, 1959, p. 355.
[11] Guar, U. and Wunderlich, G., *Macromolecules,* Vol. 13, 1980, p. 445.
[12] Browning, C., *Polymer Engineering and Science,* Vol. 18, 1978, p. 16; Ellis, et al., *Polymer,* Vol. 25, 1984, p. 664; Bauer, R. S., in *Proceedings of the 18th International SAMPE Symposium,* Vol. 18, 1986.
[13] Hiltz, J. and Keough, I. A., *Proceedings of the 20th NATAS Conference,* 1991, p. 573.
[14] Lee, B. L. and Nielsen, L. E., *Journal of Polymer Science,* Polymer Physics Edition, Vol. 15, 1977, p. 683.
[15] Chartoff, R. P., in *Polymer Composites,* B. Sedlacek, Ed., de Gruyter, Berlin, 1986, p. 89.
[16] Chartoff, R. P., *ANTEC Proceedings,* Vol. 34, Society of Plastics Engineers, 1988, p. 1143.
[17] Harrell, E. R. Jr. and Chartoff, R. P., *Journal of Macromolecular Science-Physics,* Vol. B14, 1977, p. 277.
[18] Wetton, R. E., in *Developments in Polymer Characterization,* J. V. Dawkins, Ed., Applied Science Publishers, Barking, U.K., 1986, ch. 5, p. 179.
[19] Sikka, S., *Polymer Bulletin,* Vol. 3, 1980, p. 61.
[20] Vallat, M. F. and Plazek, D. J., *Journal of Polymer Science,* Part B: Polymer Physics, Vol. 26, 1988, p. 545.
[21] Yoshino, M. and Takayanagi, M. J., *Journal of Society for Testing Materials* (Japan), Vol. 10, 1959, p. 330; see also Ref 5, p. 506.
[22] Connolly, M., Duncan, J., and Wetton, R., "Comparison of Glass Transition Temperatures in Polymer Blends via DMTA and DSC," presented at COMPALLOY'91 Conference, January 1991, New Orleans, LA (available from PL Thermal Sciences, Amherst, MA).
[23] Sircar, A. K. and Drake, M. L., in *Sound and Vibration Damping with Polymers,* ACS Symposium Series No. 424, L. Sperling and R. Corsaro, Eds., 1990, p. 132.
[24] Foreman, J., Sauerbrunn, S. R., and Marcozzi, C. L., in *Proceedings of the 22nd NATAS Conference,* 1993, p. 52.

Sue Ann Bidstrup[1] and David R. Day[2]

Assignment of the Glass Transition Temperature Using Dielectric Analysis: A Review

REFERENCE: Bidstrup, S. A. and Day, D. R., "**Assignment of the Glass Transition Temperature Using Dielectric Analysis: A Review,**" *Assignment of the Glass Transition, ASTM STP 1249*, R. J. Seyler, Ed., American Society for Testing and Materials, Philadelphia, 1994, pp. 108–119.

ABSTRACT: The use of dielectric analysis for the determination of the glass transition temperature for polymers is reviewed. Both a sharp increase in the permittivity and the dielectric loss peak have been correlated with the glass transition. Dielectric data for an epoxy resin and polyvinylchloride are presented and compared with data obtained by differential scanning calorimetry (DSC). The dielectric glass transition approaches the DSC glass transition as the frequency of the dielectric measurement is decreased. The effects of contact resistance and moisture on the dielectric measurement are also discussed.

KEYWORDS: glass transition temperature, dielectric properties, permittivity, loss factor, epoxies, polyvinylchloride

Dielectric thermal analysis is a valuable method to determine the glass transition region for polymers. This technique involves the placement of a polymer specimen between two electrodes and applying a sinusoidal voltage to one of the electrodes to establish an electric field in the specimen. In response to this field, the specimens become electrically polarized and can conduct a small charge from one electrode to another. Through measurement of the resulting current, the dielectric permittivity and loss factor for the polymer can be obtained. The dipole peak which occurs in the dielectric loss factor and a sharp transition in permittivity during a temperature scan both have been correlated with the glass transition temperature (T_g).

In this paper, the basic dielectric measurement techniques are reviewed. The use of dielectric analysis for the assignment of the glass transition temperature is discussed and compared to other thermal measurements. Issues concerning the effects of moisture interference, contact resistance, and frequency shifts are also covered.

Background

The classical dielectric measurement consists of two parallel plates, situated around the material under test [1–4]. One plate is excited with a sinusoidal electrical potential at frequencies ranging from 1 mHz to 1 MHz, and the resulting sinusoidal current passing from the second electrode to ground is monitored. The dielectric properties may then be easily calculated [1–4] based on the amplitude and phase of the measured current, and the area and

---

[1]School of Chemical Engineering, Georgia Institute of Technology, Atlanta, GA 30332-0100.
[2]Auburn International, Danvers, MA 01923.

separation of the plates. In cases where extreme accuracy is required, a guard ring may be used to eliminate current from leaking around the edges of the specimen and deposited metal electrodes may be used to ensure the best possible contact to the material under test [1,3]. To obtain high signal levels with the classical parallel plate technique, thin specimens with large areas are required, and the dielectric properties are an average of the bulk. In addition, low frequency measurements, which can reveal much about glass transition and the mechanical properties of the material, are often difficult to obtain using the parallel plate technique since the signal current is proportional to the frequency of test.

More modern dielectric measurements use two interdigitated electrodes on an insulating substrate [4–6]. This design permits a one-sided, very local measurement of dielectric properties, and is particularly well suited for thin films. Modern measurement circuitry also circumvents the traditional lower frequency limits, thus allowing dielectric analysis at frequencies as low as 0.001 Hz.

Permittivity

The dielectric response consists of two measured properties: the relative permittivity and loss factor. The relative permittivity (κ', often referred to as the dielectric constant) is related to the capacitive nature of a material or its energy storing ability. The relative permittivity is simply the absolute permittivity (ϵ') of the specimen divided by the permittivity of free space (ϵ_0 or 8.85×10^{-12} F/m)

$$\epsilon' = \frac{C_p d}{A} \tag{1}$$

$$\kappa' = \frac{\epsilon'}{\epsilon_0} \tag{2}$$

where

C_p = equivalent parallel capacitance,
d = separation between parallel plates, and
A = area of a single plate.

The dielectric permittivity of a medium measures the polarization of the medium per unit applied electric field. Since air is a very low-density material, it has a permittivity nearly equal to vacuum and, therefore, the relative permittivity is 1.0. Most organic resins have relative permittivities ranging from 2.0 to 10 although there are some that lie outside this range [1].

Loss Factor

The relative loss factor (κ'') is related to the conductive nature of the material. The relative loss factor is defined as the absolute loss factor (ϵ'') divided by the permittivity of free space

$$\epsilon'' = \frac{d}{R_p A \omega} \tag{3}$$

$$\kappa'' = \frac{\epsilon''}{\epsilon_0} \tag{4}$$

where

R_p = equivalent parallel resistance and
ω = angular frequency of applied voltage.

The dielectric loss factor arises from two sources: energy loss associated with the time-dependent polarization and bulk conduction. Since air is not conductive, it has an ideal loss factor of zero. Polymers, if well below the T_g, generally have relative loss factors less than 0.1. When heated up to the T_g and above, they can have relative loss factors as high as 1×10^9, due to ionic conduction.

Dipoles

In the presence of an electric field, electron clouds may be slightly shifted inducing a slight polarization that is aligned with the electric field. This acts to store energy and contributes to the capacitive nature of a material. The response times of the electronic shift are extremely fast so that at normal measurement frequencies the effect is always present. These induced dipoles are responsible for nonpolar or symmetrically polar polymers having permittivities of 2 or greater. Static dipoles consist of inherently polar moieties (not induced by the electric field) within the polymer such as carbonyl or alcohol groups. The static dipoles, if sufficiently mobile, may rotate in an electric field, thus also storing energy and contributing to the capacitive nature of the material. If a material, such as an epoxy resin with many polar groups, is heated so that an immobile dipole becomes mobile, then an increase in permittivity is observed as the dipole starts to oscillate in an alternating electric field. This effect is referred to as a dipole transition and has a characteristic time (τ_d) associated with it.

$$\kappa' = \left(\epsilon_u + \frac{\epsilon_r - \epsilon_u}{1 + (\omega \tau_d)}\right) \times E'_p \qquad (5)$$

$$\kappa'' = \left(\frac{\sigma}{\omega \epsilon_o} + \frac{(\epsilon_r - \epsilon_u)\omega \tau_d}{1 + (\omega \tau_d)^2}\right) \times E''_p \qquad (6)$$

where

κ' = relative permittivity,
κ'' = relative loss factor,
σ = bulk ionic conductivity,
ϵ_0 = permittivity of free space,
ω = $2\pi \times$ measurement frequency,
τ_d = dipole relaxation time,
ϵ_r = relaxed dielectric constant or low-frequency dielectric constant (relative permittivity due to induced plus static dipoles),
ϵ_u = unrelaxed dielectric constant or high-frequency dielectric constant (relative permittivity due to induced dipoles only),
E'_p = electrode polarization term for permittivity [7], and
E''_p = electrode polarization term for loss factor [7].

It should be noted that E'_p and E''_p are usually 1.0 except when ion conduction is very high [7].

Static dipoles also contribute to the loss factor as can be seen by Eq 6. This contribution arises from viscous drag as the dipole rotates through the surrounding medium. As a result of this drag, there can be a significant phase lag between maximum applied field and maximum dipole deflection. The energy lost due to this phase lag reaches a peak as ω approaches $1/\tau_d$. At higher frequencies the dipole hardly moves and so little energy is lost, while at lower frequencies the dipole can keep up with the changing field more easily, and, again, less energy is expended.

It should be noted that Eqs 5 and 6 are a form of the ideal Debye equation for a single relaxation time model [8]. In reality, polymers often contain more than one type of dipole and each of these has a distribution of relaxation times causing some differences between calculated and observed results; as the relaxation time widens and becomes skewed the loss peak might not occur at $\omega\tau_d = 1$. There has been much work at refining the model to account for distributed relaxation time systems [1,9–11]. However, the Debye model (Eqs 5 and 6) serves as a very useful template upon which dielectric data can be analyzed and interpreted.

Ionic Conduction

Ionic conduction is the result of current flow due to the motion of mobile ions within the material under test. It is often assumed that ionic conduction is insignificant due to low mobile ion concentrations. However, it has been demonstrated that concentrations well below 1 ppm are sufficient to cause significant ionic conduction levels [4,12]. From the Debye model, Eqs 5 and 6, it is seen that ionic conduction contributes only to the loss factor and does not affect the permittivity (as long as electrode polarization is negligible). It is very common to observe large loss factors in polymers when the temperature is above T_g due to the ionic conduction contribution to the loss factor.

Electrode Polarization

Electrode polarization is not due to the polymer under test alone, but is the result of the resin in combination with the electrodes used to introduce the electric field. Electrode polarization occurs when ionic conduction is extremely high, causing ions to collect at the polymer/electrode interface during one-half cycle of the oscillating electric field cycle. As ions build up at the electrode in a thin boundary layer and do not exchange their charges, a large capacitance is formed (not unlike an electrolytic capacitor). This has the effect of artificially decreasing the measured values of high loss factors and increasing the measured values of permittivity. In a Cole-Cole plot or arc diagram [1,9] electrode polarization appears much like a semicircular dipole transition. However, the measured permittivities attained through electrode polarization are usually high (>100), thus making electrode polarization easy to identify. Corrections can be made to the loss factor for electrode polarization influence if the ϵ_r value from Eq 5 is known [13]. A detailed derivation of the electrode polarization influence is given by Day et al. [7].

Dielectric Properties and the Glass Transition

Both the dipole relaxation time and ionic conductivity are related to the T_g of a polymer [1,3,4]. As a material is heated through the glass transition temperature, static dipoles gain mobility and start to oscillate in the electric field. This generally causes an increase in permittivity with a corresponding loss factor peak. At the lower temperature side of the glass transition, dipoles will only be able to respond to low frequency excitation since they do not have the higher mobility associated with higher temperatures. The frequency dispersion observed in dielectric measurements is very analogous to that observed for mechanical measurements. It has generally been observed that low-frequency dipole peaks (less than 1 Hz) correspond well with other thermal analysis T_g measurements. Materials containing static dipoles that gain mobility during the glass transition will usually exhibit permittivity changes and dipole loss peaks. However, materials with no static dipoles or non-T_g-active static dipoles will obviously exhibit little dipolar influence at the glass transition.

Charged ions also gain mobility as a material is heated through the glass transition and will start to contribute to conductive losses above the glass transition. The contribution of ion conductive loss to the measured loss factor can be enormous and can often overshadow small dipole loss contributions.

Many polymers also exhibit sub-T_g dipole relaxations which are related to beta or gamma transitions in the polymer. Just as in mechanical measurements, these are attributed to motions of polar side chain groups or any polar group than can move below the T_g. These transitions characteristically exhibit linear behavior when plotted on Arrhenius axes and activation energies can be extracted [3].

For this review paper, two polymer systems were selected to illustrate the relationship between the dielectric properties and the glass transition temperature: (1) a homologous series of diglycidyl ether of bisphenol A (DGEBA) epoxy resins and (2) polyvinyl chloride (PVC). The structural formula for the DGEBA resins is shown in Table 1. The EPON resins were supplied by Shell Chemical Company, and the DER 332 resin was obtained from Dow Chemical Company. Values of the weight average molecular weight and degree of chain extension for each of these resins were supplied by the vendors.

The temperature dependencies of the permittivity and the loss factor for one of the epoxy resins (EPON 825) in the vicinity of its glass transition ($T_g = -20°C$) at frequencies ranging from 1 to 10 000 Hz are shown in Figs. 1 and 2. At temperatures well below T_g, the

TABLE 1—*Chemical structure of a homologous series of DGEBA epoxy resins.*

RESIN	M_W	T_g (°C)[a]	n
DER 332	346	-20	0
EPON 825	354	-20	0
EPON 828	383	-17	0.2
EPON 1001F	1880	39	2
EPON 1002F	2540	54	3
EPON 1004F	3390	62	5
EPON 1007F	8430	78	12
EPON 1009F	14200	84	14

[a]Measured at 15°C/min using a Perkin Elmer DSC-4.

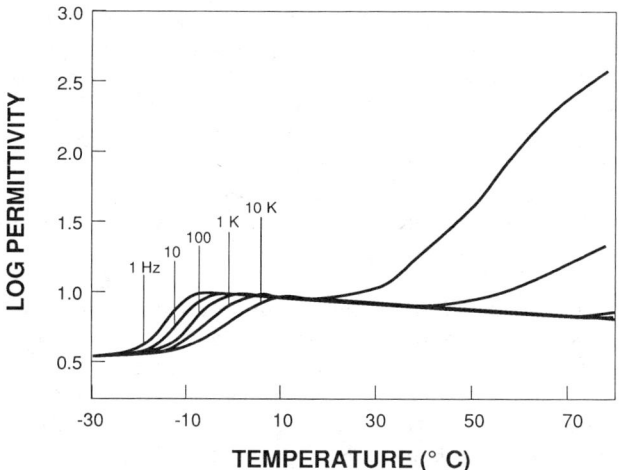

FIG. 1—*Permittivity during a thermal scan of EPON 825. The frequencies of measurement range from 1 Hz to 10 000 Hz. Note that* T_g *determined by DSC is* $-20°C$.

permittivity at all frequencies has a value of 4 (the unrelaxed permittivity), and the loss factor is below 0.1. As the temperature approaches T_g, the dipoles gain sufficient mobility to contribute to the permittivity, with evidence of this mobility increase occurring first at the lowest frequency. With a further increase in temperature, the permittivity for a given frequency levels off at the relaxed permittivity, which then decreases due to increasing temperature [14], and then abruptly increases again as a result of electrode polarization. At each frequency, a dipole peak is observed in the loss factor, which then rises continuously with temperature due to an increasing ionic conductivity. Both the frequency of maximum loss and the ionic conductivity increase by many orders of magnitude over a narrow temperature range, a characteristic of relaxation processes very close to the glass transition temperature.

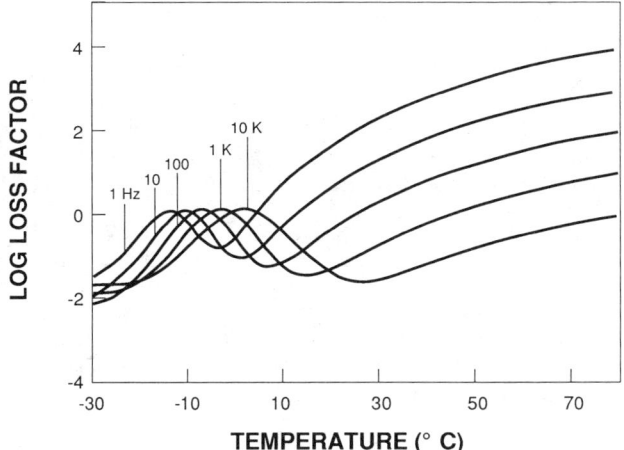

FIG. 2—*Loss factor during a thermal scan of EPON 825 over a frequency range of 1 Hz to 10 000 Hz.*

When log relaxation time or log conductivity is plotted against 1/temperature (degrees Kelvin) in an Arrhenius fashion, a curved result is usually obtained rather than a straight line with a single activation energy. This behavior of the dielectric properties is similar to that of viscosity and has been attributed to a WLF dependence. Figure 3 shows such a plot for EPON 825. Note that both the frequency of the dipole loss peak and log conductivity start falling off steeply as the glass transition temperature is approached. It is for this reason that low-frequency dipole relaxations or the appearance of ionic conductivity can often be used to track changes in T_g. At temperatures well above T_g, both dipole relaxation time and ionic resistivity often change in a similar fashion as viscosity since all three variables follow a WLF dependence [4,15–17].

For each of the epoxy resins in the homologous series shown in Table 1, the dielectric loss factor was measured during a temperature scan at frequencies ranging from 1 Hz to 10 000 Hz. The glass transition temperature for each of the resins in the homologous series was obtained using a differential scanning calorimeter (DSC); the T_g was defined as the midpoint in the inflection of the DSC heat curve scanned at 15°C/min. In Table 2, the temperature at which the maximum of the dipole peak in the dielectric loss occurred is compared with the glass transition temperature obtained using the DSC. Note that as the frequency approaches 1 Hz, the temperature of the dipole peak maximum approaches that of the glass transition temperature.

Figures 4 and 5 show the dielectric permittivity and loss factor at frequencies ranging from 1 to 10 000 Hz during the thermal ramping of PVC. The usual frequency dispersion is observed with higher dipole loss peaks occurring at higher temperatures. The frequency of the dielectric loss peaks versus inverse temperature are shown in Fig. 6. The lower frequency dipole loss peak is observed to approach the T_g of 85°C as measured by DSC (20°C/min).

This type of behavior is common for various types of polymers. Table 3 is a compilation of dielectric loss peak temperatures measured at 1 Hz [3] compared to commonly established T_g values [18]. It should be noted that the sources of T_g and dielectric loss data are different, making some of the comparisons uncertain. Additional comparisons between

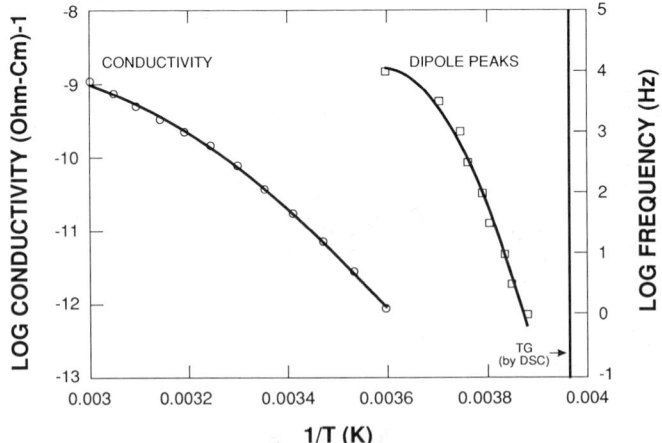

FIG. 3—*Arrhenius plot of dipole relaxation peak frequencies (Hz) and ionic conductivity $((\Omega\ cm)^{-1})$ as a function of temperature for EPON 825.*

TABLE 2—*Temperature (°C) of dipole peak maximum.*

Epoxy Resin	Hz					T_g(DSC)
	10^4	10^3	10^2	10^1	10^0	
DER 332	2	−1	−5	−10	−15	−20
EPON 825	−2	−6	−9	−12	−16	−20
EPON 828	4	2	−2	−6	−12	−17
EPON 1001F	66	59	55	51	47	39
EPON 1002F	76	69	64	60	58	54
EPON 1004F	84	78	73	68	64	62
EPON 1007F	104	96	91	87	82	78
EPON 1009F	113	105	99	96	90	84

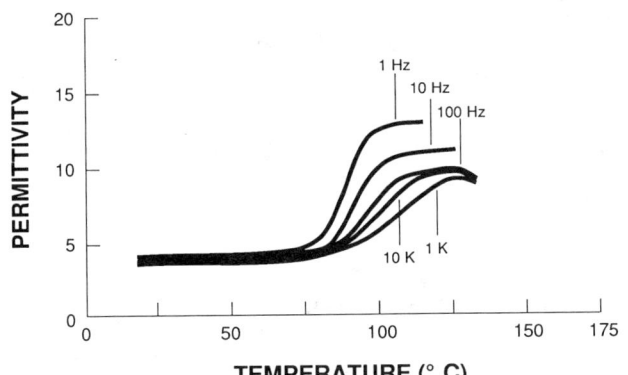

FIG. 4—*Permittivity during a thermal scan of PVC over a frequency range of 1 to 10 000 Hz.*

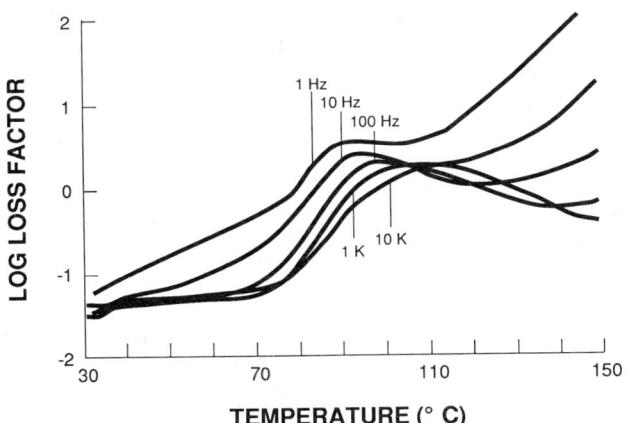

FIG. 5—*Loss factor during a thermal scan of PVC over a frequency range of 1 to 10 000 Hz.*

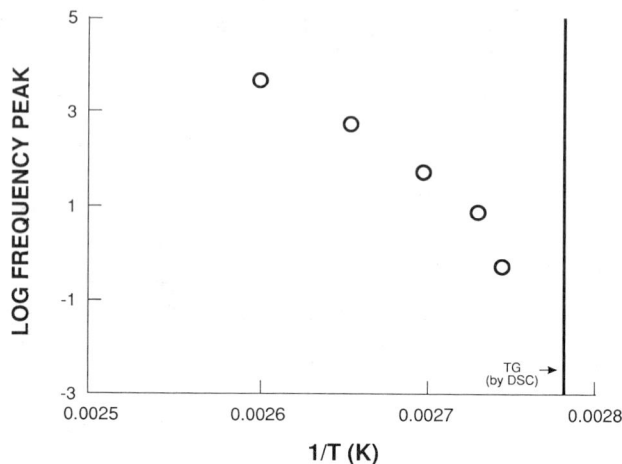

FIG. 6—*Arrhenius plot of dipole relaxation peak as a function of temperature for PVC.*

thermal scans of dielectric properties and mechanical properties can be found in McCrum, Read, and Williams [1] and Hedvig [3].

Experimental Effects

Just as in any other thermal analysis technique, care must be given to both the specimen and experimental conditions. One factor that is important for dielectric measurements is that the specimen has very good contact with the measurement electrodes. In cases of poor contact, the resulting gap between the specimen and the electrodes acts much like an insulating layer that causes severe electrode polarization. As mentioned earlier, this results in misleading values of permittivity and loss factor. The contact issue becomes especially important for thin specimens (less than 250 μm) since any small gap that exists makes up a significant portion of the volume between the electrodes. It is generally recommended that thin films that cannot be cast onto electrodes be first coated with metal electrodes (vapor or sputter deposition) to obtain the most accurate results.

Dimensional changes in the polymer specimen during heating and cooling can complicate dielectric analysis. As shown in Eqs 1 and 2, accurate information regarding the spacing

TABLE 3—*Comparison between the T_g Measured by DSC and the dipole peak temperature $(°C)^a$ 1 Hz.*

Polymer	T_g^a	Dipole Peak Temp (1 Hz)^b
Polystyrene	88	90
Polyvinylchloride	85	91
Polyethylene oxide	−53	−55
Polypropylene oxide	−63	−60
Polyoxymethylene	−68	−70
Polycarbonate	157	160
Polytetrafluoroethylene	135	130

^{a} Glass transition temperatures are from Ref *18*.
^{b} Temperatures of the dipole peaks are from Ref *3* with the exception of the PVC data.

between the electrodes is required in order to calculate the dielectric permittivity and loss factor. Therefore, the expansion or contraction of the polymer film between the parallel plate capacitors must be taken into account during thermal cycling. The alternative dielectric measurement geometry, based on a pair of planar, interdigitated electrodes, offers a distinct advantage in this regard. The metal electrodes are typically fabricated on an insulating substrate using photopatterning. The substrate is typically a ceramic or a silicon integrated circuit. Since the electrodes are rigid and are manufactured with microelectronic precision, the calibration of the interdigitated electrodes is stable with respect to temperature and pressure variations [4,5]. If the thickness of the polymer film covering the interdigitated electrodes is at least twice the spacing between the electrode fingers, then the measurement is independent of the dimensional changes that may occur in the film during thermal cycling [19].

Moisture within the specimen can also significantly influence the dielectric measurement. Since moisture can act as a plasticizer, it is possible for the T_g to shift independent of the analysis technique. In addition, moisture can contribute to the permittivity because of its large dipole strength and to the loss factor mainly due to its high ionic conductivity. However, if conditions are properly controlled, the dielectric measurement can actually be used to sense changes in moisture concentration and to determine moisture diffusion coefficients [20,21].

Acknowledgments

The authors would like to acknowledge Mr. Joseph Zellia at B. F. Goodrich for his assistance with dielectric measurements of the epoxy resins.

References

[1] McCrum, N. G., Read, B. E., and Williams, G., *Anelastic and Dielectric Effects in Polymeric Solids,* Wiley and Sons, New York, 1967.
[2] Reed, C. W., *Dielectric Properties of Polymers,* F. E. Kraus, Ed., Plenum Press, New York, 1972, p. 343.
[3] Hedvig P., *Dielectric Spectroscopy of Polymers,* McGraw-Hill, New York, 1977.
[4] Senturia, S. D. and Sheppard, N. S., Jr., "Dielectric Analysis of Thermoset Cure," *Advances in Polymer Science,* Vol. 80, No. 1, 1986.
[5] Sheppard, N. F., Day, D. R., Lee, H. L., and Senturia, S. D., "Microdielectrometry," *Sensors and Actuators,* Vol. 2, 1982, p. 263.
[6] Sheppard, N. F., Garverick, S. L., Day, D. R., and Senturia, S. D., "Microdielectrometry: A New Method for In-Situ Cure Monitoring," *Proceedings of the 26th SAMPE Symposium,* 1983, p. 65.
[7] Day, D. R., Lewis, T. J., Lee, H. L., and Senturia, S. D., "The Role of Boundary Layer Capacitance at Blocking Electrodes in the Interpretation of Dielectric Cure Data in Adhesives," *Journal of Adhesion,* Vol. 18, 1985, p. 73.
[8] Debye, P., *Polar Molecules,* Chemical Catalog Co., New York, 1929.
[9] Cole, K. S. and Cole, R. H., "Dispersion and Absorption in Dielectrics I. Alternating Current Characteristics," *Journal of Chemical Physics,* Vol. 9, 1941, p. 341.
[10] Davidson, D. W. and Cole, R. H., "Dielectric Relaxation in Glycerine," *Journal of Chemical Physics,* Vol. 18, 1950, p. 1417.
[11] Williams, G. and Watts, D. C., "Non-Symmetrical Dielectric Relaxation Behavior Arising from a Simple Empirical Decay Function," *Transactions of the Faraday Society,* Vol. 66, 1970, p. 80.
[12] Blythe, A. R., *Electrical Properties of Polymers,* Cambridge, Cambridge University Press, 1979.
[13] Day, D. R., in *Quantitative Nondestructive Evaluation,* Vol. 5B, Plenum Press, New York, 1986, p. 1037.
[14] Sheppard, N. F., Jr., "Dielectric Analysis of the Cure of Thermosetting Epoxy/Amine Systems," Ph.D. thesis, Department of Electrical Engineering, Massahusctts Institute of Technology, Cambridge, MA, 1986.
[15] Sheppard, N. F. and Senturia, S. D., "Dielectric Properties of Bisphenol-A Epoxy Resins," *Journal of Polymer Science: Polymer Physics Edition,* Vol. 27, 1989, p. 733.

[16] Gotro, J. T. and Yandrasits, M., *Polymer Engineering and Science,* Vol. 29, 1989, p. 278.
[17] Simpson, J. O. and Bidstrup, S. A., "Correlation Between Chain Segment and Ion Mobility in an Epoxy Resin System: A Free Volume Analysis," *Journal of Polymer Science: Polymer Physics Edition,* Vol. 31, 1993, pp. 609–618.
[18] Billmeyer, F. W., *Textbook of Polymer Science,* John Wiley & Sons, New York, 1971, p. 230.
[19] Lee, H. L., "Optimization of a Resin Cure Sensor," S. M. thesis, Department of Electrical Engineering, Massachusetts Institute of Technology, Cambridge, MA, 1982.
[20] Denton, D. D., Ph.D. thesis, "Moisture Transport in Polyimide Films in Integrated Circuits," Department of Electrical Engineering, Massachusetts Institute of Technology, Cambridge, MA, 1987.
[21] Day, D. R., "Microdielectrometry," *New Characterization Techniques for Thin Polymer Films,* Ho-Ming Tong and Luu T. Nguyen, Eds., John Wiley & Sons, 1990.

DISCUSSION

R. Bartnikas[3] *(written discussion)*—In the glass transition temperature region both the physical and electrical properties of the polymer undergo change. Although the glass transition temperature determined in terms of the dielectric measurements on polar materials does not in general entirely coincide with that determined using either mechanical or thermal measurement techniques, the differences are minor as concern practical applications of the result. Dielectric measurements for the determination of the glass transition point are most effective when the material under investigation contains a significant number of permanent molecular or ionic dipoles. In fact, from the dielectrics point of view, the glass transition region has always been of interest, because it has provided much information on the freedom of dipole orientation mechanism in the solid and liquid phases of electrical insulating materials. This becomes apparent if one considers the low frequency or static value of the permittivity, ϵ_s, as a function of absolute temperature, T, [1]

$$\frac{\epsilon_s - 1}{\epsilon_s + 2} \cong \frac{4\pi}{3} \sum_{j=1}^{\eta} N_j \left[\alpha_{ej} + \alpha_{aj} + \frac{\mu_j^2}{3kT} \right] \quad (1)$$

where N_j represents the number of atoms or molecules per cm^3 of the j th kind, α_{ej} and α_{aj} their respective temperature-independent electronic and atomic polarizabilities, and μ_j their permanent dipole moment. If measurements are carried out with increasing temperature at some convenient test frequency (e.g., 1, 10, or 100 kHz) sufficiently removed from the absorption region, then provided the permanent dipoles are initially rigidly fixed in the solid lattice (i.e., $\mu = 0$) to prevent their orientation, ϵ_s will exhibit an abrupt increase at T_g as soon as the dipoles are set free to rotate as the solid changes to a liquid. Often dipoles within the solid phase may not be entirely prevented from undergoing orientation, in which case the transition at T_g may not be marked by a large change in ϵ_s. Furthermore, if the dielectric material is nonpolar and free of conduction ions as well, the change in ϵ_s at T_g will be minimal, since it will reflect only density-variation effects in accordance with the Clausius-Mosotti relation.

$$\frac{\epsilon_s - 1}{\epsilon_s + 2} \cdot \frac{M}{d} = \frac{4\pi}{3} N_a(\alpha_e + \alpha_a) \quad (2)$$

[3]Institut de Recherche d'Hydro–Quebec, Varennes, Quebec, Canada J3X 1S1.

where M is the molecular weight of the material, d its density, and N_a is Avogadro's number; hence, unless the density of the nonpolar material changes abruptly, at T_g the value of ϵ_s will vary only gradually with temperature.

For materials that contain no permanent molecular dipoles but which are characterized by a finite conductivity either as a result of some intrinsic free charge carriers or contaminating ions, measurements ought to be performed at a fixed low frequency (0.01 or 0.1 Hz) in order to take advantage of space charge effects at the electrodes that are manifest by massive polarization which is accompanied by large changes in the dielectric constant and loss with temperature.

Although the approach employed by the authors to derive the T_g value in terms of dielectric measurements is somewhat unconventional, in that they perform measurements of permittivity and loss as a function of frequency with temperature as a parameter (a procedure that may lead to some interpretational difficulties), I wish nevertheless to strongly commend them for bringing to the attention the fact that dielectric measurements do constitute another important means for obtaining the T_g value for certain materials. It should be emphasized that for insulating materials that must operate under an electrical stress, the T_g value, obtained electrically, may have additional practical significance. For example, extruded polyethylene power cables are frequently operated at temperatures beyond the T_g value under emergency load conditions. In such circumstances an electrical T_g determination will not only provide the T_g value itself but will also yield valuable data on the behavior of the electrical properties within this critical regime over which the cable dielectric is adversely stressed.

Reference

[1] Bartnikas, R., "Dielectric Loss in Solids" in *Engineering Dielectrics, Vol. IIA, Electrical Properties of Solid Insulating Materials, ASTM STP 783,* R. Bartnikas and R. M. Eichhorn, Eds., American Society for Testing and Materials, Philadelphia, 1983, ch. 1.

Shu-Sing Chang[1]

Calorimetric Studies on Glasses and Glass Transition Phenomena

REFERENCE: Chang, S.-S., "Calorimetric Studies on Glasses and Glass Transition Phenomena," *Assignment of the Glass Transition, ASTM STP 1249,* R. J. Seyler, Ed., American Society for Testing and Materials, Philadelphia, 1994, pp. 120–136.

ABSTRACT: Any material possessing a wide relaxation time spectrum which is a strong function of temperature will exhibit a glass transition as the mean relaxation time crosses the time scale of observation; for example, the metastable supercooled liquid freezes into a glass state. After a review of calorimetric investigation and characterization of the glassy state, the fictive temperature concept is recommended for the assignment of glass transition temperature. The glass transition temperature (T_g) is defined as the intersecting temperature of extensive thermodynamic properties (such as enthalpy, entropy, and volume) extrapolated from temperatures above and below T_g toward the T_g. This assignment of T_g is dependent only on the conditions of glass formation and is independent of the rate of observation. Therefore, the T_g so defined may be considered as a property of the glass, free of observational artifacts. The width of the glass transition is often 30 to 50 K; therefore, a description of the width and the intensity of the transition would be helpful. By slow cooling or annealing, not only a relaxation peak shows up, but the transition also appears sharper as the width narrows to about 10 K. A continuous slow cooling procedure is preferred over the annealing procedure to locate weak glass transitions, as annealing may produce more than one relaxation peak for exceptionally wide glass transitions. For glass transitions even harder to locate, T_g may be bracketed by using the relaxation nature of the glass, by observing spontaneous exothermic adiabatic temperature drifts of quenched glasses and endothermic drifts of slow-cooled or annealed glasses.

KEYWORDS: calorimetry, differential scanning calorimetry (DSC), fictive temperature, glass transition temperature, residual entropy, vitreous state

The glass transformation occurs when a disordered state is forced out of its equilibrium and a certain configuration of the disorder becomes frozen in the time scale of observation. The high-temperature disordered state is in a thermodynamically definable equilibrium, although it may be in a metastable state. At sufficiently low temperatures below the transformation region, the glassy state is fixed in one of the myriad combinations of different orientations or configurations. As the temperature approaches the glass transformation region, relaxation times become shorter and approach the experimental time scale, and relaxations toward the equilibrium can be observed.

The most common glass transformation occurs when a liquid is cooled while the ordering process such as crystallization is being prevented. The high-temperature supercooled liquid state is in a thermodynamically defined equilibrium, although it is metastable. The glass transition phenomena of polar and nonpolar organic molecules, as well as covalent and ionic inorganic glass formers, were reviewed and classified recently [1]. Metals and alloys [2] may also be processed into glasses. However, due to their relative ease of crystallization

[1]Polymers Division, Materials Engineering and Science Laboratory, National Institute of Standards and Technology, Gaithersburg, MD 20899.

near and in the glass transition range, relatively fast heating rates in excess of 1 K/s are generally needed to observe the glass transition phenomena in metallic glasses. Calorimetry that requires the attainment of internal temperature equilibrium has an effective heating rate of less than 0.01 K/s. Thus the use of dynamic calorimetry such as differential scanning calorimetry (DSC) and flash calorimetry is indispensable in this area of research.

In a broader sense, the glass transition phenomena may be extended to the frozen disorder in plastic crystals [3] and liquid crystals, where long range order may be maintained, but the molecular orientation may be in disorder. In a rather comprehensive review of studies on various types of glasses from their laboratory, Ref 4 pointed out that more than one type of glass transition may be observed in the same material when the glasses are vitrified from different states. However, the glass transition temperatures of the liquid-glass and different plastic crystal-glass transformations appear to occur at remarkably similar temperatures, as in ethanol and cyclohexene. Multiple stages of glass transition involving the onset of different types of disorder in one material have also been observed. In aqueous clathrates, the disorder is generally associated with the host molecules. In a further extension of the glassy state concept, frozen disorder phenomena also occur in the paramagnetic-spin glass transition of magnetic materials [5], electron glass-transition in semiconductors [6] and the vortex-glass transition in superconductors [7]. As polymers seldom exist in a pure crystalline state or perfect single crystals, the existence of a glass transition is inevitable for the amorphous fraction, whether in thermoplastics or in thermosets.

The following discussion will present a brief review on the calorimetric investigations of glassy state and then concentrate on the assignment of the glass transition temperature, T_g, based on the fictive temperature concept. The fictive temperature has also been called the equilibrium temperature and the configurational temperature. The fictive temperature is defined as a point of temperature at which the extensive thermodynamic properties (e.g., enthalpy, entropy, and volume) are continuous but the temperature derivatives of these properties are different at temperatures above and below this point. Thus, the fictive temperature defines the glass transition temperature of a configurationally fixed glass existing at temperatures much below the transformation region, and therefore it depends only on the formation of the particular glass and its subsequent thermal history and is free from the artifacts of methods and rates of observation.

A Brief History of Calorimetric Investigation of Glasses

Possibly the earliest studies on the low-temperature heat capacities of molecular or organic glasses and their crystalline counterparts were reported by Nernst and Koref in Ref 8 around 1910, on the crystals and undercooled liquids of benzophenone and betol (β-naphthyl salicylate). Only three or four mean heat capacity values, each covering a range of 40 to 100 K, were reported for each of the states. Although the heat capacity behavior of glasses was not fully understood at that time, in retrospect, the salient features were already shown, as in Figs. 1 and 2, such that below the glass transition temperature specific heats of crystals and glasses are similar and that above the glass transition temperature the specific heat of the glass is higher than that of crystal. The Nernst's method of heat capacity measurements, by measuring the temperature rise after an energy input, has been used in most adiabatic calorimetry ever since. Occasionally, continuous heating has been used in adiabatic calorimetry; in those cases, the data should be analyzed in a way similar to that applicable to dynamic calorimetry.

The more systematic investigation of the glass transition behavior of organic glasses probably began with Gibson, Parks, and Latimer's study on ethyl and n-propyl alcohols reported in 1920 [9]. This paper led to the discussion on whether a glass possesses residual

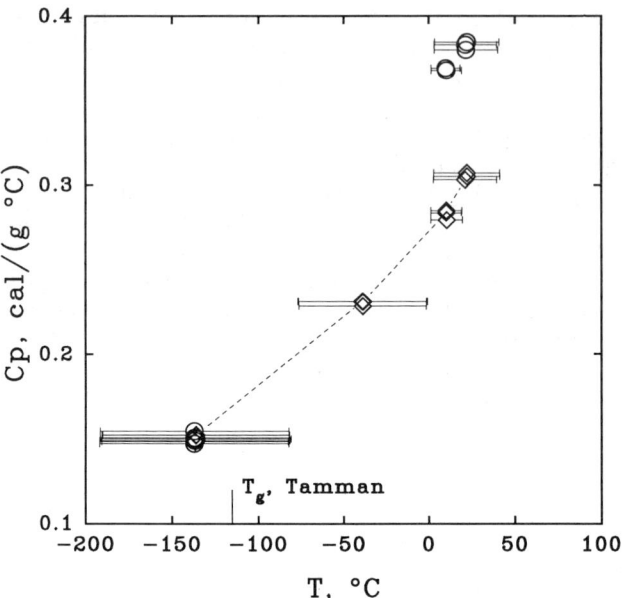

FIG. 1—*Heat capacity of benzophenone [8].* ◇—*Crystalline.* ○—*Amorphous. Error bar designates range of temperature for averaging.*

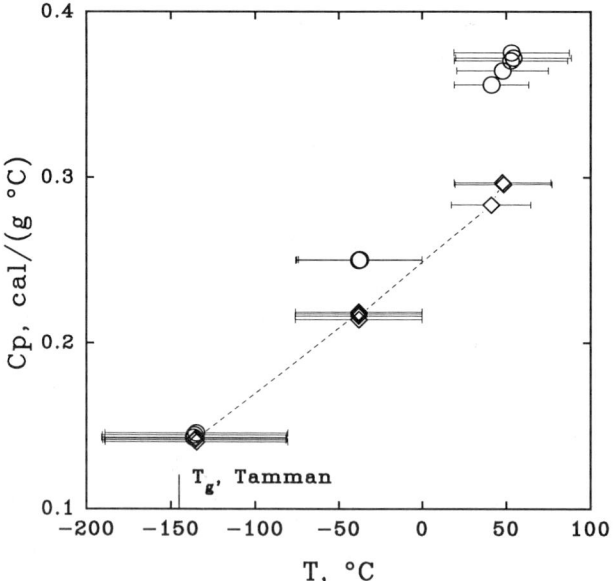

FIG. 2—*Heat capacity of betol [8].* ◇—*Crystalline.* ○—*Amorphous. Error bar designates range of temperature for averaging.*

entropy at 0 K. Although measurements from 70 or 100 K up yielded indications that residual entropies of glasses are probably nonzero, this question was settled in the late 1920s following the availability of more precise measurements and extending the temperature range of measurement from liquid air or nitrogen temperatures to liquid hydrogen or helium temperatures. The first low temperature study of organic glasses to 10 K was reported for glycerol in 1926 by Simon and Lange [10], and below the boiling point of helium by Ahlberg [11] in 1937. More refined low temperature data on glycerol were reported later by Craig et al. [12] and Leadbetter and Wycherley [13]. However, the argument of the validity of the residual entropy concept continued well into the late 1930s. Until the mid-1940s, only two examples of residual entropies were determined in reasonable precision, i.e., ethanol and glycerol. Three more compounds, 1-butene, 2-methylpentane, and sulfuric acid trihydride, were added to the list before the mid- 1960s.

The residual entropies of glasses formed from liquid are remarkably similar in all disordered systems at a value near $(R\ln2)/2$, or about 3 J/(K bead), if one adopts the concept of the configurational significant unit of "beads" [14] in a chain. The bead is loosely defined as an atom other than hydrogen, or a group of rigidly bounded atoms such as ethylene or benzene groups, in a chain. The assignment of the number of beads in Table 1 is not without ambiguity. Nevertheless, the residual entropies for configurational disorder per bead appear to fall within a relatively narrow band, much better than the originally proposed constancy for configurational heat capacity changes, ΔC_p, at T_g. The residual entropies are observed only for systems that can exist in both ordered and disordered solid states. The configurational entropy theory [15] has been quite successful in explaining the occurrence of the glass transition. Although the differences in the configurational entropies at T_g as compiled [16] may be approximated as the residual entropies, heat capacities should be measured to much below 20 K in order to yield the residual entropy in reasonable accuracy. Only observations extending to 20 K or below are collected in Table 1; this is to minimize the uncertainties due to the extrapolation of heat capacity to 0 K. The original assignment of one bead for the larger end groups such as ethylene or phenyl groups appeared to underrate the contributions of these groups; therefore, two beads were assigned for these groups in Table 1. A variation in the residual entropies due to the state of the glass, whether quenched or annealed, persists. The heat capacity of the less dense and disordered glass is in general only slightly higher than its crystalline counterpart at temperatures above 30 to 50 K. However, at lower temperatures, the heat capacity is often more than double that of the crystalline counterpart. Accordingly, the extrapolation of Debye temperature from results of heat capacity measurements at temperatures above 20 K may yield rather erroneous values due to the neglected large non-Debye contributions at lower temperatures.

The early studies in the 1920s of organic glasses were mainly on associated polar molecules, e.g,. mono- and polyalcohols. Beginning in 1930, nonpolar simple glasses, such as selenium [42] and heptanes [43], were studied.

Many plastic crystals can also have their high temperature form supercooled below their transition temperatures and into a glassy crystalline state. The first study of glassy crystals was probably Kelley's work [3] in 1929 on cyclohexanol. Only much later were other glassy crystals, such as cis-1,2-dimethylcyclohexane [44], being studied. Haida, Suga, and Seki [34] showed that the early studies on ethanol glasses [9,19,21] were really on glasses of supercooled plastic crystals, rather than on glasses of supercooled liquid. A large number of glassy crystals and glassy liquid crystals was studied by Suga, Seki, et al., e.g. Ref 4. The residual entropies on these supercooled crystals are generally rather small, only in the order of 2-3 J/(K mole), where one mole contains several beads.

The first polymeric material to have its heat capacity measured was probably rubber, reported as a single averaged value in 1889 [45]. Ruhemann and Simon in 1928 [46] were

TABLE 1—*Residual entropies of glasses.*

Substance	S_0 J/(K mol)	Beads	S_0 J/(K bead)	Reference
Ethanol	8.9	3	3.0	[9,19,21,34]
n-Propanol	11.3	4	2.8	[9,20,21,26]
Glycerol	19.2	6	3.2	[10,11,17,18]
1-Butene	12.8	4	3.2	[22,41]
2-Methylpentane	19.2	6	3.2	[23]*
Sulfuric acid trihydride	24.7	8	3.1	[24]
Diethyl Phthlate	22	9	2.4	[25]
iso-Butanol	9	5	1.8	[26]
iso-Pentane	14	5	2.8	[27]
Methanol	7.1	2	3.5	[28]
2-Butanol	9.6	5	1.9	[29]
o-Terphenyl	15	5	3.0	[30]
i-Propylbenzene	12	5	2.4	[31]
Selenium	3.8	1	3.8	[32]
3-Methyl-l-butanethiol	15.9	6	2.7	[33]
N-(β-trimethylsilylethyl) trimethylenimine	18.7	8	2.3	[35]
Triacetin	40	12	3.3	[36]
Dibutyl Phthlate	39	13	3.0	[37]
Di(2-ethylhexyl) Phthlate	59	21	2.8	[37]
Dimethyl Phthalate	16.8	7	2.4	[38]
Triethylbismuth	17.5	7	2.5	[39]
Dimethylzinc/dimethyl-tellurium complex	23	6	3.8	[40]

*Recalculated from original data

the first to show the glass transition in various types of rubbers. Besides the early studies on ketone resin [17] and rosin (colophony) [42] before 1930, only a few studies on the heat capacities of polymers appeared before 1950: cis-polyisoprene [47], polyisobutylene [48], polystyrene [49], polyethylene [50], and a few papers with rather poor data, enthalpy measurements, or a single average value over a wide temperature range. In each of the following decades, the number of polymers studied increased to about 10 and 20, respectively. After the development and popularization of commercialized differential scanning calorimeters in the 1960s, thermal studies on polymers and plastics, concentrating on the effects of compositions and various treatments, proliferated. Although the precision in the heat capacity measurement from dynamic calorimetry is in the order of 1 to 5% with sound practices and the temperature range is limited to 100 K and above, the speed of analytical experimentation and the small specimen size really give rise to its popularity. The large volume of heat capacity data on linear polymers has been reviewed extensively by a series of publications by Wunderlich et al. from 1981 [51] through 1991 [52]. Cross-linked polymers were much less studied.

The effect of thermal history on glass formation was well known in the silicate glass manufacturing industry. Tool et al. in a series of papers beginning with Refs 53 and 54 showed "abnormal" absorption of heat below the deformation temperature in silicate glasses using differential thermal analysis (DTA) techniques. This large absorption of heat was due to the heat capacity changes in the glass transition region. Effects due to quenching and annealing were also shown. The effect of thermal history on heat capacities of organic glasses was probably first studied by Parks and Huffman in 1927 [55] on propyl glycol glasses; the effect on the enthalpy of glycerol and glucose glasses was studied later by the reverse drop or mixture method [56,57].

However, the effect of thermal history of glass formation and treatment on heat capacities and enthalpies was generally not described in the literature until much later, perhaps coinciding with the popularity of adiabatic calorimeters with automatic shield temperature controllers, or later to commercially available differential scanning calorimeters. Small peaks, whether due to annealing or artifacts of measurement techniques, may be noticed for a few glasses in an early study by adiabatic calorimetry [55]. A pronounced peak was produced using a radiation calorimeter for continuous measurement in both heating and cooling [58], or by a slow continuous heating method [59]. Continuous heating method was employed later by many authors, e.g., a series of papers beginning with Ref 60 and a differential calorimeter [61]. These modes of observation are probably the predecessors of the modern DSC.

The relaxation of a glass toward the equilibrium supercooled liquid at temperatures below its fictive temperature may be observed by the loss in the enthalpy in reference to the supercooled liquid. The concept of fictive temperature was generally attributed to Tool [62], as a result of dilatometric relaxation studies. This concept has also been termed the equilibrium temperature or the configurational temperature [63]; however, the term "fictive temperature" seemed to have won wider acceptance. The rate of the enthalpy loss, which can be monitored by the spontaneous temperature drifts under adiabatic conditions, depends not only on real time and temperature, but also on the configuration of the glass, e.g., its fictive temperature and the deviation of the real temperature from the fictive temperature. Thus, for quenched glasses, i.e., the cooling rate is much greater than the heating rate used for observation, the relaxation process produces large spontaneous temperature increases as the glass transition region is approached. The annealing process was called the stabilization process. A well-annealed or slow-cooled glass does not show significant relaxation before the fictive temperature. However, the annealed glass can easily be overheated beyond its fictive temperature until its relaxation time approaches the experimental time, then the overheated glass relaxes toward the liquid configuration by absorbing heat. This process causes negative spontaneous temperature drifts and a relaxation peak by the continuous heating method. These drift and relaxation phenomena were generally ignored in most calorimetric investigations for simple molecular glasses. Early studies on glasses generally quoted a long equilibration time in the glass transformation range. For polymers, qualitative directions and magnitudes of spontaneous temperature drifts versus thermal treatments were probably first reported for rubbers in 1942 [64,65], and in a series of papers following Ref 66. Qualitative drifts were also noted in a few later papers, e.g., on polypropylene [67,68], polycarbonate [68], cis- and trans-polybutadiene [69], often observed together with crystallization phenomena. The time constants of the approach toward equilibrium were used in relaxation time studies in 1953 [70] and later in Ref 71. The graphical representations of the drift observations were reported in a series of papers following Ref 25. Drifts due to glass transition and crystallization may be differentiated. The adiabatic drift observations were summarized and used for locating the weak glass transition of polyethylene [72], and the weak glassy crystal transition in ice [73]. Typical maximum drifts are observed at around 10 mK/min or greater.

Characteristics of a Glass

A glass is expected to show all following common characteristics:

(a) exhibition of a glass transition as observed by a broad discontinuity in heat capacity (C_p) and in the coefficient of thermal expansion (α), or by a change in the slope of extensive thermodynamic properties, such as enthalpy (H), entropy (S), and volume (V) with temperature;

(b) relaxation toward equilibrium state in the glass transformation range;
(c) large excess (non-Debye contribution) in the low-temperature heat capacity (<10 K) and a slight excess configurational heat capacity near and in the glass transition region; and
(d) excess enthalpy and residual entropy at 0 K.

C_p or α discontinuities are most often used to locate the glass transition region. Detailed discussion is presented in the next section.

Relaxation toward the equilibrium state is often studied by dielectric and mechanical experiments. Long-term relaxations are studied in physical aging experiments. Thermal relaxations have often been observed in the past couple of decades with thermal analytical instruments. In the quenched glass, where the rate of observation is much slower than the rate of glass formation, the glass relaxes toward the equilibrium supercooled liquid state by spontaneously releasing heat just before the T_g. On the other hand, the annealed or slow-cooled glass can be superheated beyond the T_g; such glasses relax toward the supercooled liquid state by spontaneously absorbing heat. Aged or slow-cooled glass would thus enhance the observation of the glass transition by the presence of the relaxation peak and a narrower transition region above the T_g.

The last two characteristics, (c) and (d), are common to all disordered structures. The residual entropies have been discussed in the previous section. Heat capacities of amorphous materials at low temperature (e.g., less than 20 K) characteristically exhibit large deviations from an ordered hard solid which obeys the Debye continuum theory. Thus, the most dramatic example may be found in the behavior of various forms of carbon, including the hardest material, diamond. C_p of diamond [74] is rather small due to its high Debye temperature around 2200 K. C_p of glassy carbon is one to two orders of magnitude greater than that of diamond, and is slightly higher than that of graphite [75]. C_p of recently discovered C_{60} buckminsterfullerene [76] is almost another order magnitude greater than that of glassy carbon. Because of the large differences, specific heats of these materials are shown in a logarithmic plot (Fig. 3). The characteristic hump in the C_p/T^3 plot and the linear dependency below 1 to 2 K for C_{60}, as shown in Fig. 4, are also seen in molecular glasses and polymers, e.g., atactic polystyrene [77]. Other recent examples include SiO_2 aerogel [78], glass ceramics [79] and orientational glasses [80]. Typically in this temperature region, heat capacities of molecular solids and polymers in the amorphous form are often several times greater than their crystalline counterpart. These low temperature characteristics are generally attributed to low frequency vibrations from large molecules and "soft" modes of vibration from defective or disordered structures. New models to explain the occurrence of these universal phenomena in glasses have been proposed recently [81-83].

Furthermore, many glasses appear to have a region of temperature in which C_p is proportional to temperature. Tarasov [84] considered that these were due to the linear high polymer-like structure for inorganic glasses. This led to the treatment by many authors of low temperature heat capacities of polymers, by fitting the data through a combination of linear, quadratic, and cubic dependencies for linear, sheet-like, and network structures and their interactions. Because the heat capacity reported for cubic As_2O_3 has a much longer region of temperature proportionality than the monoclinic (ribbon-like structure) or the vitreous modification, Tarasov applied his theory in reverse, reaching a conclusion that the earlier heat capacity measurements on As_2O_3 [85] was probably reported for wrong forms of crystals. Later measurements [86] confirmed that the earlier measurements were made on the most common form of cubic As_2O_3 (globular dimer molecules), and that both monoclinic crystals (ribbon-like structure) and the vitreous modification have much shorter regions, if any, of linear temperature proportionality. Guttman [87] calculated that, as long

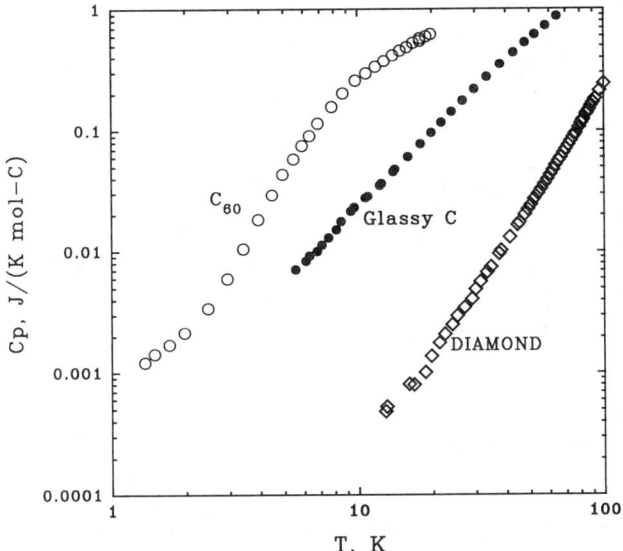

FIG. 3—*Heat capacity of various forms of carbon.* ◇—*Diamond* [74]. ○—*Glassy carbon* [75]. ●—C_{60} [76].

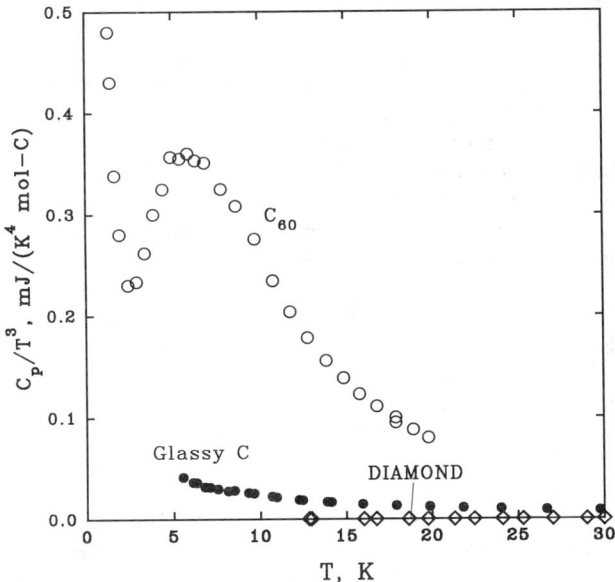

FIG. 4—*Low temperature heat capacity of various forms of carbon.* ◇—*Diamond* [74]. ○—*Glassy carbon* [75]. ●—C_{60} [76].

as there are evenly distributed vibrational frequencies to about 100 cm^{-1}, the vibrational heat capacity will be linear in temperature for a wide range.

For most glassy systems, C_p of the glass is somewhat greater than that of the crystalline counterparts. The relative difference is greatest below about 5 K, where C_p of glasses can be several times greater that of the crystal. The difference, $C_{p,\text{glass}} - C_{p,\text{crystal}}$, generally reaches a maximum around 20 to 40 K. Guttman [88,89] has attributed the heat capacity difference mainly to the density difference. However, the difference is generally minimized at temperatures above around 50 K to approximately 50 K below the T_g. In some systems, C_p of glass may even be smaller than that of the crystal over a short temperature interval. In the glass transition region, the heat capacity for the quenched glass is always slightly greater than that of the annealed glass by 1 to 2%, and the heat capacity of the glasses is always greater than that of the crystals by about 5 to 10%.

Assignment of the Glass Transition Temperature

Although the glass transformation occurs over a relatively wide temperature region, over at least 10 to 20 K in less sensitive measurements and over 50 K in more sensitive techniques, there appears a need to use a single temperature to represent the transition, mainly to compare techniques or to compare trends of changes. However, regardless of the methodology used, it is unlikely that this single temperature may be defined to better than 1 or 2 K.

The first assignment of a transition temperature, T_g, based on the idea that below which the enthalpy of glass becomes equal to that of the crystal, was probably by Tammann in 1930 [90] for a list of five substances. This definition is similar to the assignment of the hypothetical second order transition temperature, T_2, of Gibbs and DiMarzio [91], as the temperature where an idealized equilibrium glass losses all its configurational entropy. Simon in 1931 [92] used the entropy differences between the amorphous and crystalline glycerine to locate the break point, T_e. This also paved the way for the configurational entropy concept of glass transition.

The definition of the T_g varies with different techniques, because of the different frequency domains being applied by the technique. Even with the same technique, T_g may be assigned by different definitions to different regions of the phenomena by different investigators. As the glass transformation is kinetic in nature, the temperature at which an apparent transformation between a liquid-like behavior and a solid-like behavior occurs depends on the frequency of the observation. Not only is the mean relaxation time a function of temperature, but also the relaxation time spectrum and the activation energy spectrum [93]. Both spectra widen as the temperature is lowered. The so-called heating rate dependency of the glass transition temperature seen in DSC measurements is partly a result of the frequency dependence; such a discussion is outside the scope of the discussion here. The following discussion will be limited to the rather conventional observation of the glass transition by calorimetry at frequencies much lower than 1 Hz, or with the relaxation time in the order of around an hour, which corresponds to the definition that the glass transition occurs at a temperature where the viscosity of the supercooled liquid reaches 10^{12} to 10^{13} Pa·s (10^{13} to 10^{14} poise).

Calorimetry has been the most popular and simple method in locating the glass transition region. In both adiabatic and differential scanning calorimetry, the glass transition phenomena are observed as a jump or discontinuity in C_p. A list of the middle point of the transition region for nine compounds was given by Parks et al. [94]. In the absence of a relaxation peak, the choosing of the midpoint of the ΔC_p, or where the rate of C_p rise is at its maximum, does appear to be a consistent technique and corresponds to the temperature where a change in the slopes of the glass-liquid enthalpy occurs. Although the midpoint of the transition region is often taken as T_g, the glass transition temperature is better defined by the fictive temperature concept [62], similar to the temperature at which the volume of the glass and the liquid intersects. This concept has been widely accepted in dilatometry, as well as in aging or relaxation studies. This definition of T_g depends only on the history of the glass formed and is independent of the rate of observation. The heating curves at different heating rates for glasses of different cooling histories are shown in Fig. 5. It was often quoted that if the cooling rate and the heating rate are the same, the cooling curve and the heating curve should follow the same course. The cooling curve can never be reproduced by a heating curve, because the configuration of the glass is constantly changing in an irreversible fashion toward the lower energy states during both cooling and heating paths. Although the superheated annealed glass may be temporarily stable at temperatures higher than the fictive temperature for a short while, the annealed glass is not to be assigned with a higher T_g than the quenched glass. Conceptually, the configurational theory of glass formation suggested that the leveling off of the configurational entropy should be used to define the glass transition; the similar breaks in other extensive thermodynamic properties, such as enthalpy and volume, versus temperature are much easier to observe experimentally.

Rands et al. in 1943 [65] was probably first to plot the cumulative heat input in adiabatic calorimetry versus temperature, minus the enthalpy of the liquid, to locate the T_g. The difference in the fictive temperature due to thermal treatments was also demonstrated. Karasz et al. [95] later used both excess enthalpy and entropy plots to define T_g. However, the practice of summing enthalpy inputs was generally not followed, and the effect of thermal history was not studied, in most adiabatic calorimetric studies. The heat capacity values of glasses reported for the glass transition region are subject to the data reduction

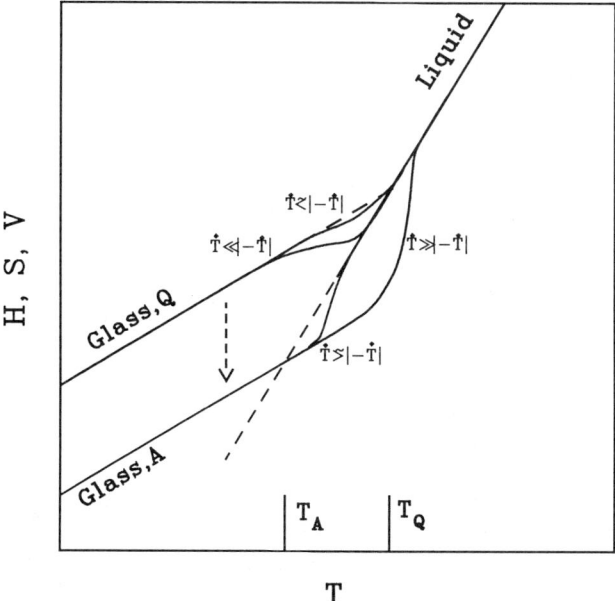

FIG. 5.—*Schematic heating curves of extensive thermodynamic properties in the glass transformation region. $\dot{T}$—Heating rates. $-\dot{T}$—Cooling rates. A—Annealed. Q—Quenched.*

procedures in Nernst's method of calorimetry. If each determination is made by waiting for a sufficiently long time to reach an "equilibrium" value with most of the effects of relaxation included, then the integration of the resultant C_p may yield a reasonable enthalpy diagram. On the other hand, if determinations were aimed at "instantaneous" lattice heat capacity with minimal relaxation effects included, then the enthalpy diagram should be obtained by the summation of all energy inputs. The "instantaneous" lattice heat capacity may include thermal effects that may be completed within a short period of time, e.g., 10 min. In a series of papers following Ref 25, the practice was to obtain both the "instantaneous" heat capacities and the enthalpy diagrams for glasses with various thermal histories. The diagrams were obtained by summation of each energy input minus the enthalpy absorbed by the container. By assigning the enthalpy of the supercooled liquid above the glass transition region to a singular definable equation of state, the difference in the enthalpies and in the glass transition temperatures of quenched and annealed glass is easily observed. Enthalpy changes during a measurement on poly(vinyl chloride) [96] are shown in Fig. 6 as an example.

Similarly, glass transition temperatures in DSC can also be defined by the fictive temperature concept. In the absence of a relaxation peak, the most common way of assigning the glass transition temperature is to take the temperature of at the midpoint of the C_p discontinuity. For materials that show relaxation peaks due to annealing or quenching exothermic

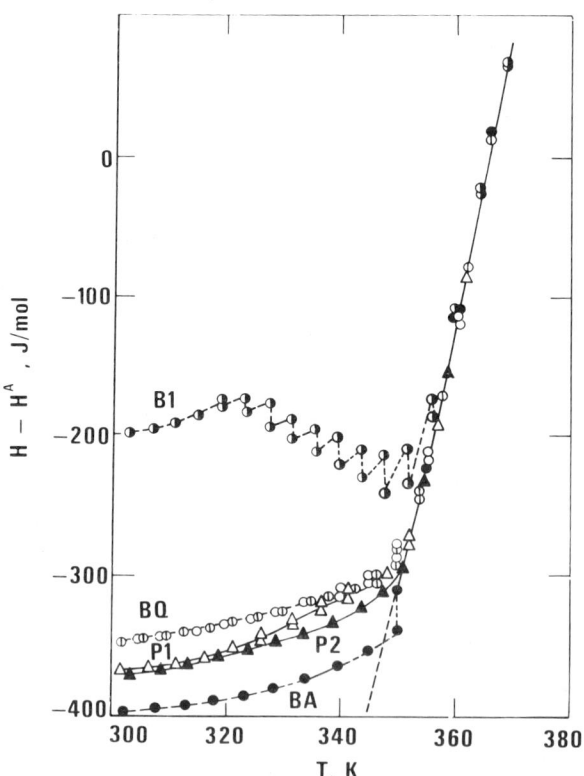

FIG. 6—*Enthalpy changes of poly(vinyl chloride) [96] in the glass transformation region.* H^A—*Extrapolated enthalpy of the annealed glass.* ⊙: *B1—Pelletized (first heating);* ○ ⊙ *BQ—Quenched;* ● *BA—Annealed.* Δ: *P1 and* ▲: *P2—First and second heating of powder.*

heat releases, the procedures [97–99] to reduce the observations to the fictive temperature should be followed. Although the actual integration of C_p to enthalpy might not have been performed in these procedures, at least the energy involved in the relaxation peak was compensated. Using the jump in the C_p curve alone will yield an erroneously high value for the relaxed material over that of a quenched material. In the dynamic methods, the defining of temperature is, however, much easier with DTA where the thermocouple may be in intimate contact with the specimen. DTA was also used in early days for locating the glass transition region of a number of compounds [100,101]. The so-called abnormal absorption of heat of glasses near the deformation temperature as noted by Tool et al. in a series of papers beginning with Refs 53 and 54 is essentially the DTA curve of a glass transition, due mainly to the heat capacity rise plus quenching and annealing effects.

The advantages of the dynamic calorimetry are its ready availability, ease of operation, speed of the measurement, and small specimen size requirement. However, calibration errors and the effects of thermal lags in the specimen and in the thermal analytical equipments at different heating rates and operational conditions may not be completely compensated. The advantages of the adiabatic calorimetry lie not only in its much greater accuracy, but also in its greater sensitivity. Adiabatic calorimetry provides definitive thermodynamic functions, including residual entropy measurements and accurate integration of enthalpy increments between any two temperatures. The drift techniques for glasses of different thermal histories as mentioned in the previous section are useful in detecting and confirming weak glass transitions, such as that occurred in polyethylene [72] (Fig. 7), and broad glass transitions, as in poly(chlorotrifluoroethylene) [102], in weak glassy crystalline transitions in ice [73] and in acetone clathrate hydrate [103]. The difference in the T_g's of

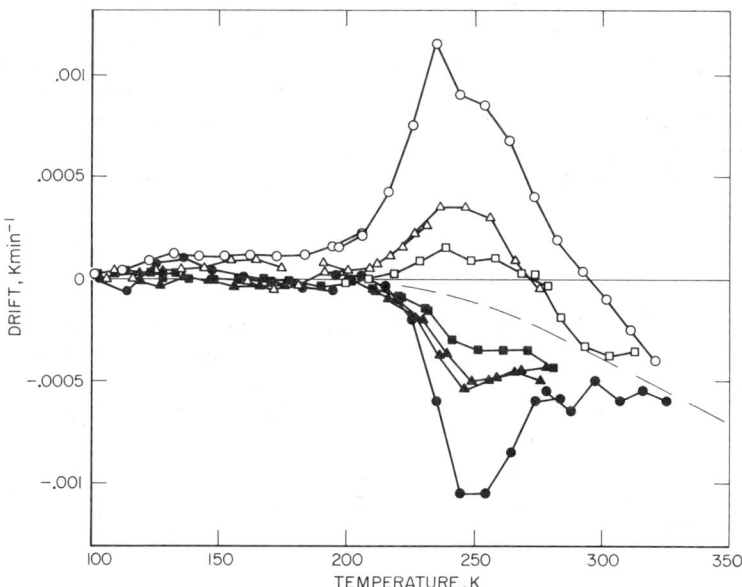

FIG. 7—*Spontaneous adiabatic temperature drifts of linear polyethylene [72] in the glass transformation region. Open symbols—Quenched. Filled symbols—Annealed. Crystallinities:* ○, ●—71%. △, ▲—88%. □, ■—96%.

acetone clathrate hydrate, 90 K from ΔC_p and 85 K from drift, is probably due to the way C_p was measured.

The uncertainty of the assignment of the midpoint of the transition is at best about 1 to 2 K, as the shape of the glass transition region is unsymmetrical about the midpoint, in addition to the uncertainties due to the extrapolation of configurationally fixed heat capacity of glass to the midpoint of the transition region. The onset of the glass transition region is ambiguous, as it depends on the sensitivity of the measurement. Therefore, it is rather meaningless to define an onset temperature for the glass transition. The liquefaction point, where the relaxation time of the liquid is well within the experimental time scale, is much easier to define. It is thus possible to approximately indicate the width of the transition region.

For highly cross-linked polymeric systems, the glass transition region is often very broad with relatively small discontinuity in C_p. Depending on the curing history and the network structure, higher T_g appears to associate with smaller ΔC_p [104]. The weak glass transition may be made more visible if the glass is cooled slowly through the glass transformation region. The slow cooling process is preferred over annealing, as annealing procedures may produce multiple relaxation peaks in a wide transformation range [105].

Due to the relaxation nature of the glass transition, attempts have been made to study the so-called complex heat capacity or thermal diffusivity [106], or the frequency-dependent specific heat via specific heat spectroscopy [107,108]. On a similar approach, a low-frequency sinusoidal temperature excursion in the order of 0.01 to 0.04 Hz is superimposed on either isothermal or scan temperature profile during a DSC run [109]. This technique is said to be able to separate the reversible part of lattice vibration heat capacity from the nonreversing and frequency-dependent part of contributions. It would be interesting to use these newer techniques to study the effects of annealing on glass transition, by separating the so-called annealing peak from the lattice heat capacity. However, similar to that mentioned previously for careful adiabatic measurements, where spontaneous thermal effects with long-time constants may be separated from the instantaneous lattice heat capacity, the resolved "reversible" heat capacity curve may also show C_p of overheated annealed glass beyond its T_g. Hence, the jump in C_p would also occur at a temperature higher than T_g. Therefore, an enthalpy diagram obtained by integrating the power consumed in the heating process is still required to locate the T_g by locating the intersection temperature of the configurational fixed glass at much below the T_g with that of the equilibrium supercooled liquid.

Conclusion

Calorimetric studies on the glasses and glass transition phenomena since the days of W. Nernst have been reviewed. The assignment of the glass transition temperature is only an initial step in describing a glass. The fictive temperature concept, at least, makes an attempt to define the transition temperature of a configurationally fixed glass, free from artifacts and dependency on the rate of observation. By the singular value itself, there is no information about previous history of the glass and other qualities of the glass transition, such as the relaxation spectrum and its function of temperature. It is rather difficult to describe the relaxation time spectrum, and its function as a temperature, in any simple terms. At a minimum, qualitative, or semi-quantitative description of the glass transition, such as the approximate width of the transitional region and the intensity of the glass transition as may be indicated by the magnitude of the C_p jump, should be noted.

For weak transitions, the detection can be enhanced by heating glasses produced by slow continuous cooling. The glass transition of annealed glass or slow-cooled glass appears sharper and often associates with a relaxation peak. For even weaker transitions, spontaneous adiabatic drifts in the glass transition range may be used to locate the transition.

References

[1] Angell, C. A., *Pure and Applied Chemistry*, Vol. 63, 1991, p. 1387.
[2] Chen, H. S. and Turnbull, D., *Applied Physics Letters*, Vol. 10, 1967, p. 248; *Journal of Chemical Physics*, Vol. 48, 1967, p. 2560.
[3] Kelley, K. K., *Journal of the American Chemical Society*, Vol. 51, 1929, p. 1400.
[4] Suga. H. and Seki, S., *Journal of Non-Crystalline Solids*, Vol. 16, 1974, p. 171.
[5] Chowdhury, D., *Spin Glasses and Frustrated Systems*, Princeton University Press, 1986.
[6] Baranovskii, S. D., Thomas, P., and Vaupel, H., *Philosophical Magazine B*, Vol. 65, 1992, p. 685.
[7] Fisher, M. P. A., *Physical Review Letters*, Vol. 62, 1989, p. 1415.
[8] Nernst, W., Koref, F., and Lindemann, F. A., Sitzungsberichte der Deutschen Akademie der Wissenschaften zu Berlin, 1910, p. 247.
[9] Gibson, G. E., Parks, G. S., and Latimer, W. M., *Journal of the American Chemical Society*, Vol. 42, 1920, p. 1542.
[10] Simon, F. and Lange, F., *Zhurnal der Physik*, Vol. 38, 1926, p. 227.
[11] Ahlberg, J. E., *Journal of Chemical Physics*, Vol. 5, 1937, p. 539.
[12] Craig, R. S., Massena, C. W., and Mallya, R. M., *Journal of Applied Physics*, Vol. 36, 1965, p. 108.
[13] Leadbetter, A. J. and Wycherley, K. E., *Journal of Chemical Thermodynamics*, Vol. 2, 1970, p. 855.
[14] Wunderlich, B., *Journal of Physical Chemistry*, Vol. 64, 1960, p. 1052.
[15] Adams, G. and Gibbs, J. H., *Journal of Chemical Physics*, Vol. 43, 1963, p. 139.
[16] Chang, S. S., Bestul, A. B., and Horman, J. A., *Comptes Rendus 7th International Congress on Glass*, Brussels, Sess. I.3.1, No. 26, 1965, p. 1.
[17] Simon, F., *Annalen der Physik*, Vol. IV 68, 1922, p. 241.
[18] Gibson, G. E. and Giauque, W. F., *Journal of the American Chemical Society*, Vol. 45, 1923, p. 93.
[19] Parks, G. S., *Journal of the American Chemical Society*, Vol. 47, 1925, p. 338.
[20] Parks, G. S. and Huffman, H. M., *Journal of the American Chemical Society*, Vol. 48, 1926, p. 2788.
[21] Kelley, K. K., *Journal of the American Chemical Society*, Vol. 51, 1929, p. 779.
[22] Aston, J. G., Finke, H. L., Bestul, A. B., et al., *Journal of the American Chemical Society*, Vol. 68, 1946, p. 52.
[23] Douslin D. R. and Huffman, H. M., *Journal of the American Chemical Society*, Vol. 68, 1946, p. 1704.
[24] Kunzler, J. E. and Giauque, W. F., *Journal of the American Chemical Society*, Vol. 74, 1952, p. 797.
[25] Chang, S. S., Horman, J. A., and Bestul, A. B., *Journal of Research of the National Bureau of Standards*, Vol. 71A, 1967, p. 293.
[26] Counsell, J. F., Lees, E. B., and Martin, J. F., *Journal of the Chemical Society,* Vol. 1968 (A), 1968, p. 1819.
[27] Sugisaki, M., Adachi, K., Suga, H., and Seki, S., *Bulletin of the Chemical Society of Japan,* Vol. 41, 1968, p. 593.
[28] Sugisaki, M., Suga, H., and Seki, S., *Bulletin of the Chemical Society of Japan,* Vol. 41, 1968, p. 2586.
[29] Andon, R. J. L., Connett, J. E., Counsell, J. F., et al., *Journal of the Chemical Society*, Vol. 1971 (A), 1971, p. 661.
[30] Chang, S. S. and Bestul, A. B., *Journal of Chemical Physics*, Vol. 56, 1972, p. 503.
[31] Kishimoto, K., Suga, H., and Seki, S., *Bulletin of the Chemical Society of Japan,* Vol. 46, 1973, p. 3020.
[32] Chang, S. S. and Bestul, A. B., *Journal of Chemical Thermodynamics*, Vol. 6, 1974, p. 325.
[33] Messerly, J. F., Finke, H. L., and Todd, S. S., *Journal of Chemical Thermodynamics*, Vol. 6, 1974, p. 635.
[34] Haida, O., Suga, H., and Seki, S., *Journal of Chemical Thermodynamics*, Vol. 9, 1977, p. 1133.
[35] Lebedev, B. V., Rabinovich, I. B., and Evstropov, A. A., *Journal of Chemical Thermodynamics*, Vol. 9, 1977, p. 101.

[36] Rabinovich, I. B., Khlyustova, T. B., Mochalov, A. N., and Kokurina, N. N., *Zhurnal Fizicheskoi Khimi*, Vol. 57, 1983, p. 2867; *Russian Journal of Physical Chemistry*, Vol. 57, 1983, p. 1736.
[37] Rabinovich, I. B., Novoselova, N. V., Moseeva, E. M., et al., *Zhurnal Fizicheskoi Khimi*, Vol. 59, 1985, p. 2127; *Russian Journal of Physical Chemistry*, Vol. 59, 1985, p. 1266.
[38] Rabinovich, I. B., Novoselova, N. V., Moseeva, E. M., et al., *Zhurnal Fizicheskoi Khimi*, Vol. 60, 1986, p. 545; *Russian Journal of Physical Chemistry*, Vol. 60, 1986, p. 325.
[39] Nistratov, V. P., Rabinovich, I. B., Sheiman, M. S., et al., *Zhurnal Fizicheskoi Khimi*, Vol. 63, 1989, p. 1779; *Russian Journal of Physical Chemistry*, Vol. 63, 1989, p. 981.
[40] Lebedev, B. V. and Kulagina, T. G., *Journal of Chemical Thermodynamics*, Vol. 22, 1990, p. 21.
[41] Takeda, K., Yamamura, O., and Suga, H., *Journal of Physics and Chemistry of Solids*, Vol. 52, 1991, p. 607.
[42] Tammann, G. and v. Gronow, H. E., *Zeitschrift für Anorganische und Allgemeine Chemie*, Vol. 192, 1930, p. 193.
[43] Huffman, H. M., Parks, G. S., and Thomas, S. B., *Journal of the American Chemical Society*, Vol. 52, 1930, p. 3241.
[44] Huffman, H. M., Todd, S. S., and Oliver, G. D., *Journal of the American Chemical Society*, Vol. 71, 1949, p. 584.
[45] Gee, W. W. and Terry, H. L., *British Association of Advancement of Science Report*, Vol. 1889, 1889, p. 516; *Memoirs and Proceedings, Manchester Literary and Philosophical Society*, [4], Vol. 4, 1891, p. 34.
[46] Ruhemann, M. and Simon, F., *Zeitschrift für Physikalische Chemie*, Vol. 138A, 1928, p. 1.
[47] Bekkedahl, N. and Matheson, H., *Journal of Research of the National Bureau of Standards*, Vol. 15, 1935, p. 503.
[48] Ferry, J. D. and Parks, G. S., *Journal of Chemical Physics*, Vol. 4, 1936, p. 70.
[49] Boyer, R. F. and Spencer, R. S., *Journal of Applied Physics*, Vol. 15, 1944, p. 398.
[50] Raine, H. C., Richards, R. B., and Ryder, H., *Transactions of the Faraday Society*, Vol. 41, 1945, p. 56.
[51] Guar, U., Shu, H. C., Mehta, A., and Wunderlich, B., *Journal of Physical and Chemical Reference Data*, Vol. 10, 1981, p. 1.
[52] Varma-Nair, M. and Wunderlich, B., *Journal of Physical and Chemical Reference Data*, Vol. 28, 1991, p. 349.
[53] Tool, A. Q. and Valasek, J., *Bureau of Standards Paper No. 358*, 1920.
[54] Tool, A. Q. and Eichlin, C. G., *Journal of Optical Society of America*, Vol. 4, 1920, p. 341.
[55] Parks, G. S. and Huffman, H. M., *Journal of Physical Chemistry*, Vol. 31, 1927, p. 1842.
[56] Oblad, A. G. and Newton, R. F., *Journal of the American Chemical Society*, Vol. 59, 1937, p. 2495.
[57] Nelson, E. W. and Newton, R. F., *Journal of the American Chemical Society*, Vol. 63, 1941, p. 2178.
[58] Thomas, S. B. and Parks, G. S., *Journal of Physical Chemistry*, Vol. 35, 1931, p. 2091.
[59] Parks, G. S. and Thomas, S. B., *Journal of the American Chemical Society*, Vol. 56, 1934, p. 1423.
[60] Kanda, E., Otsubo, A., and Haseda, T., *Science Reports of the Research Institute, Tohoku University*, Vol. A2, 1950, p. 9.
[61] Huffman, J. D., *Journal of the American Chemical Society*, Vol. 74, 1952, p. 1696.
[62] Tool, A. Q., *Journal of Research of the National Bureau of Standards*, Vol. 37, 1946, p. 73; *Journal of the American Ceramic Society*, Vol. 29, 1946, p. 240.
[63] Jones, G. O., *Journal of the Society of Glass Technology*, Vol. 32, 1948, p. 381; "Glass," Methuen & Co. Ltd., London, 1956.
[64] Bekkedahl, N. and Scott, R. B., *Journal of Research of the National Bureau of Standards*, Vol. 29, 1942, p. 87.
[65] Rands, R. D. Jr., Ferguson, W. J., and Prather, J. L., *Journal of Research of the National Bureau of Standards*, Vol. 33, 1943, p. 63.
[66] Furukawa, G. T., McCoskey, R. E., and King, G. J., *Journal of Research of the National Bureau of Standards*, Vol. 49, 1952, p. 273.
[67] Dainton, F. E., Evans, D. M., Hoare, F. E., and Melia, T. P., *Polymer*, Vol. 3, 1962, pp. 286, 297.
[68] O'Reilly, J. M., Karasz, F. E., and Bair, H. E., *Journal of Polymer Science C*, Vol. 6, 1964, p. 109.
[69] Abu-Isa, I., Crawford, V. A., Haly, A. R., and Dole, M., *Journal of Polymer Science C*, Vol. 6, 1964, p. 149.

[70] Davies, R. O. and Jones, G. O., *Advanced Physics*, Vol. 2, 1953, p. 370; *Proceedings of the Royal Society*, Vol. A217, 1953, p. 26.
[71] Adachi, K., Suga, H., and Seki, S., *Bulletin of the Chemical Society of Japan*, Vol. 45, 1972, p. 1960.
[72] Chang, S. S., *Journal of Polymer Science Symposium*, Vol. 43, 1973, p. 43.
[73] Haida, O., Suga, H., and Seki, S., *Proceedings of the Japan Academy*, Vol. 49, 1973, p. 191.
[74] Desnoyers, J. E. and Morrison, J. A., *Philosophical Magazine*, Vol. 8, No. 3, 1958, p. 42.
[75] Takahashi, Y. and Westrum, E. F. Jr., *Journal of Chemical Thermodynamics*, Vol. 2, 1970, p. 847.
[76] Beyermann, W. P., Hundley, M. F., Thompson, J. D., et al., *Physical Review Letters*, Vol. 68, 1992, p. 2046.
[77] Chang, S. S., *Journal of Polymer Science, Polymer Symposium*, Vol. 71, 1984, p. 59.
[78] Sleator, T., Bernasconi, A., Felder, E., et al., *Physica B*, Vol. 165, 1990, p. 907.
[79] Cahill, D. G., Olsen, J. R., Fisher, H. E., et al., *Physical Review B*, Vol. 44, 1991, p. 12226.
[80] Berret, J.-F., Meissner, M., and Mertz, B., *Zeitschrift für Physik B—Condensed Matter*, Vol. 87, 1992, p. 213.
[81] Grannan, E. R., Randeria, M., and Sethna, J. P., *Physical Review B*, Vol. 41, 1990, pp. 7784, 7799.
[82] Westrum, E. F., Jr., *Pure and Applied Chemistry*, Vol. 64, 1992, p. 137.
[83] Malinovsky V. F. and Novikov, V. N., *Journal of Physics: Condensed Matter*, Vol. 4, 1992, p. L139.
[84] Tarasov, V. V., "New Problems in the Physics of Glass," OTS 63-11004, Office of Technical Services, 1963; translated from "Novye Voprosy Fiziki Stekla," Moskva, 1959.
[85] Anderson, C. T., *Journal of the American Chemical Society*, Vol. 52, 1930, p. 2296.
[86] Chang, S. S. and Bestul, A. B., *Journal of Chemical Physics*, Vol. 55, 1971, p. 933.
[87] Guttman, C. M., Private communications.
[88] Guttman, C. M., *Journal of Chemical Physics*, Vol. 56, 1972, p. 627.
[89] Mopsik, F. I. and Guttman, C. M., *Journal of Research of the National Bureau of Standards*, Vol. 83, 1978, p. 283.
[90] Tammann, G., *Zeitschrift für Anorganische und Allgemeine Chemie*, Vol. 190, 1930, p. 48.
[91] Gibbs, J. H. and DiMarzio, E. A., *Journal of Chemistry Physics*, Vol. 28, 1958, p. 373.
[92] Simon, F., *Zeitschrift für Anorganische und Allgemeine Chemie*, Vol. 203, 1931, p. 219.
[93] Tauke, J., Litovitz, T. A., and Macedo, P. B., *Journal of the American Ceramics Society*, Vol. 51, 1968, p. 158.
[94] Parks, G. S., Thomas, S. B., and Light, D. W., *Journal of Chemical Physics*, Vol. 4, 1936, p. 64.
[95] Karasz, F. E., Bair, H. E., and O'Reilly, J. M., *Journal of Physical Chemistry*, Vol. 69, 1965, p. 2657.
[96] Chang, S. S., *Journal of Chemical Thermodynamics*, Vol. 9, 1977, p. 189.
[97] Flynn, J. H., *Thermochimica Acta*, Vol. 8, 1974, p. 69.
[98] Moynihan, C. T., Easteal, A. J., DeBolt, M. A., and Tucker, J., *Journal of the American Ceramic Society*, Vol. 59, 1976, p. 12.
[99] Richardson M. J. and Savill, N. J., *British Polymer Journal*, Vol. 11, 1979, p. 123.
[100] Tammann, G. and Tampke, R., *Zeitschrift für Anorganische und Allgemeine Chemie*, Vol. 162, 1927, p. 1.
[101] Parks, G. S., Huffman, H. M., and Cattoir, F. R., *Journal of the American Chemical Society*, Vol. 32, 1928, p. 1366.
[102] Chang, S. S., *Journal of Research of the National Institute of Standards and Technology*, Vol. 97, 1992, p. 341.
[103] Kuratomi, N., Yamamuro, O., Matsuo, T., and Suga, H., *Journal of Chemical Thermodynamics*, Vol. 23, 1991, p. 485.
[104] Chang, S. S., *Polymer*, Vol. 33, 1992, p. 4768.
[105] Kong, E. S., *American Chemical Society Symposium Series*, Vol. 211, 1983, p. 171.
[106] Shelley, D. L., *Proceedings of the 10th Thermal Conductivity Conference*, Boston, MA, 1970.
[107] Birge, N. O. and Nagel, S. R., *Physical Review Letters*, Vol. 54, 1985, p. 2674.
[108] Dixon, P. K., *Physical Review B*, Vol. 42, 1990, p. 8179.
[109] Sauerbrunn, S., Crowe, B., and Reading, M., *The TA Hotline*, Vol. 2, 1992, p. 6.

DISCUSSION

B. Wunderlich[2] (written discussion)—
1. Determination of the calorimeter drift after heating or cooling steps to evaluate the glass transition is a very useful technique. The value of 240 K that you report for polyethylene fits well with the half-vitrification point established from extrapolated heat capacity measurements to the completely amorphous polymers.[3]
2. The values of the residual entropies of small organic molecules of 1.9 to 3.5 J/(K mol of beads) can be compared to the collection of linear macromolecules in our *ATHAS* data bank where for 56 polymers an average of 4.5 ± 2.0 J/(K mol of beads) is found for S_0. In addition, an average of 11.1 ± 2.5 J/(K mol of beads) is observed for the increase in heat capacity at the glass transition temperature.[4]

*S.-S. Chang (author's closure to Discussion #2)—*The reason that residual entropies of polymers were not included in Table 1 is that we chose only those materials or substances with both of their crystalline and glassy heat capacities measured at least down to 10 to 20 K. Reported residual entropies of polymers generally contain relatively large error, although they may be good educated guesses. The reasons are: (1) only a small number of polymers have their low temperature heat capacities measured; (2) it is difficult to produce completely amorphous phase and completely crystalline phase of the same polymer; and (3) the heat of fusion and entropy of fusion, and the deconvolution of the crystalline and amorphous contributions to the heat capacity, were estimated based on their crystallinities estimated elsewhere. Therefore, reported values of residual entropies for polymers are most likely the estimated configurational entropies of the amorphous phase near the glass transition temperature.

[2]University of Tennessee, Department of Chemistry, Knoxville, TN 37996-1600.
[3]Wunderlich, B., 237 K, *Journal of Chemical Physics,* Vol. 37, 1962, p. 1203.
[4]ATHAS Data Bank; see, for example, Wunderlich, B., *Thermal Analysis,* Academic Press, Boston, 1990.

John R. Saffell[1]

Analysis of DSC Thermal Curves for Assigning a Characteristic Glass Transition Temperature, Dependent on Either the Type or Thermal History of the Polymer

REFERENCE: Saffell, J. R., "**Analysis of DSC Thermal Curves for Assigning a Characteristic Glass Transition Temperature, Dependent on Either the Type or Thermal History of the Polymer,**" *Assignment of the Glass Transition, ASTM STP 1249,* R. J. Seyler, Ed., American Society for Testing and Materials, Philadelphia, 1994, pp. 137–150.

ABSTRACT: DSC was used to study the thermal curve characteristics of four classical polymer glasses: polycarbonate (PC), anionic polystyrene (aPS), polysulfone, and polymethyl methacrylate (PMMA). The onset, peak, and fictive temperatures as well as endotherm characteristics were studied over two decades of heating and cooling rates. Assuming that the fictive temperature is independent of heating rate, the usefulness of T_{onset} and ambiguity of T_{peak} are demonstrated. The useful [heat/cool] ratio allows use of T_{onset} to calculate the effective cooling rate of the sample (which is easier to calculate than $T_{fictive}$). Comparison with results from annealed samples are shown. The fictive and onset temperatures can be used together to improve the reliability of glass temperature characterization and results are surprisingly constant between the four different types of glasses, leading to a method of providing a single temperature to characterize the thermal history/morphology using a DSC thermal curve. The peak temperature is nearly independent of thermal history, and so provides a characterization parameter that is insensitive to manufacturing and molding procedures.

KEYWORDS: glass transition, onset temperature, peak temperature, fictive temperature, cooling exotherm, differential scanning calorimetry (DSC), DSC heat/cool ratio, DSC heating rate, DSC cooling rate

Differential Scanning Calorimetry (DSC) is the easiest method of gathering information about the glass transition, so there is a clear advantage in specifying a DSC-type glass transition measurement. However, for such a parameter to be specified reliably it must: (1) be insensitive to the manufacture or type of the DSC, (2) apply generally to all polymers/glasses, and (3) be either sensitive or insensitive to thermal history, as required.

Equipment and Materials Characterization

A DuPont 990 with DSC cell was used to obtain the thermal curves. Also, results obtained from a Perkin-Elmer DSC II are compared where relevant. The four polymers were: Pressure Chemicals anionic polystyrene (PS) ($\bar{M}n = 2 \times 10^5$, $\bar{M}w/\bar{M}n = 1.06$); ICI Perspex (PMMA) ($\bar{M}w = 1.5 \times 10^7$, $\bar{M}w/\bar{M}n = 3.1$); Bayer Makrolon 3201 (polycarbonate (PC), high-viscosity extruded sheet); and injection-molded Union Carbide Udel (polysulfone).

[1] Technical director, Solomat Ltd., Heathpark Industrial Estate, Honiton, Devon, U.K. EX14 8SQ.

Experimental Technique

Thermal Curve Characteristics

Three characteristic temperatures of a glass transition were measured as shown in Fig. 1. T_{onset} is the extrapolated intersection of the baseline below the glass transition and the leading edge of the peak. T_{peak} is the maximum of the endotherm and $T_{fictive}$, as explained by Flynn [1–3], reconstructs the peak area into an ideal second-order transition.

The peak area on heating was also measured, assuming a flat baseline extrapolated from the baseline above the glass transition. The two characteristics of the thermal curve measured during cooling are also shown: T_{onset} is the extrapolated deviation from the baseline above the glass transition and T_x is the extrapolated intersection of the baseline below the glass transition.

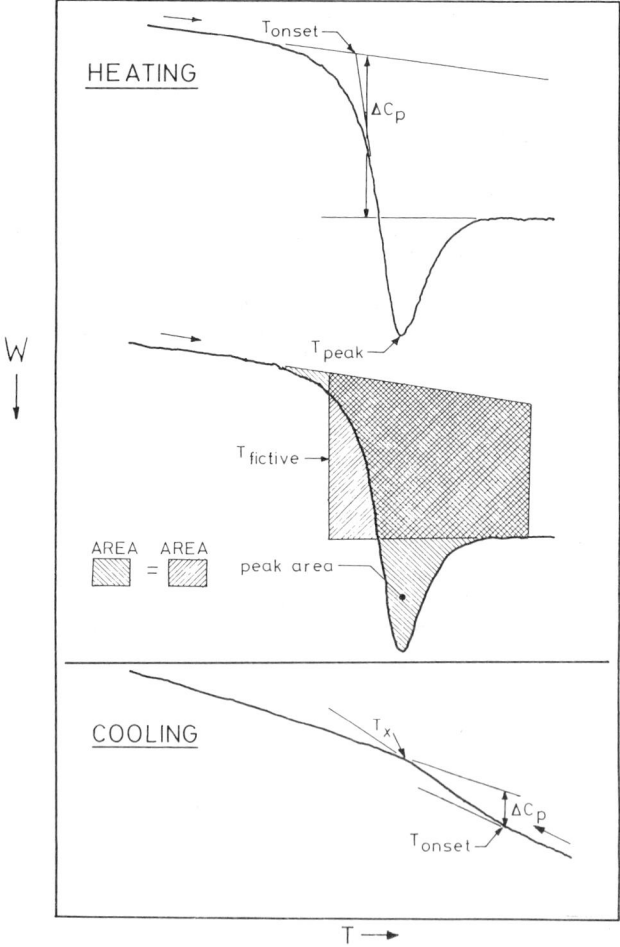

FIG. 1—*Determining characteristic temperatures from glass transition thermal curves.*

Thermal Lag Correction

Reference Material Variations—It is vital that DSC instrument variations, aging, and sample effects are eliminated from thermal lag calculations in order to establish reproducible thermal curve characteristics for the glass transition. Historically this technique has been done using either indium or benzoic acid and monitoring the onset temperature of the melting exotherm. However, variations between different DSCs and oxidation of DSC platforms can lead to incorrect calculations of the thermal lag (Fig. 2). Using indium or benzoic acid corrects for changes in the sample pan-to-DSC platform thermal resistance but incorrectly specifies the specimen-to-pan thermal resistance, specimen mass, and specimen thermal conductance.

The more correct method for calibrating the thermal lag for polymers is to use Flynn's method [1–3] of the fictive temperature (Figs. 2 and 3), which accounts for specimen-to-pan thermal resistance, specimen mass, and specimen thermal conductance.

Calibration Using Fictive Temperature—This method of thermal lag calibration assumes that the fictive temperature, which is a rearrangement of the glass transition endotherm into

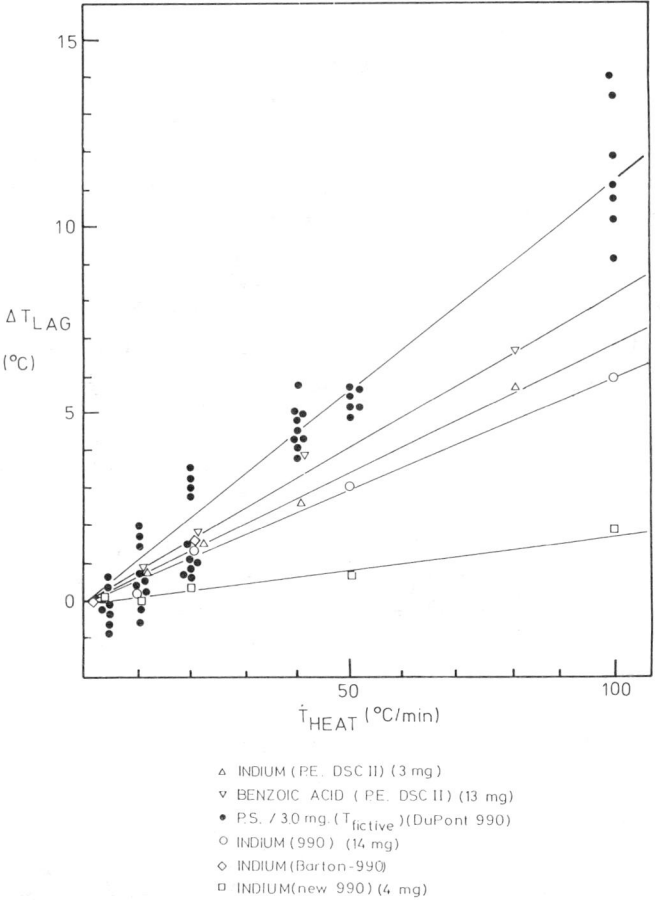

FIG. 2—*Measured thermal lag using melting point and fictive temperature.*

140   ASSIGNMENT OF THE GLASS TRANSITION

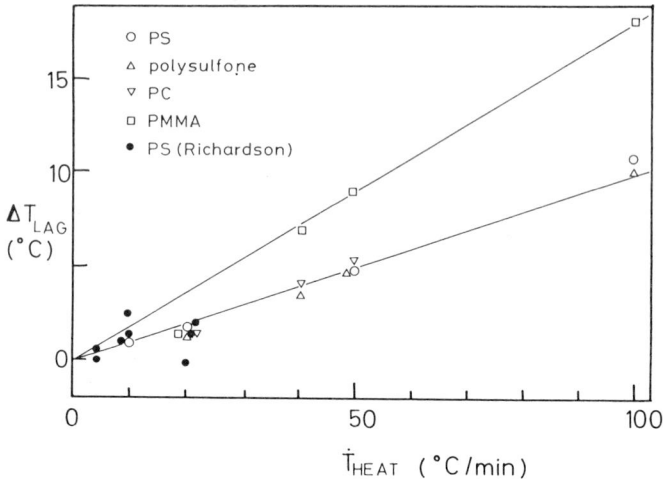

FIG. 3—*Measured thermal lag using fictive temperature for four noncrystalline polymers.*

an ideal second-order transition, is independent of the heating rate. Calibration of thermal lag uses a single specimen (3 to 10 mg) that is cooled at a constant rate, then reheated at a family of heating rates between the minimum and maximum heating rates of the DSC. The plot of the fictive temperature as a function of heating rate (Fig. 3) is the thermal lag calibration curve. Note that three of the polymers were repeatable; PMMA, due to its high viscosity above the glass transition, shows a different specimen-to-pan thermal resistance, requiring a different thermal lag calibration.

Error Histograms

The repeatability of the measurement of the three characteristic temperatures of the glass transition is shown in Fig. 4. The top three histograms show the repeatability from 1° to 100°C/min, after thermal lag correction. The bottom two lines of histograms show the

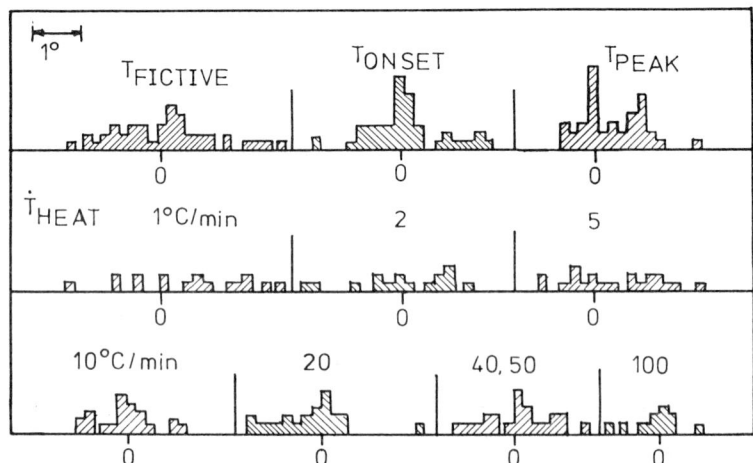

FIG. 4—*Error histograms for the three characteristic temperatures of the glass transition at various heating rates.*

repeatability of fictive, onset, and peak temperatures at different heating rates. The 10°C and 20°C/min heating rates are a good compromise between repeatability and rapid testing. Note that although the peak temperature is the most repeatable, it also has the fewest data points and avoids the large temperature spread at the slower heating rates where a peak temperature was not observed. The onset and fictive temperatures show about 1°C repeatability with improved repeatability at higher heating rates.

Typical Thermal Curves

It is worthwhile to remind ourselves that the well-defined endotherm of a carefully preheated and annealed sample is not the typical result when measuring incoming batches of polymer or looking for effects due to rapid quenching or postmolding treatments. Figure 5 shows typical glass transition thermal curves for "as-received" PC. We must be careful not to specify a repeatability for as-received specimens that is not attainable in the real world.

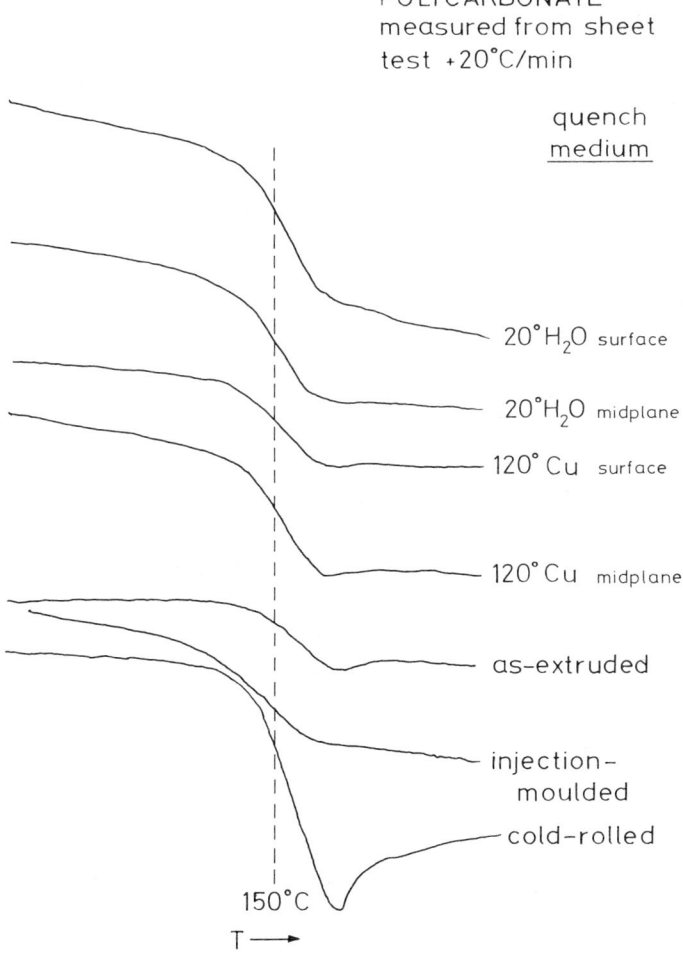

FIG. 5—*Actual as-received thermal curves for PC with different thermal histories and processing conditions.*

The testing technique of careful sample preparation and testing to obtain a thermal curve followed by controlled quenching at a fixed rate and then reheating to provide a reference thermal curve that is free of thermal history effects is the most repeatable technique for measuring as-received specimens. The second thermal curve provides a reference fictive temperature, while the first thermal curve provides a fictive temperature that is representative of the specimen's thermal history.

Results

Cooling Rate Results

The fictive temperature is a measure of the "excess enthalpy" of the material. In a noncrystalline polymer this excess enthalpy will increase (and the fictive temperature proportionately decrease) by changing the thermal history which includes either changing the cooling rate or annealing after cooling. Other treatments such as annealing above the glass transition and orientation also change the excess enthalpy and hence the fictive temperature.

The fictive temperature linearly decreases with the logarithm of the cooling rate (Fig. 6) for all four polymers tested. This simple rule will be complicated when working with semicrystalline polymers since the cooling rate will also affect the % crystallinity. But for simple glasses, this rule is followed.

Further proof that the fictive temperature is a good parameter for monitoring thermal history can be seen in Fig. 7. The fictive temperature is used to monitor the excess enthalpy of PC following annealing treatments between 100 and 140°C.

The classical sigmoidal decrease in T_{fictive} (when plotted against (log) annealing time) can

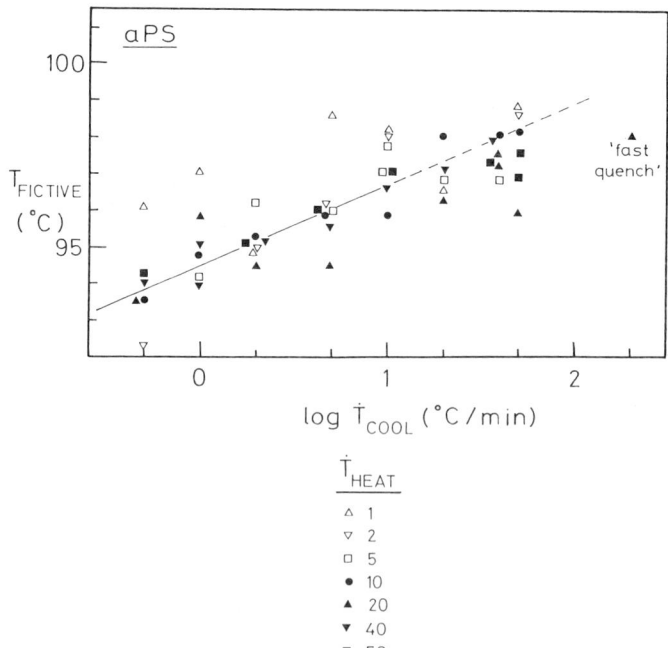

FIG. 6—*Fictive temperature as a function of cooling rate for aPS.*

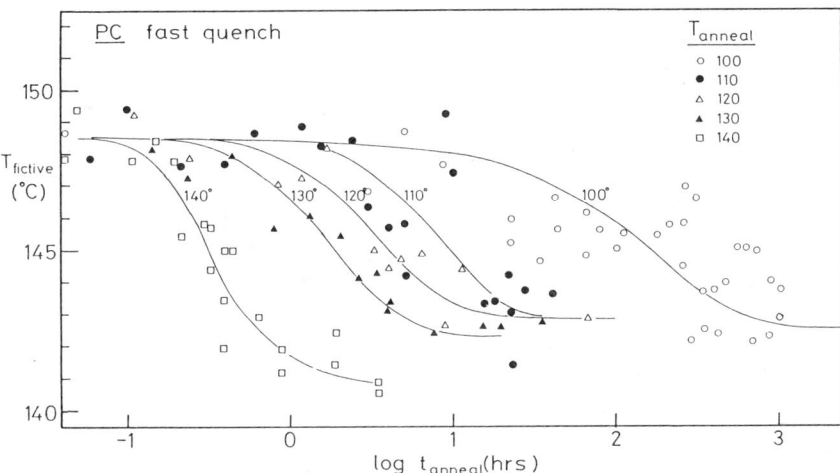

FIG. 7—*Fictive temperature can be used to monitor the excess enthalpy for PC after annealing.*

be seen (except at 100°C anneal, where although a sigmoidal curve could be inferred, it is not obvious). These results compare well to the change in material density and tanδ, as reported in the literature. The shift of inflection time with annealing temperature shows approximately the same slope on an Arrhenius plot for DSC, DMA, and density experiments, although the exact (inflection) time/anneal temperature (i.e., Arrhenius plot offset) varies with method.

Heating Rate Results

The onset and peak temperatures for anionic polystyrene are plotted as a function of heating rate from 0.5 to 50°C/min (Fig. 8). The fictive temperature is assumed to be independent of heating rate as per Flynn's assumption (Fig. 6); this thermal lag correction was applied to the raw data before plotting in Fig. 8. The peak temperature is dependent only on heating rate for aPS and shows only a very weak dependence on cooling rate for polysulfone and PC; as a first approximation the peak temperature is dependent only on the heating rate. The onset temperature is dependent on both heating rate and cooling rate as shown. This trend is true for all noncrystalline polymers tested. Note the linear dependence of the onset temperature with the (log) heating rate.

It is worthwhile to consider further the relation between heating rate and cooling rate on the onset temperature.

Heat/Cool Ratio

When looking for the link between heating rate and cooling rate it is interesting to plot a family of thermal curves with similar heat rate to cool rate ratios (Fig. 9). Note how the endotherms show the same shape with just a variation in the size of the endotherm, as expected for different heating rates.

If we replot Fig. 8 and re-identify the points as those with the same heat/cool rate ratio, we now see that the onset temperature is constant for a constant heat-to-cool ratio (Fig. 10).

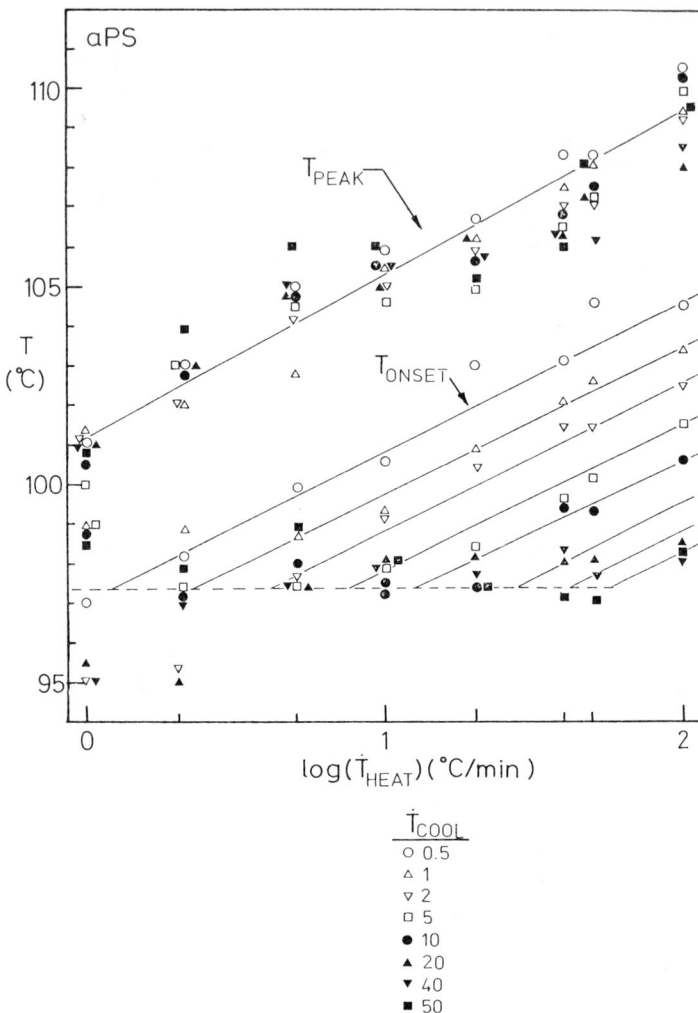

FIG. 8—*Peak and onset temperatures for aPS at various heating and cooling rates.*

Although shown only for aPS, this simple rule is true for all four noncrystalline polymers that were studied.

One would suspect that the peak area would follow the trend of the fictive temperature, i.e., be dependent only on cooling rate. However when we plot the (log) peak area against the (log) heat/cool ratio (Fig. 11) we then obtain a simple linear law. Therefore the peak area of the glass transition endotherm is a function of heat-to-cool ratio, like T_{onset}, and not of the excess enthalpy.

Glass Transition During Cooling

By careful experimentation one can monitor the glass transition during cooling. There is no characteristic peak, but there are two characteristic temperatures: T_{onset} and T_x (Fig. 1).

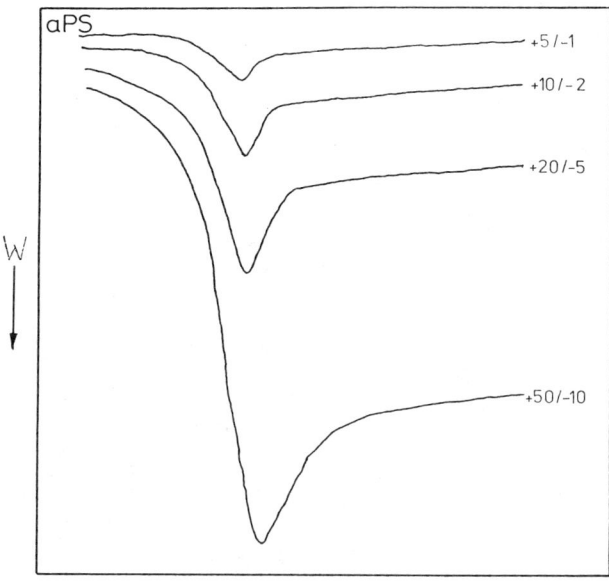

FIG. 9—*A family of thermal curves for aPS showing the same characteristic shape. Heating rate is five times the cooling rate.*

Likewise $T_{fictive}$ could also be calculated but since the glass transition trace is always a simple linear segment between T_x and T_{onset}, $T_{fictive}$ is approximately halfway between T_{onset} and T_x.

Cooling transitions were measured for PS (Fig. 12), polysulfone, and PC. All three showed the same behavior: T_x is independent of cooling rate and T_{onset} increases with cooling rate. During cooling the thermal lag correction is added to, rather than subtracted from, the uncorrected data. At very fast cooling rates T_{onset} appears to level off which may be due to the low thermal conductivity of the material rather than actually reaching a plateau of behavior. The effect of cooling rate on T_{onset} is not as marked as the effect of heating rate on T_{peak} or T_{onset}.

Discussion

The three characteristic temperatures and endotherm characteristics (peak height and peak area) can be used to define a material parameter that is independent of its thermal history, or to define a material thermal history parameter or to back-calculate a cooling rate for processed material. However, although DSC is a simple technique, it is seen that correct calibration of thermal lag and careful sample preparation for as-received samples are paramount to obtaining accurate information with the required repeatability. A simple technique to ensure accurate measurement of thermal/processing history on a repeatable basis is to first heat the sample at a constant heating rate (such as 10°C/min or 20°C/min), cool at a fixed rate (e.g., 10°C/min) and then reheat at the same heating rate.

This heat/reheat experiment provides a repeatable baseline result on the second run that adds reliability to glass transition measurements.

We have seen how the fictive, onset, and peak temperatures are each dependent on or

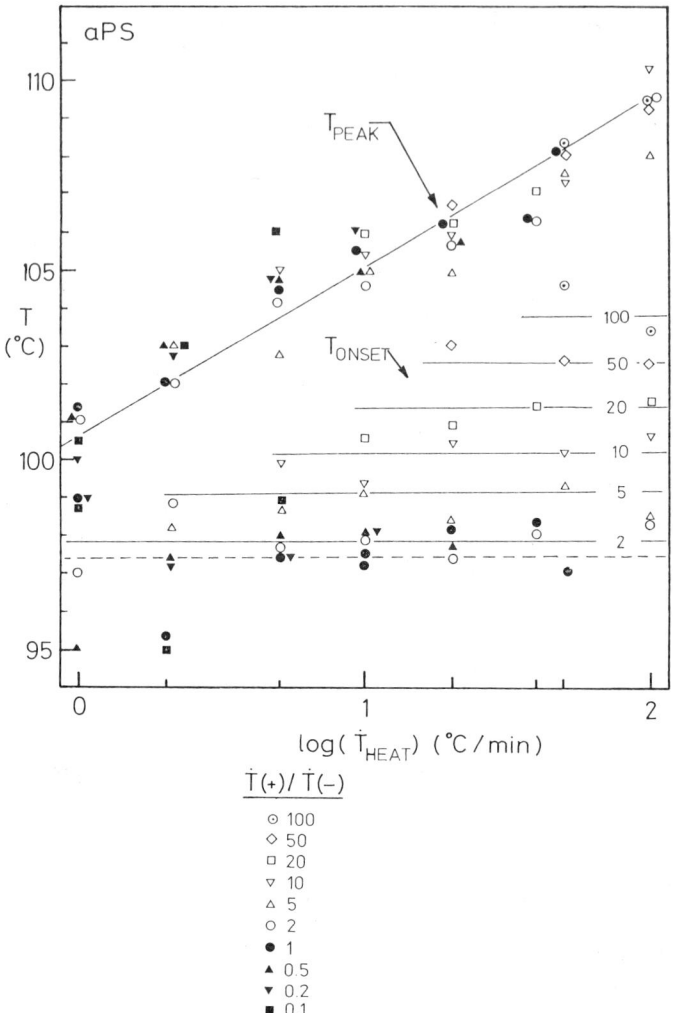

FIG. 10—*A replot of Fig. 8, plotted at constant heat/cool rate ratios.*

independent of the three parameters: heating rate, cooling rate, and heat rate/cool rate ratio. Figure 13 shows each of these three temperatures plotted against these three variables and we now see the trends for these characteristic temperatures.

Thermal History/Processing Effects

If one wishes to use the characteristics of the glass transition temperatures to characterize (with a single parameter) the thermal history or the processing of a sample then the fictive temperature is the best choice.

However, since the onset temperature is dependent on the heat/cool ratio and since the heating rate is known, then the onset temperature can also be used to confirm the fictive

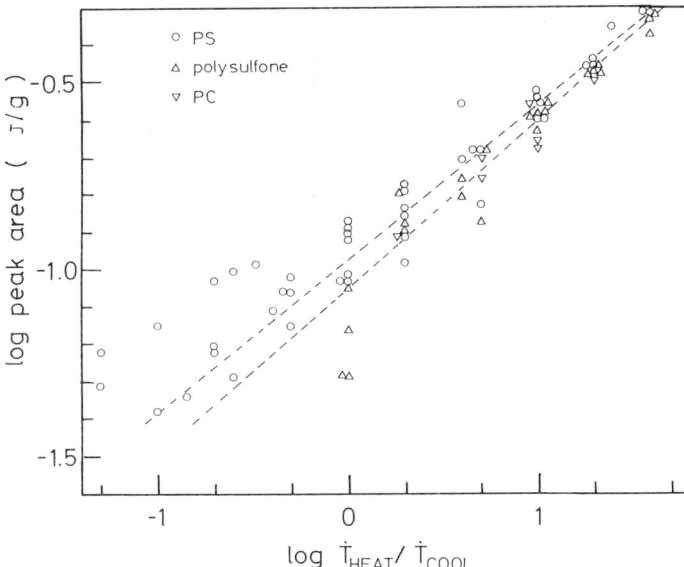

FIG. 11—*The (log) peak area plotted against (log) heat/cool rate ratio for three noncrystalline polymers. The upper dashed line is best fit for PS and lower dashed line is best fit for three polymers.*

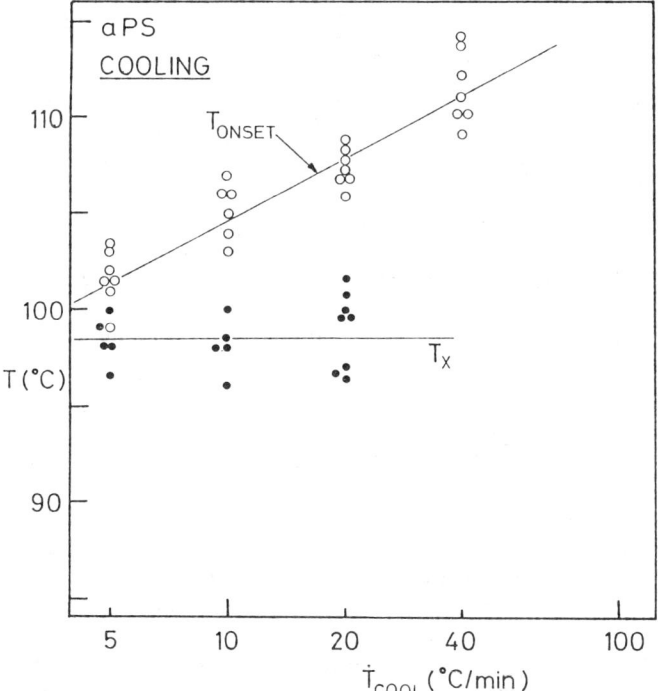

FIG. 12—*The two characteristic temperatures measured during cooling through the glass transition for aPS.*

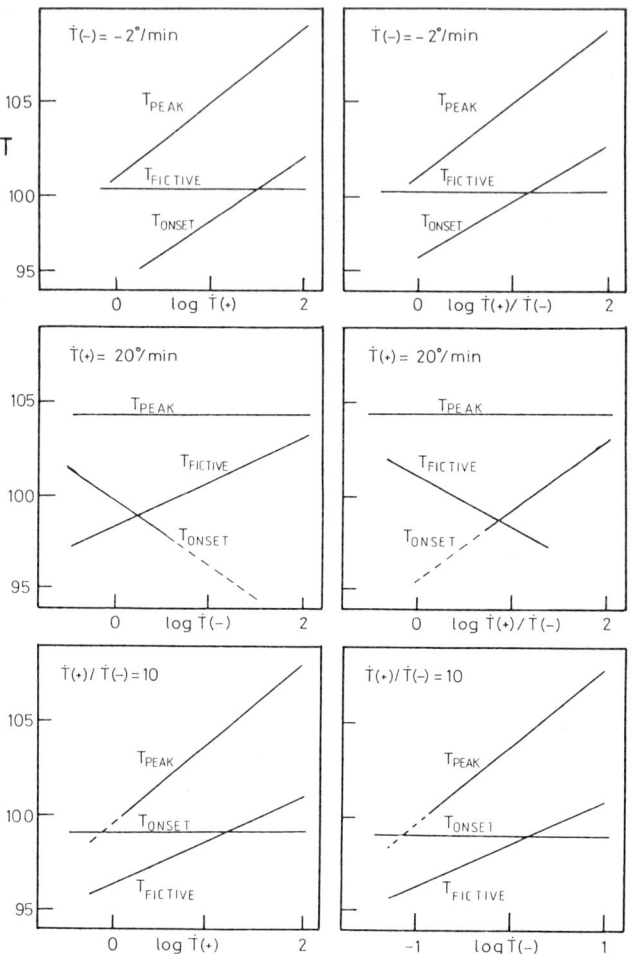

FIG. 13—*Replots of the three characteristic temperatures of the glass transition for aPS. These three characteristic temperatures are plotted against heating rate, cooling rate, and heat/cool rate ratio. (+) is heating rate, (−) is cooling rate, and (+)/(−) is heat/cool rate ratio.*

temperature measurement. The onset temperature is also easier to measure than the fictive temperature.

Each polymer performs similarly, but when trying to correlate a new polymer it is necessary to run a set of heat/cool experiments to obtain the effect of cooling rate on fictive temperature and the effect of heat/cool ratio on the onset temperature. After these simple tests, one can calculate the cooling rate for a material with a simple thermal history from either the onset or fictive temperature measurements.

Checking both temperatures would confirm that there was a simple thermal history to the material or, if the two temperatures do not correlate to the expectations, then perhaps annealing or orientation had also occurred which affected the fictive temperature more than the onset temperature. This could be the subject of further work.

Glass Transition Characteristic, Independent of Thermal History

As a characteristic temperature of a polymer that is independent of thermal history but dependent on molecular weight and molecular number, the peak temperature is the best choice. This experiment must be done at reasonable heating rates (e.g., 20°C/min or 40°C/min) especially if the material is rapidly quenched; otherwise there will be no characteristic endotherm by which to identify T_{peak} for simple cool/heat thermal histories (an endotherm peak will form if a specimen is annealed before heating).

Arrhenius-type Analysis

Since the glass transition is a cooperative transition with a wide distribution of relaxation times [5–7], it should not be analyzed using simple Arrhenius graphing. However, it is still useful to plot results on a relaxation map (log heating rate versus $1/T$) to study the various slopes of the characteristic temperatures; if the characteristic temperatures have the same slope then they are all equally valid for characterization of the glass transition. This plot is shown in Fig. 14 for aPS; this "relaxation map" is similar for the other three noncrystalline polymers.

The peak and onset temperatures behave similarly (except for the onset temperature measured during cooling). Note that the fictive temperature has a different "activation energy" than the other measurements; this is because the fictive temperature is actually a characteristic of the excess enthalpy and not an easily extrapolated temperature from the thermal curve. The onset temperature slope (as a function of cooling rate) is reversed, which is consistent with the independence of T_{onset} when measured at a constant heat/cool rate ratio.

Conclusions

1. The fictive temperature can be used to calibrate the thermal lag of polymers; this method is superior to using crystalline calibration material such as benzoic acid and indium.

2. The fictive temperature is representative of the "excess enthalpy" and can be used to identify the cooling rate for a polymer with simple thermal history, or to characterize the state of annealing and cooling rate for a polymer that has undergone a more complex thermal history.

3. The peak temperature is nearly independent of the cooling rate and thermal history and so can be used as a characteristic temperature of the polymer (at a fixed heating rate), independent of its thermal history/processing.

4. The onset temperature, peak height, and peak area are dependent on the heat/cool rate ratio and can be used to either calculate the cooling rate having known the heating rate, or to further verify and qualify the fictive temperature characteristic.

5. The fictive temperature is independent of DSC instrument and heating rate, and hence represents the best characteristic temperature to specify the glass transition. However, since the fictive temperature is more difficult to determine than the onset temperature, the onset temperature can also be used to characterize the thermal history/processing of a sample if all testing is performed at a constant heating rate.

6. Figures 6 and 12 show the fictive temperature for polystyrene as a function of cooling rate and T_{onset} and during cooling. The fictive temperature, if measured from a cooling thermal curve, is the midway point between T_x and T_{onset}.

7. The heat/cool rate ratio controls the onset temperature, peak area, and peak height, reminding us that the glass transition is kinetically controlled.

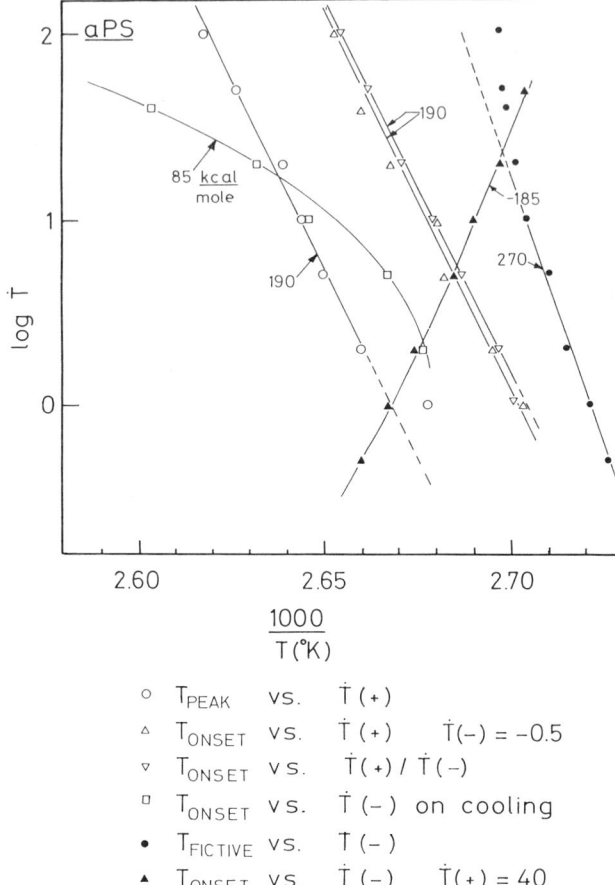

FIG. 14—*Relaxation map of the various characteristic temperatures. The actual "activation energies" do not have physical significance but show that the characteristic temperatures behave in a similar fashion.*

References

[1] Petrie, S. E. B., "Thermal Behavior of Annealed Organic Glasses," *Journal of Polymer Science: Part A-2*, Vol. 10, 1972, pp. 1255–1272.
[2] Flynn, J. H., "Thermodynamic Properties from Differential Scanning Calorimetry by Calorimetric Methods," *Thermochimica Acta*, Vol. 8, 1974, pp. 69–81.
[3] Moynihan, C. T., Easteal, A. J., DeBolt, M. A., and Tucker, J., "Dependence of the Fictive Temperature of Glass on Cooling Rate," *Journal of the American Ceramic Society*, Vol. 59, No. 1-2, 1976, pp. 12–16.
[4] Richardson, M. J. and Savill, N. G., "Derivation of Accurate Temperatures by Differential Scanning Calorimetry," *Polymer*, Vol. 16, 1975, pp. 753–757.
[5] Hutchinson, J. M. and Kovacs, A. J., "Effects of Thermal History on Structural Recovery of Glasses During Isobaric Heating," *Polymer Engineering and Science*, Vol. 24, October 1984, No. 14, pp. 1087–1103.
[6] Gourari, A., Bendaoud, M., Lacabanne, C., and Boyer, R. F., "Influence of Tacticity on Tbeta, Tg and Tl,l in PMMAs by the Method of Thermally Stimulated Current (TSC)," *Journal of Polymer Science, Polymer Physics Edition*, Vol. 23, No. 5, May 1985, pp. 889–916.
[7] Ibar, J. P., "On the Use of Compensation Law to Describe Cooperative Relaxation Kinetics in Thermally Stimulated Processes: a New View," NATAS symposium, Cambridge, MA, 1990.

Instrumental Techniques

Jay W. Grate[1]

Sensing Glass Transitions in Thin Polymer Films on Acoustic Wave Microsensors

REFERENCE: Grate, J. W., "**Sensing Glass Transitions in Thin Polymer Films on Acoustic Wave Microsensors,**" *Assignment of the Glass Transition, ASTM STP 1249*, R. J. Seyler, Ed., American Society for Testing and Materials, Philadelphia, 1994, pp. 153–164.

ABSTRACT: Acoustic wave microsensors are capable of sensing transition behaviors of homogeneous amorphous polymers applied as thin films to the surface of the sensor. The device and wave characteristics of four such microsensors, and examples of their uses in monitoring polymer properties, are described. The devices are the thickness-shear mode (TSM), surface acoustic wave (SAW), flexural plate wave (FPW), and shear horizontal acoustic plate mode (SH-APM) devices. Using the FPW device, the change in the polymer coefficient of thermal expansion at the static glass transition temperature is sensed as a change in slope of the frequency-temperature plot. This behavior reflects the fact that the frequency measures acoustic velocity, which decreases as the polymer modulus decreases. Modulus decreases as the polymer volume increases. Frequency-dependent relaxation properties are detected with a sigmoidal change in slope of the frequency-temperature plot and a minimum in signal amplitude. The minimum in signal amplitude is analogous to the maximum in the loss tangent in dynamic mechanical analysis, or the maximum in sound damping in bulk-wave ultrasonic studies.

KEYWORDS: acoustic wave, microsensor, thin film, glass transition, thickness-shear mode (TSM), surface acoustic wave (SAW), flexural plate wave (FPW), shear horizontal acoustic plate mode (SH-APM), ultrasonic

The term "glass transition" refers generally to the transition of a polymer from a glassy material to a rubbery material as the temperature of the sample increases. There are a variety of polymer properties that change at the glass transition, and a variety of physical methods to measure the glass transition, including dilatometry, dynamic mechanical analysis (DMA), and differential scanning calorimetry (DSC) [1,2]. Conventional bulk wave acoustic methods measuring sound velocities and damping through polymer samples are also informative [3–5]. These methods are generally applied to bulk polymer samples. This paper will describe the observation of polymer glass transition phenomena in thin films on surfaces, using a number of acoustic wave microsensors.

Two types of transitions will be examined, both of which are referred to as glass transitions. Changes in polymer thermal expansion rates, which are frequency-independent, are associated with a transition sometimes referred to as the static or dilatometric glass transition [3,4,6–8]. The temperature of this transition will be referred to as the T_g. Transitions associated with polymer relaxation processes under dynamic conditions are referred to as dynamic glass transitions. The transition temperatures associated with relaxation processes are highly frequency-dependent [1–4]. T_α will be used to denote the transition

[1]Pacific Northwest Laboratory,* Richland, WA 99352.
*Operated for the U.S. Department of Energy by Battelle Memorial Institute under Contract DE-AC06-76RLO 1830.

temperature associated with the primary relaxation of polymer chain segments (the α relaxation). At low frequencies (i.e., a few Hz or less) T_α is close to T_g. As the measurement frequency is increased, T_α increases while T_g remains the same, giving rise to two separate transition temperatures.

Acoustic Wave Devices

The acoustic wave devices considered in this paper are shown in Fig. 1. They are the thickness-shear mode (TSM), surface acoustic wave (SAW), flexural plate wave (FPW), and shear horizontal acoustic plate mode (SH-APM) devices. The sensing surfaces of these devices are typically 1 cm² or less in area. Each device consists of a piezoelectric plate with one or more metal transducers that convert radio frequency electrical energy to mechanical energy in the form of acoustic waves at ultrasonic frequencies. The wave characteristics of each device are summarized in Table 1. When viscoelastic polymers are present on the surfaces of these devices, the polymer chains are forced to move in response to the wave motion at the surface. As a result, the polymer stores and dissipates wave energy, measurably influencing both the wave velocity and attenuation (or damping).

The acoustic waves generated in TSM devices (Fig. 1) are bulk waves that travel in a direction perpendicular to the plate surfaces [9–12]. The motions of particles on the surface are entirely in-plane. In this regard, the device could be considered a shear micromotor [13]. Most TSM sensors are constructed on quartz discs, and are commonly known as quartz crystal microbalances, or QCMs. The plate thickness determines the wavelengths of the fundamental and harmonic modes: the fundamental mode has a wavelength of twice the plate thickness, and its frequency is typically between 5 and 10 MHz.

The interdigital transducers (IDTs) on a SAW delay line generate Rayleigh waves whose energy is largely confined to a zone at the surface approximately one acoustic wavelength thick [14–16]. The waves travel the length of the device from one transducer to another with wavefronts parallel to the lengths of the IDT fingers. Surface particles move in elliptical

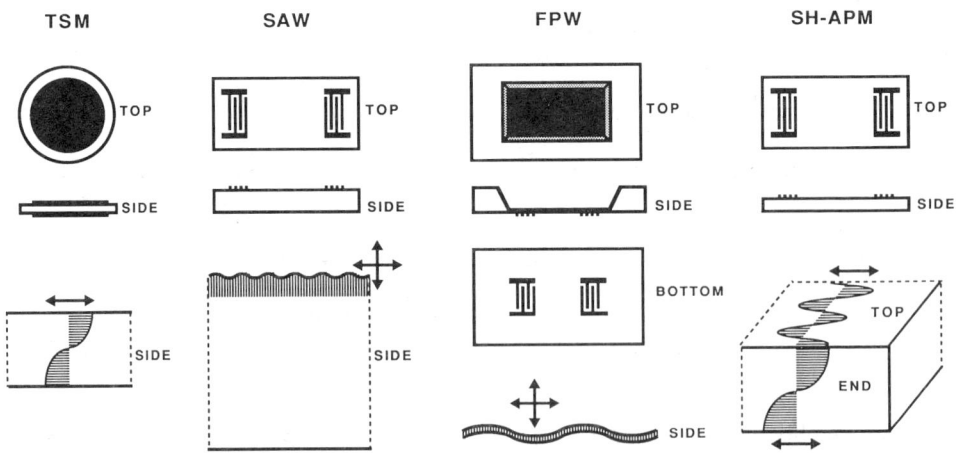

FIG. 1—*Views illustrating the structures and acoustic waves of TSM, SAW, FPW, and SH-APM devices. Side views are in cross-section. Double-headed arrows indicated the directions of surface particle displacements, while shaded areas help to illustrate the wave motion or the depth of wave penetration in the plate.*

TABLE 1—*Comparisons of acoustic wave microsensors*

Device	Wave Type	Particle Displacement Relative to Propagation Direction	Particle Displacement Relative to the Sensing Surface	Typical Frequency (MHz)	Example V = velocity λ = wavelength d = plate thickness
TSM (thickness shear mode)	bulk	transverse	in-plane	5 to 10	6 MHz V = 3330 m/s λ = 555 μm d = 277 μm
SAW (surface acoustic wave)	surface	transverse[a]	surface-normal[a]	30 to 300	158 MHz V = 3160 m/s λ = 20 μm d = 760 μm
FPW (flexural plate wave)	plate	transverse[a]	surface-normal[a]	2 to 7	5.5 MHz V = 550 m/s λ = 100 μm d = 3.5 μm
SH-APM (shear-horizontal acoustic plate mode)	plate	transverse	in-plane	25 to 200	101 MHz V = 5060 m/s λ = 50 μm d = 203 μm

[a]Also has a component of particle displacement in the direction of wave propagation that is in-plane.

paths with both surface-normal and in-plane components. Typical plate materials for SAW devices are quartz and lithium niobate. The plate is much thicker than an acoustic wavelength, and typical device frequencies (determined by the spacing of the fingers in the IDTs) are 30 to 300 MHz.

IDTs on piezoelectric plates are also capable of launching a variety of other waves, depending on the crystal orientation and plate thickness. On plates oriented for Rayleigh wave generation, most of the energy input by the IDT is transmitted as Rayleigh waves provided that the plate thickness is much greater than the acoustic wavelength (e.g., at least >10λ). In a semi-infinite medium, only the Rayleigh wave can exist. On thin plates, IDTs can generate two types of plate waves: lamb waves and shear horizontal acoustic plate waves.

The FPW device generates Lamb waves in a micromachined structure with a very thin (ca. 3 μm) composite plate [*17–20*]. The waves are excited by IDTs in contact with a zinc oxide piezoelectric layer, setting the entire thickness of the plate in motion. Because the plate is so thin, just a few percent of an acoustic wavelength, it is sometimes referred to as a membrane. It is supported by the surrounding silicon substrate. Particle motions on the FPW device surface are similar to those on the SAW device: elliptical with both surface-normal and in-plane components, where the surface-normal component is shear relative to the direction of wave propagation. Typical device frequencies are 2 to 7 MHz.

SH-APM devices are essentially identical in appearance to SAW delay line devices, except that their plates are thinner [*21–25*]. Typical plates are a few acoustic wavelengths thick. In addition to generating a Rayleigh wave, the IDTs on these thinner plates generate shear horizontal waves that propagate in the bulk at angles to the surface, reflecting between the plate surfaces as they progress down the plate between the IDTs. The superpositions of these waves give rise to a series of plate modes ($n = 0,1,2 ...$), each having a slightly different wave propagation velocity in the plate. SH-APM devices and their associated electronics are designed to tune in on a particular plate mode. Shear horizontal plate waves generate surface particle motions that are almost entirely in-plane. Typical SH-APM devices are constructed with quartz plates and their frequencies are 25 to 200 MHz.

Various methods exist for operating acoustic wave microsensors. In each case, it is necessary to generate the acoustic waves by supplying rf electrical energy to the transducers, and to have a method of measuring wave characteristics. The most common methods are the oscillator, which generates a frequency proportional to the wave velocity, the source/vector voltmeter combination, which permits measurement of both wave velocity and attenuation, and the network analyzer, which also permits measurement of both wave velocity and attenuation, and various other characteristics of the device [10,12,14,16,25-27].

Acoustic wave characteristics can be altered by a variety of physical changes at the surface or in thin films on the surface, including mass changes, polymer modulus changes, and acoustoelectric effects. In the experiments below involving polymer films on acoustic wave device surfaces, the surface mass does not change and the films are insulating; the effects observed as the temperature is changed are due to changes in the polymer modulus. Various theoretical and experimental studies have established that acoustic waves are sensitive to film modulus changes [13,14,21,28-30].

When discussing modulus in the context of ultrasonic devices, it is essential to understand that the measurement of modulus is highly frequency-dependent [2]. To the probing high-frequency waves, films of rubbery polymers on ultrasonic devices appear to have moduli that are more characteristic of a glassy material. (This result is a consequence of relaxation effects to be described below.) It should be emphasized, however, that the state of the material is unchanged by such probing. If these films were probed simultaneously by a low-frequency method, the measured modulus would be that of a rubber, as expected. For reference, the typical modulus of a rubbery polymer is 10^6 N/m^2, whereas that of a glassy polymer or a rubbery polymer at high frequency is usually about 10^9 N/m^2. At room temperature and frequencies above a megahertz, nearly all rubbery polymers have measured moduli near 10^9 N/m^2 [31].

Polymer modulus is also temperature- and volume-dependent. At temperatures below T_α, polymer moduli decrease continuously with temperature as the polymer expands. This small continuous change should not be confused with the large sudden drop in modulus that is associated with the dynamic glass transition at T_α. Temperature/volume/modulus effects are well-known in conventional bulk-wave ultrasonics, where the sonic velocity through a polymer sample decreases with increasing temperature [3,4]. For reference, shear sound speed, V_s, is directly proportional to the square root of the shear storage modulus, G', and inversely proportional to the square root of the density, ρ: $V_s = (G'/\rho)^{1/2}$. Since both modulus and density decrease with temperature, the observed decreases in sonic velocity indicate that the modulus is the dominant factor influencing the sonic velocity. However, density and volume do influence sonic velocity indirectly, since the modulus is strongly volume-dependent. Increasing volume decreases the chain-chain interactions, which decreases the modulus. This volume effect, *via its influence on modulus*, is so important that polymer volume can be considered a fundamental influence on acoustic velocities [4]. The same principles apply to polymer thin films on planar acoustic devices. Increasing temperature increases the polymer volume, which decreases the modulus, which decreases acoustic wave velocities, which decreases the observed frequencies.

Studies on Polymer Thin Films

Early Studies on SAW devices

The use of SAW devices in chemical sensing applications was first reported in a series of three papers by Wohltjen and Dessy in 1979 [32-34]. The third of these publications discussed methods for observing glass transitions of polymer samples in contact with the

SAW device surface [33]. These investigators used 30 MHz SAW devices and measured signal amplitudes to observe the attenuation of the waves by the polymer sample. Polymers such as poly(ethylene terephthalate), polycarbonate, and polysulfone were investigated as thin samples pressed against the SAW device surface. Raising the temperature caused a sharp drop in amplitude when the static T_g of the sample was reached. This effect was not reversible. These results were explained by noting that the surface waves were not efficiently coupled into the polymer material at the beginning of the experiment. When the experiments were assembled, microscopic roughness limited contact between the glassy polymer films and the SAW device surfaces. When the polymer samples softened at the T_g, they were pressed into more intimate contact with the SAW surface and acoustic losses into the polymers increased. Each result obtained by this method was in agreement with the known static T_g of the polymer and with DMA at a low frequency (3.5 Hz).

A single polymer, poly(methyl methacrylate), was investigated as a film cast directly on the device surface to create intimate contact between the polymer film and the SAW surface throughout the experiment. The film thickness was ca. 2.5 μm. A reversible drop in signal amplitude was observed at 150°C, which is above the static T_g of this polymer at ca. 105°C. This result was interpreted as indicating that the surface waves were coupled into the polymer film, inducing periodic strains in the polymer, and increasing the measured temperature of the transition according to the time-temperature superposition principle. One troubling aspect of this result was that the α relaxation of poly(methyl methacrylate) is known to occur at temperatures above 200°C at frequencies above 1 MHz, [35] suggesting that the polymer sample in this particular study may have contained some plasticizer.

Investigations of polymer samples on 30 MHz SAW devices were continued by Groetsch and Dessy, who also measured signal amplitudes [36]. Polystyrene, polysulfone, poly(methyl methacrylate), and polycarbonate were examined as cast films. Results on SAW devices were compared with T_g values determined by DSC and loss peaks observed by DMA at 1 Hz. With one exception, poly(methyl methacrylate), the temperatures of the SAW amplitude minima were similar to the loss peaks observed by DMA at 1 Hz. These results are difficult to reconcile with the expectation that the high-frequency surface waves should couple into the polymer film and increase the temperature at which the dynamic transition is observed. If this were the case, then one would expect transition temperatures measured at 30 MHz on the SAW device to be significantly higher than those determined at only 1 Hz by DMA. The temperatures of the SAW minima for poly(methyl methacrylate) and polystyrene reported by these authors are also not consistent with known dynamic relaxation behavior of these polymers at high frequencies. The poly(methyl methacrylate) results were similar to those reported previously by Wohltjen [33]. The position of the polystyrene SAW minimum reported at 110°C is much lower than the temperatures of the known transitions of polystyrene at high frequencies: at frequencies above 1 MHz, the α relaxation of this polymer occurs at temperatures above 150°C [1]; and recent bulk-wave ultrasonic studies indicate transition temperatures of 154 and 163°C for polystyrene samples at 1.75 MHz [5].

More recently, Ballantine and Wohltjen attempted to observe the dynamic glass transition by following SAW device frequency [37], which monitors the velocities of the surface waves. 158 MHz SAW devices were used with polymer films of ca. 50 to 100 nm thickness. It was assumed that the frequency of the SAW would raise polymer transition temperatures. The experiments did not investigate whether SAW frequency changes would occur in the range of the static T_g. Of the four polymers examined, three were examined only at temperatures well above their static T_g, while a fourth was examined at temperatures beginning quite near the T_g. The results of these SAW frequency investigations were difficult to interpret. The actual dynamic transition temperatures of the selected polymers at high frequencies (as determined by independent methods such as DMA or bulk-wave ultrasonic studies) were not known.

These early investigations left a number of questions unanswered. First, clear distinctions between the frequency-independent T_g and the frequency-dependent T_α were not made, and in some cases the discussions were confusing because both these phenomena are referred to as glass transitions. The static glass transition was not observed in cast films, nor was it known if or how it should be observed. Results claimed to be due to the dynamic glass transition were not correlated with known properties of the polymers at high frequencies. In one study frequency appeared to have little effect on the transition phenomena claimed to be observed. So it was not entirely clear if the periodic motion at the device surfaces would cause the observed temperature of the polymer glass transition in a thin film to rise with frequency, as would be expected from conventional dynamic mechanical analysis and considerations of polymer relaxation phenomena. The full effects of such a transition on wave velocity and attenuation, and hence sensor signals, were not elucidated.

Transitions Observed on FPW Devices

In 1991, Grate, Wenzel, and White reported that both the static and dynamic glass transitions could be observed in thin films on the FPW device at 5 MHz [*17*]. Full experimental details were presented in 1992 [*18*]. Polymer samples were applied as thin films (ca. 0.5 μm) by a spray-coating technique. Changes in wave velocities were indicated by measurements of the frequency of the device in an oscillator circuit (recording frequency measurements by computer at one-minute intervals while the temperature was ramped at 10 to 20°C per h), and changes in wave attenuation were monitored by following the oscillator signal amplitude with an oscilloscope (amplitude data were recorded manually). The polymer-coated devices were maintained under an atmosphere of dry nitrogen during these experiments. Results were reported for poly(vinyl acetate), poly(t-butyl acrylate), poly(vinyl propionate), and poly(isobutylene). In every case, the observed results were correlated with the properties of the polymers known from conventional methods for measuring glass transition phenomena.

Results for poly(t-butyl acrylate), illustrating the observation of the static glass transition at T_g, are shown in Fig. 2. The T_g of this polymer occurs at ca. 43°C [*38–40*]. In Fig. 2, it is seen that the frequency-temperature plot for the poly(t-butyl acrylate)-coated FPW device is linear below and above T_g with a clear change in slope at T_g. These results are similar to

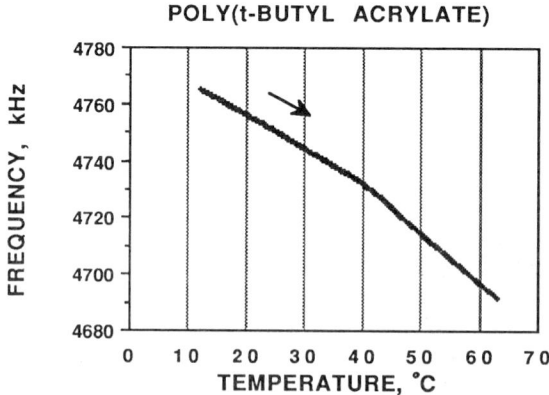

FIG. 2—*The observation of the static glass transition of poly(t-butyl acrylate) at ca. 43°C as a change in the slope of the frequency-temperature plot of a poly(t-butyl acrylate)-coated FPW device at 5 MHz.*

those obtained in dilatometric and conventional bulk wave ultrasonic studies of polymers at the T_g. In dilatometric studies, the polymer volume is monitored as function of temperature, where the slope indicates the coefficient of thermal expansion. This slope changes at the T_g. Similarly, in conventional bulk wave ultrasonic studies, the acoustic velocity through a polymer decreases with increasing temperature, and this decrease becomes steeper at T_g. (Acoustic velocities decrease as the modulus decreases.) It should be noted that acoustic studies of polymers are dynamic experiments where the acoustic wave motion induces strains in the polymer at the test frequency. Nevertheless, the change in the rate of polymer thermal expansion occurs at the static glass transition temperature. Thus, the transition in thermal expansion rates is frequency-independent.

The static glass transition of poly(vinyl acetate) was also observed as a change in slope of the frequency-temperature plot of a poly(vinyl acetate)-coated FPW device, as shown in Fig. 3. The change in slope occurs in the range of 30 to 40°C, compared to the known T_g of this polymer of ca. 33°C. This polymer was further investigated at temperatures up to 110°C, following both frequency and signal amplitude to see if the dynamic glass transition (at T_α) could be observed. Data compiled by McCrum et al. in a plot of T_α as a function of temperature and frequency indicate that T_α falls between 90 to 120°C in the 1 to 10 MHz range [1]. Thurn and Wolf show plots of sound velocity and absorption at 2 MHz with a loss peak at 93°C [41]. On the FPW device, as shown in Fig. 3, there is a minimum in signal

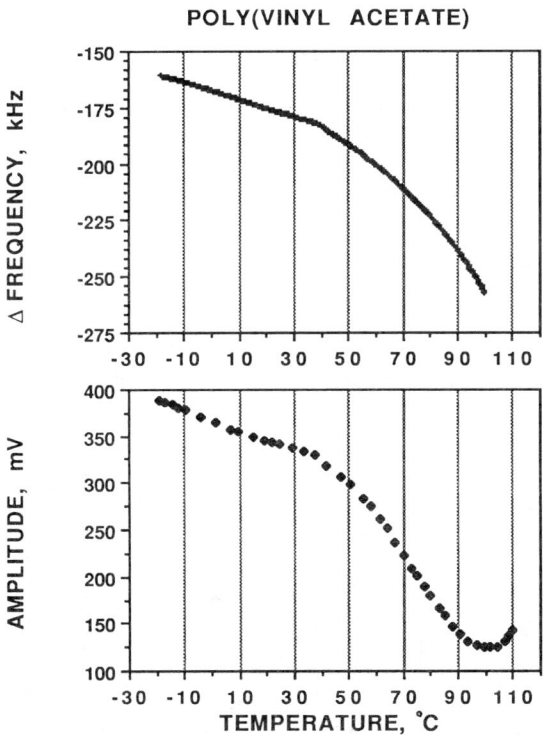

FIG. 3—*Static and dynamic glass transitions in a film of poly(vinyl acetate) on a 5 MHz FPW device. The static glass transition is indicated by the change in slope of the frequency-temperature plot in the 30 to 40°C range. They dynamic glass transition is indicated by the minimum in amplitude and drop in frequency at ca. 100°C.*

amplitude at 100°C, indicating a maximum in wave attenuation by the polymer sample. At the same time, the frequency curves downward as the temperature approaches 100°C. These results are very similar in appearance to the plots shown by Thurn and Wolf, where sound velocity drops sigmoidally in the region of the dynamic glass transition, and a peak in sound damping occurs. These are also the types of results that might be expected by comparison with DMA. At the dynamic glass transition, the measured elastic modulus drops from a high value typical of a glass (or a rubber at high frequency), usually ca. 10^9 N/m^2, to a lower value typical of a rubber, usually ca. 10^6 N/m^2, and a peak in the lost tangent is observed.

Results for poly(isobutylene) are shown in Fig. 4. A previous paper by Ivey et al. described conventional bulk-wave ultrasonic studies on a cured sample of butyl rubber, and showed plots of both sound velocity and sound absorption. These studies were conducted at various frequencies, including 3 MHz and 10 MHz [42]. Broad peaks of sound absorption with maxima at ca. 25°C at 3 MHz and 40°C at 10 MHz indicated the occurrence of the dynamic glass transition. Sound velocity curves in the same temperature regions were nonlinear. Butyl rubber is actually a copolymer of isobutylene with a few percent of isoprene. Results obtained with this rubber may not be identical to those of a pure uncrosslinked poly(isobutylene), but they are representative of the type of behavior to be expected. The viscoelastic behavior of poly(isobutylene) is described by McCrum et al. [1]. A plot of T_α as a function of temperature and frequency in this reference suggests that the temperature at which the α relaxation is observed in the 1 to 10 MHz frequency domain should be in between 0 to 20°C. For comparison, the T_g of poly(isobutylene) is -76°C.

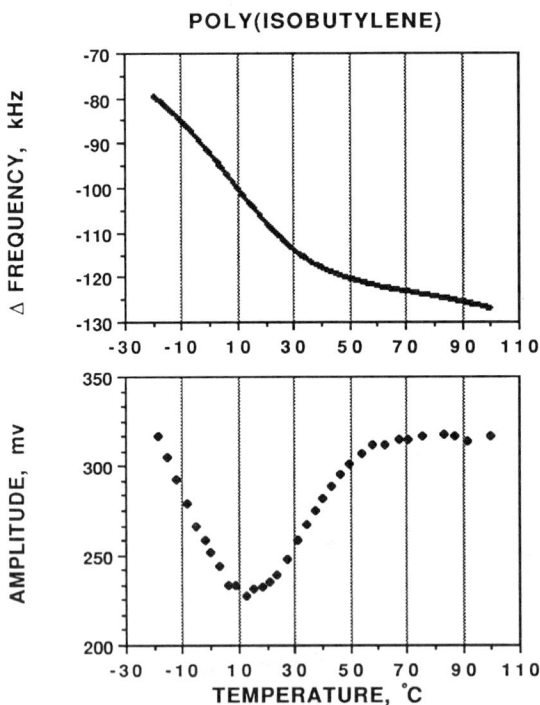

FIG. 4—*The observation of the dynamic glass transition in a film of poly(isobutylene) on a 5 MHz FPW device, indicated by the minimum in amplitude and drop in frequency at ca. 15°C.*

The frequency-temperature plot for the poly(isobutylene)-coated FPW device in Fig. 4 is highly curved. It is very similar in appearance to the sound velocity curves published by Ivey et al. where the slope changes in a sigmoidal fashion. The FPW frequency results also appear to be sigmoidal in shape, although more data at lower temperatures would be required to fully confirm this. The plot of the signal amplitude clearly shows a broad minimum, indicating a maximum in signal attenuation around 15°C which is associated with the sigmoidal changes in frequency. This maximum in signal attenuation occurred at slightly lower temperatures than Ivey et al. [42], but consistent with the compilation in McCrum's book [1]. The overall behavior of poly(isobutylene) was clearly in accord with the known dynamic properties of this polymer.

The dynamic glass transition process is associated with the relaxation of polymer chain segments as they respond to mechanical strains. The results obtained in these studies of polymer-coated FPW devices demonstrated unambiguously that relaxation processes in polymer thin films on an acoustic wave device can be observed, and that the temperatures at which the α relaxation processes are observed (T_α) are shifted to temperatures higher than the polymer T_g values. The acoustic waves are therefore coupled into the polymer film and strain the polymer. The results with poly(vinyl acetate) graphically demonstrated that separate transitions occur at both T_g and T_α.

These transitions can be observed because acoustic wave devices are sensitive to the modulus of the medium in contact with their surfaces. At temperatures below T_α the polymer modulus decreases continuously with increasing temperature as the polymer expands. At the T_g, there is a change in the rate of thermal expansion, so there is a corresponding change in the rate of modulus change with temperature, and this can be observed as a change in the rate of wave velocity decrease with temperature. (In other words, there is not a modulus drop at the static T_g, but rather an increase in the rate of change of modulus with temperature which is related to the change in thermal expansion coefficient.) As the temperature continues to rise, a regime is reached where the characteristic relaxation time of the polymer sample is similar to the wave period. In this region there is a sudden drop in the modulus as it is perceived by the high-frequency waves that is associated with the dynamic glass transition, resulting in a sigmoidal drop in wave velocity. In addition, wave energy is efficiently coupled into the polymer material and dissipated in this regime, causing the peak in wave attenuation. These results are analogous to those seen in conventional bulk wave acoustic studies.

Modulus Measurements and Film Resonance Effects on the TSM Device

Since the acoustic wave devices are sensitive to polymer modulus, it follows that it may be possible to measure absolute modulus values. The theory for this is somewhat difficult for SAW and FPW devices, but is simpler for TSM devices. On TSM devices, the surface motion is entirely in-plane, whereas SAW and FPW devices have surface-normal wave motion. In 1991 Martin and Frye reported a model describing the response of a quartz TSM device to changes in the modulus of a polymer film on its surface, and experimentally verified this model in studies of polymer-coated TSM devices [13,29]. Using a network analyzer to measure admittance as a function of frequency, and an equivalent circuit model to evaluate the data, they were able to determine polymer elastic shear modulus G' and loss modulus G'' from acoustic wave sensor data [13,29]. Plotting these values as a function of temperature, the characteristics of a polymer relaxation process were observed for a poly(isobutylene) film (15.6 μm thick). Thus, the quartz TSM device can be used to monitor intrinsic polymer properties.

These TSM experiments and models also demonstrated an additional viscoelastic effect, referred to as film resonance, that can occur in relatively thick films. The film resonance effect depends on device frequency, film thickness, the polymer complex modulus, and the polymer density. It is observed experimentally by ramping the temperature of a polymer-coated device in order to vary the modulus. At lower temperatures where the polymer modulus is high and the film is rigid, the entire thickness of the film moves synchronously with the device surface. As the temperature increases and the polymer modulus decreases, the displacement at the upper surface of the film begins to exceed that at the polymer/device interface, where the polymer adheres and moves synchronously. This nonsynchronous motion, or phase lag, induces strain in the film. Continuing decreases in modulus with increasing temperature increase the phase lag and the strain. When the phase lag reaches $\pi/2$, a condition of film resonance is reached. The frequency suddenly increases to values exceeding the initial frequency, and the wave energy is highly damped. For a given polymer, the temperature at which film resonance occurs decreases with increasing device frequency or with increasing film thickness. Although these film resonance effects are simplest to understand on a TSM device where the surface motion is entirely in-plane, they also occur on SAW devices if relatively thick films are used [29]. It should be noted that both the dynamic glass transition and the film resonance effect produce peaks in the damping or attenuation of wave energy. The two effects can be distinguished by determining the effects of the device frequency and film thickness on the temperature of the loss peak.

Polymer Curing on the SH-APM Device

SH-APM devices have been used to monitor the photopolymerization of a negative photoresist [21]. Resist films were spin-coated onto a 100 MHz quartz SH-APM device and solvent was evaporated at 85°C (nominal film thickness: 1.1 μm). The acoustic wave characteristics of the coated device were measured using an oscillator to generate frequency signals proportional to wave velocities, and a vector voltmeter to follow wave attenuation. Cross-linking was initiated with 380-nm light from a xenon lamp. As the reaction proceeded, wave velocity increased, while the attenuation decreased, indicating that the elastic modulus was increasing and the loss modulus was decreasing. By measuring curing rates as a function of optical wavelength, action spectra for the photopolymerization reaction could be determined.

Final Remarks

Acoustic wave microsensors are capable of observing a variety of phenomena in polymer thin films, including the static and dynamic glass transitions. The acoustic waves are sensitive to changes in the polymer's modulus because the waves couple into the polymer film, where wave energy is stored and dissipated. Changing polymer properties are observed as variations in the acoustic wave velocity and attenuation (or damping). The results obtained with polymer thin films on the FPW device were essentially similar in form to those that are observed in conventional bulk wave ultrasonic studies, and are related to dilatometric and dynamic mechanical studies. In addition, acoustic measurements on thin films on surfaces can give rise to the phenomenon of film resonance, which must be distinguished from the observations of intrinsic polymer properties such as glass transitions.

Acoustic wave microsensors are also capable of observing phase transitions in a variety of other thin film materials, including thin liquid crystalline layers, Langmuir-Blodgett films, and multilayer films [43,44]. Cross-linking in resist layers can be monitored [21], and it may be possible to monitor paint drying. The latter potential application would require that the device operate with an initially liquid film applied. The FPW, TSM, and SH-APM devices can all operate in contact with liquids, whereas the SAW device cannot. Viscosity sensors based on FPW and SH-APM devices can sense viscosities in samples as small as 10 μL [24,25].

Acoustic wave microsensors offer the advantages of very small sample size, the capability to make measurements on thin films, and instrumentation that is well suited to automation and interfacing with digital electronics. Both the static and dynamic glass transitions can be observed in one measurement. On devices such as the TSM where harmonic modes are available, one can make measurements at various frequencies on a single device and polymer film. In addition, acoustic wave microsensors allow direct measurements at high frequencies.

Acknowledgments

The author is happy to acknowledge collaborations and discussions with Richard M. White, Stewart W. Wenzel, Stephen J. Martin, and Gregory C. Frye on acoustic wave microsensors, film resonance effects, and polymer transitions.

References

[1] McCrum, N. G., Read, B. E., and Williams, G., *Anelastic and Dielectric Effects in Polymeric Solids*, John Wiley and Sons, New York, 1967.
[2] Ferry, J. D., *Viscoelastic Properties of Polymers*, John Wiley and Sons, Inc., New York, 1980.
[3] Massines, R., Piche, L., and Lacabanne, C., "Ultrasonic Characterization of Polymer Viscoelasticity. Application to Polystyrene," Makromolecular Chemie, Macromolecular Symposium, Vol. 23, 1989, pp. 121–137.
[4] Hartman, B., "Acoustic Properties," in *Encyclopedia of Polymer Science and Engineering, 2nd Ed., Vol. 1*, John Wiley and Sons, Inc., New York, 1984, pp. 131–160.
[5] Piche, L., "Phenomena Related to the Propagation of Ultrasound in Polymers," *Proceedings of the IEEE Ultrasonics Symposium*, 1989, pp. 599–608.
[6] Kwan, S. F., Chen, F. C., and Choy, C. L., "Ultrasonic Studies of Three Fluoropolymers," *Polymer*, Vol. 16, 1975, pp. 481–488.
[7] Wada, Y. and Yamamoto, K., "Mechanical Dispersion and Transition Phenomena in Semicrystalline Polymers," *Journal of the Physical Society of Japan*, Vol. 11, 1956, pp. 887–892.
[8] Work, R., "On the Discontinuity of the Temperature Coefficient of the Velocity of Ultrasonic Waves in Polymer Materials," *Journal of Applied Physics*, Vol. 27, 1956, pp. 69–72.
[9] Alder, J. F. and McCallum, J. J., "Piezoelectric Crystals for Mass and Chemical Measurements," *Analyst (London)*, Vol. 108, 1983, pp. 1169–1189.
[10] Buttry, D. A. and Ward, M. D., "Measurement of Interfacial Processes at Electrode Surfaces with the Electrochemical Quartz Crystal Microbalance," *Chemical Reviews*, Vol. 92, 1992, pp. 1355–1379.
[11] McCallum, J. J., "Piezoelectric Devices for Mass and Chemical Measurements: An Update," *Analyst (London)*, Vol. 114, 1989, pp. 1173–1189.
[12] Lu, C.-S. and Czanderna, A. W., *Applications of Piezoelectric Quartz Crystal Microbalances*, Elsevier, New York, 1984.
[13] Martin, S. J. and Frye, G. C., "Polymer Film Characterization Using Quartz Resonators," *Proceedings of the IEEE Ultrasonics Symposium*, 1991, pp. 393–398.
[14] Wohltjen, H., "Mechanism of Operation and Design Considerations for Surface Acoustic Wave Device Vapour Sensors," *Sensors and Actuators*, Vol. 5, 1984, pp. 307–325.
[15] Venema, A., Nieuwkoop, E., Vellekoop, M. J., et al., "Design Aspects of SAW Gas Sensors," *Sensors and Actuators*, Vol. 10, 1986, pp. 47–64.
[16] Frye, G. C. and Martin, S. J., "Materials Characterization Using Surface Acoustic Wave Devices," *Applied Spectroscopy Reviews*, Vol. 26, 1991, pp. 73–149.
[17] Grate, J. W., Wenzel, S. W., and White, R. M., "Flexural Plate Wave Devices for Chemical Analysis," *Analytical Chemistry*, Vol. 63, 1991, pp. 1552–1561.
[18] Grate, J. W., Wenzel, S. W., and White, R. M., "Frequency-Independent and Frequency-Dependent Polymer Transitions Observed on Flexural Plate Wave Ultrasonic Sensors," *Analytical Chemistry*, Vol. 64, 1992, pp. 413–423.
[19] Wenzel, S. W. and White, R. M., "A Multisensor Employing an Ultrasonic Lamb-wave Oscillator," *IEEE Transactions, Electron Devices*, Vol. 35, 1988, pp. 735–743.

[20] White, R. M., Wicher, P. J., Wenzel, S. W., and Zellers, E. T., "Plate-mode Ultrasonic Oscillator Sensors," *IEEE Transactions, Ultrasonics, Ferroelectrics, and Frequency Control*, Vol. UFFC-34, 1987, pp. 162–171.
[21] Martin, S. J. and Ricco, A. J., "Monitoring Photo-Polymerization of Thin Films Using SH Acoustic Plate Mode Sensors," *Sensors and Actuators*, Vol. A21-A23, 1990, pp. 712–718.
[22] Martin, S. J., Ricco, A. J., Niemczyk, T. M., and Frye, G. C., "Characterization of SH Acoustic Plate Mode Liquid Sensors," *Sensors and Actuators*, Vol. 20, 1989, pp. 253–268.
[23] Hughes, R. C., Martin, S. J., Frye, G. C., and Ricco, A. J., "Liquid-Solid Phase Transition Detection with Acoustic Plate Mode Sensors: Application to Icing of Surfaces," *Sensors and Actuators*, Vol. A21-A23, 1990, pp. 693–699.
[24] Ricco, A. J. and Martin, S. J., "Acoustic Wave Viscosity Sensor," *Applied Physics Letters*, Vol. 50, 1987, pp. 1474–1476.
[25] Martin, S. J. and Ricco, A. J., "Effective Utilization of Acoustic Wave Sensor Responses: Simultaneous Measurement of Velocity and Attenuation," *Proceedings of the IEEE Ultrasonics Symposium*, 1989, pp. 621–625.
[26] Kipling, A. L. and Thompson, M., "Network Analysis Method Applied to Liquid-Phase Acoustic Wave Sensors," *Analytical Chemistry*, Vol. 62, 1990, pp. 1514–1519.
[27] Martin, S. J., Granstaff, V. E., and Frye, G. C., "Characterization of a Quartz Crystal Microbalance with Simultaneous Mass and Liquid Loading," *Analytical Chemistry*, Vol. 63, 1991, pp. 2272–2281.
[28] Bartley, D. L. and Dominguez, D. D., "Elastic Effects of Polymer Coatings on Surface Acoustic Waves," *Analytical Chemistry*, Vol. 62, 1990, pp. 1649–1656.
[29] Martin, S. J. and Frye, G. C., "Dynamics and Response of Polymer-coated Acoustic Devices," *Proceedings of the 1992 Solid State Sensor and Actuator Workshop*, 1992, pp. 27–31.
[30] Costello, B. J., Wenzel, S. W., Wang, A., and White, R. M., "Gel-coated Lamb Wave Sensors," *Proceedings of the IEEE Ultrasonics Symposium*, 1990, pp. 279–283.
[31] Lewis, A. F., "The Frequency Dependence of the Glass Transition," *Polymer Letters*, Vol. 1, 1963, pp. 649–654.
[32] Wohltjen, H. and Dessy, R. E., "Surface Acoustic Wave Probe for Chemical Analysis. I. Introduction and Instrument Description," *Analytical Chemistry*, Vol. 51, 1979, pp. 1458–1464.
[33] Wohltjen, H. and Dessy, R. E., "Surface Acoustic Wave Probe for Chemical Analysis. III. Thermomechanical Polymer Analyser," *Analytical Chemistry*, Vol. 51, 1979, pp. 1470–1475.
[34] Wohltjen, H. and Dessy, R. E., "Surface Acoustic Wave Probe for Chemical Analysis. II. Gas Chromatography Detector," *Analytical Chemistry*, Vol. 51, 1979, pp. 1465–1470.
[35] Phillips, D. W. and Pethrick, R. A., "Ultrasonic Studies of Solid Polymers," *Journal of Macromolecular Science—Reviews of Macromolecular Chemistry*, Vol. C16, 1977–78, pp. 1–22.
[36] Groetsch, J. A. III and Dessy, R. E., "A Surface Acoustic Wave (SAW) Probe for the Thermomechanical Characterization of Selected Polymers," *Journal of Applied Polymer Science*, Vol. 28, 1983, pp. 161–178.
[37] Ballantine, D. S. and Wohltjen, H., "Elastic Properties of Thin Polymer Films Investigated with Surface Acoustic Wave Devices," in *Chemical Sensors and Microinstrumentation, ACS Symposium Series 403*, American Chemical Society, Washington, DC 1989, pp. 222–236.
[38] Shetter, J. A., "Effect of Stereoregularity on the Glass Temperatures of a Series of Polyacrylates and Polymethacrylates," *Journal of Polymer Science, Part B*, Vol. 1, 1963, pp. 209–219.
[39] Kine, B. B. and Novak, R. W., "Acrylic and Methacrylic Ester Polymers," in *Encyclopedia of Polymer Science and Engineering, 2nd Ed.*, Vol. 1. John Wiley and Sons, Inc., 1984, pp. 257.
[40] Lewis, O. G., *Physical Constants of Homopolymers*, Springer-Verlag, New York, 1968.
[41] Thurn, H. and Wolf, K., "Vergleichende dielektrische und ultraschall-messungen bei 2 megahertz an polyvinylestern, polyacrylestern und polyvinylathern (Comparative Dielectric and Ultrasound Measurements at 2 MHz of poly(vinyl esters), poly(vinyl acrylates), and poly(vinyl ethers))," *Kolloid—Zietschrift.*, Vol. 148, 1956, pp. 16–30.
[42] Ivey, D. G., Mrowca, B. A., and Guth, E., "Propagation of Ultrasonic Bulk Waves in High Polymers," *Journal of Applied Physics*, Vol. 20, 1949, pp. 486–492.
[43] Muramatsu, H. and Kimura, K., "Quartz Crystal Detector for Microrheological Study and Its Application to Phase Transition Phenomena of Langmuir-Blodgett Films," *Analytical Chemistry*, Vol. 64, 1992, pp. 2502–2507.
[44] Okahata, Y. and Ebato, H., "Application of a Quartz-Crystal Microbalance for Detection of Phase Transitions in Liquid Crystals and Lipid Multibilayers," *Analytical Chemistry*, Vol. 61, 1989, pp. 2185–2188.
[45] Martin, B. A., Wenzel, S. W., and White, R. M., "Viscosity and Density Sensing with Ultrasonic Plate Waves," *Sensors and Actuators*, Vol. A21-A23, 1990, pp. 704–708.

Mark L. O'Neill[1] and Y. Paul Handa[2]

Plasticization of Polystyrene by High Pressure Gases: A Calorimetric Study*

REFERENCE: O'Neill, M. L. and Handa, Y. P., "**Plasticization of Polystyrene by High Pressure Gases: A Calorimetric Study,**" *Assignment of the Glass Transition, ASTM STP 1249*, R. J. Seyler, Ed., American Society for Testing and Materials, Philadelphia, 1994, pp. 165–173.

ABSTRACT: A Tian-Calvet type heat-flow calorimeter has been modified for use at pressures up to 400 bar, and has been used for studying the glass transitions in the systems polystyrene (PS)-methane (CH_4), -ethylene (C_2H_4), and -carbon dioxide (CO_2). All three gases plasticize PS leading to lowering of the glass transition temperature T_g. The low-solubility gas CH_4 induces somewhat weak plasticization; the T_g decreases gradually, with a dT_g/dP value of about -0.2 K bar^{-1}, for pressures up to 150 bar after which the T_g remains almost constant for pressures up to 350 bar. At higher gas pressures, CH_4 acts more like a hydrostatic-pressure generator than as a plasticizer such that the hydrostatic effect, which tends to raise the T_g, cancels out any additional plasticization effect. The moderately soluble gases, C_2H_4 and CO_2, show rather strong plasticization effects with a dT_g/dP value of about -0.9 K bar^{-1} for both gases. The highest pressures at which the glass transition in the two-phase polymer-gas systems PS-C_2H_4 and PS-CO_2 could be measured were 88.0 and 60.0 bar, respectively. For the PS-CO_2 system where gas solubility data are available, the measured T_g's agree very well with the values calculated from the model proposed by Chow [25].

KEYWORDS: high-pressure calorimetry, polymer plasticization, glass-transition temperature depression, polystyrene-methane, polystyrene-ethylene, polystyrene-carbon dioxide

There are many applications in which polymers come into contact with high-pressure gases or supercritical fluids [1]. These include extraction of unreacted monomers or other species from the polymer matrix [1,2], incorporation of additives into the polymer matrix [3], gas separations using polymeric membranes [4], and production of microcellular plastics [5] and foams [6]. The solubility of a gas in a polymer depends largely on how much the gas is above its critical temperature and, somewhat, on the polymer-gas interactions. Under elevated pressures, certain gases like carbon dioxide (CO_2) can dissolve to appreciable levels, as high as 30 to 40% by mass. In this respect, the dissolved gas will be expected to plasticize the polymer in the same way as sorbed liquids or vapors do and, thus, to lead to a lowering of the glass-transition temperature, T_g, of the polymer. In fact, CO_2 has been shown to be a very effective plasticizer that can lower the T_g of polymers by a few tens of degrees [7–9], and a recent paper [10] has reported a depression of 226 K in the T_g of poly(2,6-dimethyl phenylene oxide) induced by CO_2 at 60 bar.

Though the solubilities of compressed gases in polymers have been studied quite extensively for over 40 years, there have been only a few studies on the plasticization effect of the

---

[1]Ph.D. candidate, Department of Chemistry, Carleton University, Ottawa, Ontario, Canada K1S 5B6.
[2]Senior research officer, Institute for Environmental Chemistry, National Research Council of Canada, Ottawa, Ontario, Canada K1A 0R6; to whom all correspondence should be addressed.
*Issued as NRCC No. 37550.

dissolved gas. Assink [11] measured the NMR relaxation times of poly(dimethyl siloxane) under high-pressure gases and found that helium, a low-solubility gas, acted only as a pressure-generating medium and tended to raise the T_g whereas argon and nitrogen, slightly soluble gases, tended to lower the T_g. Wang et al. [7] in their study of the polystyrene (PS)-CO_2 system found that the T_g decreased with pressure P rather sharply at first but at high enough pressures it started to increase, concluding thereby that the plasticization and hydrostatic effects are operative at all pressures and that the competition between the two effects leads to a minimum in the T_g-P profile. They measured Young's modulus and static creep compliance to obtain the T_g-P profiles. This technique was also used by Wissinger and Paulaitis [9] to study the plasticization of PS and poly(methyl methacrylate) (PMMA) by methane (CH_4) and CO_2. DSC measurements on polymer samples pretreated with high-pressure gas [8] and on sealed polymer-gas systems [10] have also been attempted. These techniques give reasonable signatures for the glass transition though the thermodynamic state of the polymer-gas system is not well-defined. The onset of plasticization has also been inferred from observations of the anomalous permeability [12–14], loss of permselectivity [14,15], anomalous sorption [16,17] and dilation [16,18] behavior demonstrated by polymeric gas-separation membranes under elevated pressures.

Recently, we developed a high-pressure calorimetric technique for measuring directly the variation in T_g as a function of the gas pressure. A preliminary report on this technique and some results on the PS-ethylene (C_2H_4) system have been presented elsewhere [19]. In this paper, we describe the high-pressure calorimeter and report the T_g measurements for the systems PS-CH_4, -C_2H_4, and -CO_2.

Experimental

Methane, ethylene, and carbon dioxide with specified minimum purities of 99.9% were obtained from Matheson Canada. Polystyrene, Styron 680, was supplied by Dow Chemicals of Canada. It contained no additives, was in the form of 2-mm by 3-mm cylindrical pellets, and had M_w = 193 700 and M_w/M_n = 2.45.

A Tian-Calvet heat-flow calorimeter (Setaram, model BT) was modified to allow for in-situ measurements of the glass transition in polymer-gas systems. A schematic of the setup used in the present work is given in Fig. 1. The calorimeter as purchased was suitable for work at ambient pressures only. The details of the calorimeter, data acquisition and control system (DA/C), and the operating procedure can be found elsewhere [20,21]. Briefly, the heart of the calorimeter is a massive aluminum block with a heating coil on the outside and two identical cylindrical cavities located symmetrically about the center into which are installed a reference and a measurement cell. Each cell is surrounded by a thermopile, and the two thermopiles are connected in opposition to give the differential heat-flow signal. The computer through the DA/C regulates the power applied, via the power supply, to the heater coil in order to maintain a steady heating rate. It also makes, via the digital voltmeter (DVM), four-wire resistance measurements of the platinum resistance thermometers located in between the two cells and in the heating block. The thermopile signal via the nanovoltmeter (NV) and the voltage output from the pressure transducer (T) are also recorded by the computer-DA/C system.

The major modification to the apparatus was in developing high-pressure cells and the associated system, and in developing the procedure for making measurements under high pressures. Two sets of matched pairs of reference and sample cells were fabricated. One set was made from 316SS for use at pressures to 250 bar, and the second set was made from 410SS for use at pressures to 400 bar. The mass of the 316SS cell was about 48 g and that of the 410SS cell was about 66 g. The internal volume of the cells was about 8 cm^3. The lower

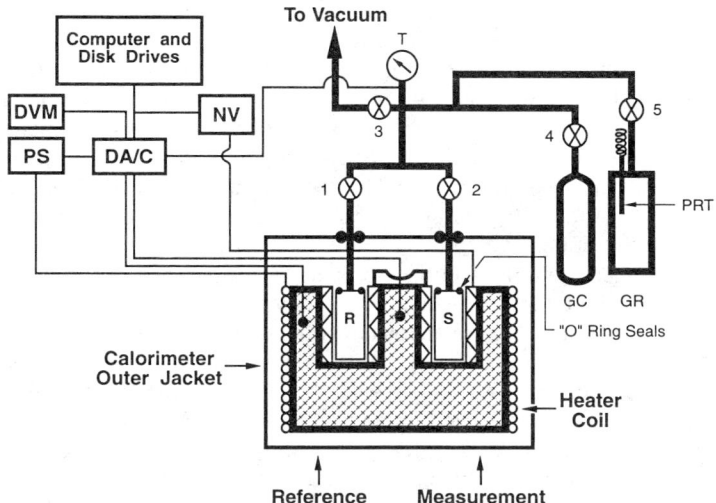

FIG. 1—*Block diagram of high-pressure calorimeter (see text for details).*

pressure cells were sealed using aluminum O-rings whereas the higher pressure cells were sealed using copper O-rings. The sample cell was frequently removed to replace the polymer contents; the reference cell was permanently installed in the calorimeter. Both cells were connected to a high-pressure manifold, which in turn was connected to a vacuum pump, the pressure transducer (T), and a high-pressure gas reservoir (GR) of approximately 500 cm^3 internal volume. The gas reservoir was equipped with a platinum resistance thermometer (PRT) and was connected directly to the desired gas cylinder (GC). The total volume including the cells, the manifold, and the gas reservoir was about 550 cm^3. This large volume served as a pressure sink and ensured that measurements were essentially isobaric; as, without the buffer volume, increases of up to 30% in the system pressure were experienced during the calorimeter scan from room temperature to about 373 K. With the buffer volume on line, the system pressure rose only about 2% during a scan. This translated into a rather small pressure increase during the actual transition which was usually spread over about 20 K. The top of the calorimeter was controlled at 303 or 323 K, depending on the gas. The temperature of the cells was always kept above the condensation temperature of the gas. This ensured that the gas did not condense anywhere inside the calorimeter. The rest of the gas system outside the calorimeter was thermostatted at 313 K in the case of CO_2 or was kept at (295 ± 1) K in the case of CH_4 and C_2H_4, again to avoid any condensed phase of the gas. Dry helium was used as the heat exchange fluid during annealing and cooling sequences, while dry nitrogen was used during measurements.

The sample cell was charged with about 2.5 g of PS, installed in the calorimeter, and the whole system evacuated for at least 30 min. The gas reservoir was then filled with an appropriate amount of gas so as to generate the desired pressure in the entire system. Pressures higher than gas cylinder pressure were achieved by cooling the reservoir with liquid nitrogen and condensing into it the required amount of gas. Both reference and measurement cells were pressurized, and during measurements valves 1, 2, and 5 were open. This ensured that perturbations on the heat-flow signal were minimized, as about the same amount of heat leak occurred on the reference and the measurement side. In the case of the 410SS cells, about 0.05 bar of helium was also added to the cells before pressurizing with the gas. This improved the heat flow into both cells and considerably sharpened the glass-transition signal. Once the whole system was pressurized, the sample was scanned at 10 K h^{-1} to a temperature about 5 K above the estimated T_g of the plasticized polymer, and the polymer annealed at this temperature for at least 2 h. Annealing above T_g allowed for rapid equilibration of the polymer with the gas and also destroyed any conformational strains due to previous thermal history. After annealing, the system was cooled at about 5 K h^{-1} to ensure that the polymer remained essentially in equilibrium with the gas as the temperature was lowered through the transition region. The cooling rate was increased once the calorimeter temperature was sufficiently below the transition temperature of the polymer-gas mixture. Complete equilibration time for a sample was at least 20 h.

Once the polymer-gas solution had been properly prepared, the calorimeter was scanned at 7 or 10 K h^{-1} from a temperature about 2 K above the condensation temperature of the gas or from room temperature. Simultaneous measurements of the sample temperature, heat-flow signal, and pressure were made as described before [20,21]. The choice of the starting temperature for a scan depended on the gas and was such that no condensed phase of the gas was encountered during the scan. The sample was scanned to at least 20 K above T_g. It took about 1 h for the calorimeter to reach a steady state and establish a baseline. Thus, the lowest T_g that could be determined was about (scan rate + starting temperature of the scan). For measurements at different pressures, either a fresh sample was used at each pressure or the same sample was used at increasingly higher pressures.

Results and Discussion

The calorimeter output from a scan on pure PS is shown by curve 1 in Fig. 2. In spite of the fact that the thermal mass of the cell was much larger than that of the polymer sample, a fairly sharp glass transition is observed. On first heating the polymer in the presence of gas, a rather sluggish transition is observed, curve 2a. However, on cooling the sample back and reheating it, the transition becomes quite sharp, curve 2b. Thus, it appears that the plasticization starts at the surface of the pellet and progresses inwards as the temperature is raised. At the end of the first heating, the pellet is fully plasticized. Any further change in temperature

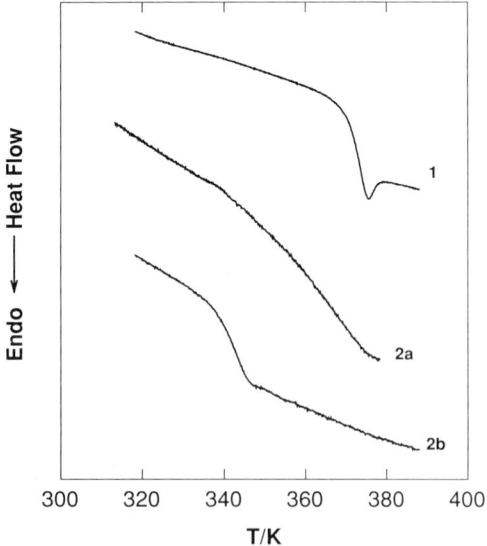

FIG. 2—*Calorimeter output from scans on pure PS (curve 1); first heating of PS under 26.3 bar of C_2H_4 (curve 2a); second heating of PS under 26.3 bar of C_2H_4 (curve 2b).*

only results in changing the solubility of the gas in the polymer, otherwise the polymer-gas interface is maintained within the entire pellet. This behavior was observed in all the PS-gas systems studied in this work. The glass transition temperature was established from the second scan, and was taken as the onset temperature. The glass transition temperature is generally taken as the temperature corresponding to 50% conversion of the glass into liquid, and is normally identified with the temperature at the midpoint of the transition. However, due to sluggishness of the transition at higher pressures (see below), this temperature was found to be less reproducible than the onset temperature and, thus, was not the favored choice as T_g. Moreover, since we are interested in the relative shift in T_g as a function of pressure, the use of the reproducible onset temperature as T_g is quite justified. For the same reason, the T_g of pure PS was also taken as the onset temperature.

The calorimeter outputs at various pressures for the system PS-C_2H_4 are shown in Fig. 3. Similar scans were obtained for the other systems. For the pure polymer, a small overshoot in the heat capacity is seen in the curve labeled 0 bar. As mentioned above, the cooling rate after the first heating was quite slow. This provides some thermal annealing of the polymer as it cools below its T_g. The enthalpy relaxed during this annealing process is then gained back as the overshoot during the second heating. On the other hand, the enthalpy of the polymer is raised due to sorption of the gas. At low pressures, the increase in enthalpy is not large enough to compensate for the relaxed enthalpy. As a result, the overshoot is still observed at low pressures, as seen in the curve labeled 5.6 bar. However, at higher pressures the overshoot disappears due to larger gains in the enthalpy. The solubility of the gas in the polymer increases with pressure. This raises the enthalpy of the glass to increasingly higher levels, and the difference between the enthalpies of the glass and the liquid becomes increasingly smaller. Consequently, the approach to the glass transition becomes somewhat diffused, as seen by the scans at 27.0 and 50.0 bar. The scans at 91.2 and 304 bar are shown in Fig. 4. At 91.2 bar, the polymer still seems to undergo a glass transition though the initial part of the transition could not be established. The critical temperature T_c and critical pressure of C_2H_4 are 282.4 K and 50.4 bar, respectively. It is possible that at 91.2 bar the T_g

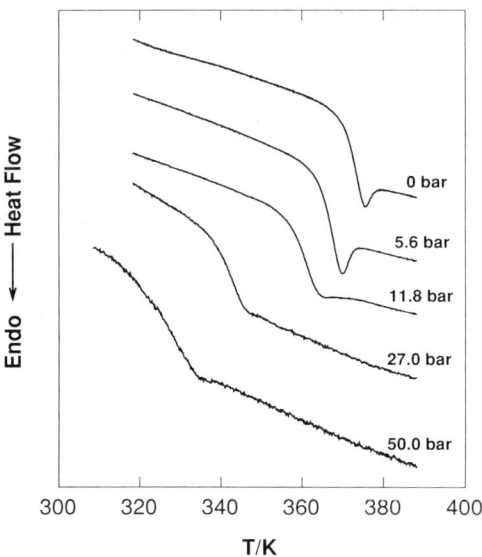

FIG. 3—*Some representative calorimeter-outputs for PS-C_2H_4, scanned under various pressures.*

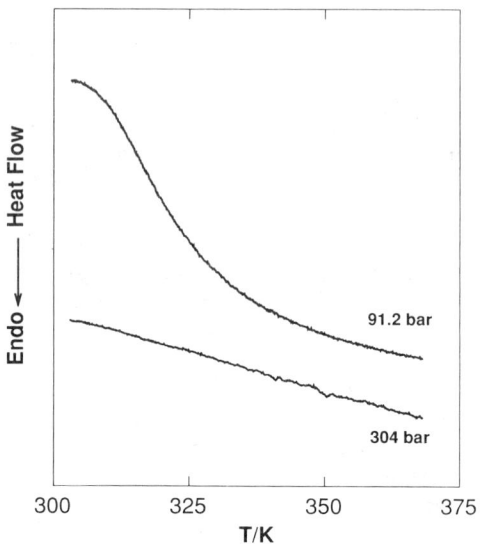

FIG. 4—*Calorimeter outputs for PS-C_2H_4 at 91.2 and 304 bar.*

falls below T_c of C_2H_4. However, the measurements could not be extended to lower temperatures due to condensation of the gas. At 304 bar, no transition is observed indicating that the polymer is already in the liquid state at the lowest temperature from which the scan could be started. The highest pressure studied at which a glass transition in PS-C_2H_4 could be established unambiguously was 88.0 bar. For PS-CO_2, this pressure was 60.0 bar.

The glass transition temperatures at various pressures for the three systems are given in Table 1 and plotted in Fig. 5. For the low-solubility gas CH_4, the plasticization effect is quite weak with dT_g/dP of about -0.2 K bar^{-1} at pressures up to about 150 bar beyond which the $T_g - P$ curve tends to level off. As reported by Assink [11] and Wang et al. [7], the gas exerts both plasticization and hydrostatic effects. Whereas plasticization tends to lower the T_g due to an increase in the free volume of the polymer, the hydrostatic effect tends to decrease the free volume and, thus, to limit the gas solubility, and also to raise the T_g. For PS, limiting values for the hydrostatic effect dT_g/dP_{hyd} in the range 0.034 to 0.046 K bar^{-1}

TABLE 1—T_g's for PS-gas systems at various gas-pressures.

PS-CH_4		PS-C_2H_4		PS-CO_2	
P/bar	T_g/K	P/bar	T_g/K	P/bar	T_g/K
0.0	369.5	0.0	369.5	0.0	369.5
9.9	366.0	5.6	363.6	5.0	365.3
27.0	361.6	11.8	354.5	9.4	360.4
50.2	356.2	27.0	337.9	14.9	355.8
82.4	349.9	34.4	332.6	22.6	347.5
126.6	342.7	50.0	318.1	35.8	337.2
206.1	338.0	79.3	302.9	51.8	323.8
238.8	338.0	88.0	293.5	60.0	314.2
267.5	331.6				
321.8	335.0				
358.6	328.5				

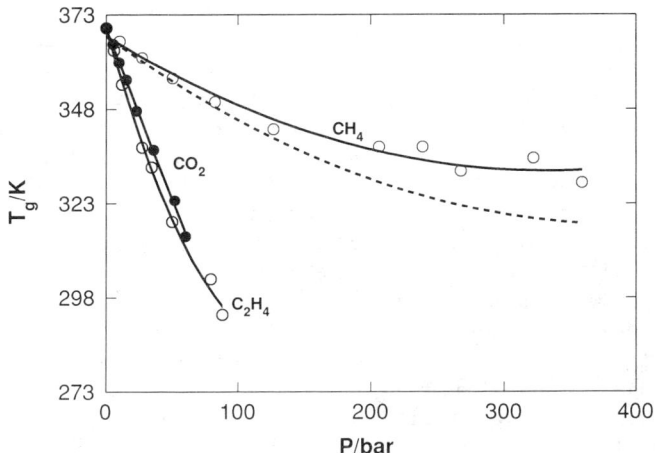

FIG. 5—*Glass transition temperatures for the various PS-gas systems plotted against the gas pressure. Solid curves are drawn to aid the eye; the dashed curve is for the system PS-CH_4 after correcting for the hydrostatic-pressure effect induced by the high-pressure CH_4.*

have been reported in the literature. At high pressures, the increase in T_g due to the hydrostatic effect starts to compensate, and eventually overtakes, the decrease in T_g due to plasticization. This leads to the $T_g - P$ profile for PS-CH_4 as shown in Fig. 5. Wang et al. [7] reported a similar phenomenon for PS-CO_2, whereas no such behavior was observed in the present work. For the PS-CH_4 system, the experimental T_g values were corrected for the hydrostatic effect assuming a value of 0.04 K bar^{-1} for dT_g/dP_{hyd}. The T_g values due to *plasticization* alone are shown by the dashed curve in Fig. 5. Unlike the systems PS-CO_2 and PS-C_2H_4, the lower limit in T_g is not imposed by the fact that the T_g becomes smaller than the T_c of CH_4 but by the limiting solubility of CH_4. In gas separation studies using polymeric membranes, CH_4 has often been assumed to be a nonplasticizing gas [23], though it has been proposed that at high enough pressures CH_4 should induce the same kind of plasticization effect as observed for CO_2 [24]. The gas-separation membranes are normally operated at ambient temperatures, and the present work shows that CH_4 depresses the T_g only by few tens of degrees which may not be enough to bring down the T_g of the polymer to the operating temperature. In this context, CH_4 can essentially be regarded as a nonplasticizing gas.

A model based on the Gibbs-DiMarzio quasi-lattice theory for amorphous polymers has been proposed by Chow [25] to correlate the depression in T_g due to the sorbed component. The basic equation of the model is

$$\ln(T_g/T_{g0}) = \beta[(1-\theta)\ln(1-\theta) + \theta \ln \theta]$$

$$\beta = \frac{zR}{M_p \Delta C_p}$$

$$\theta = \frac{M_p}{zM_d} \frac{\omega}{1-\omega}$$

where T_g and T_{g0} are the glass transition temperatures for the polymer-gas system and the pure polymer, respectively, M_p is the molar mass of the polymer repeat unit, M_d is the molar mass of the gas, R is the gas constant, ω is the gas solubility in the polymer, ΔC_p is the heat capacity change associated with the glass transition of the pure polymer, and z is the lattice coordination number that depends on the sizes of the gas and the polymer repeat unit. Chow [25] suggested a value of $z = 2$ for polystyrene-liquid systems whereas Chiou et al. [8] found a value of $z = 1$ to give better predictions for the polymer-gas systems. The solubility of CO_2 has been reported at various temperatures and pressures. These were used to estimate the solubility of CO_2 at the temperatures and pressures reported in Table 1. The calculated T_g's using $z = 1$, $M_p = 104.2$ g mol^{-1}, $\Delta C_p = 0.259$ J K^{-1} g^{-1} are shown by the solid curve in Fig. 6. The agreement with the experimental values is quite satisfactory. Also shown in Fig. 6 are the literature values obtained from DSC [8] and static creep compliance measurements [7,9]. The agreement among the various data sets is quite reasonable, considering the different kinds of technique used.

For reasons mentioned above, the $T_g - P$ curves for PS-CO_2 and PS-C_2H_4 (Fig. 5) terminate at temperatures and pressures just below which a condensed phase of the gas appears. The solubility of C_2H_4 in PMMA has been reported to be about half that of CO_2 [15,22], and C_2H_4 has been reported to be a nonplasticizer for PMMA [15]. As seen in Fig. 5, both C_2H_4 and CO_2 show a rather strong plasticization effect on PS with dT_g/dP of about -0.9 K bar^{-1}. Furthermore, the depressions in T_g induced by C_2H_4 and CO_2 are almost the same indicating that their solubilities in PS may also be the same. Work on solubility of C_2H_4 in PS and the plasticization of PMMA by various gases is now being pursued to further understand the interactions in polymer-gas systems.

Acknowledgments

The authors thank Dow Chemicals Canada for supplying the polystyrene sample, and the Natural Sciences and Engineering Research Council of Canada for a post-graduate scholarship to Mr. O'Neill.

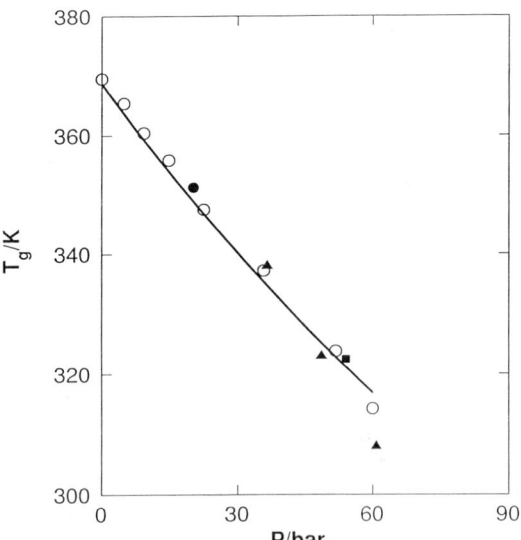

FIG. 6—*Glass transition temperatures for PS-CO_2 plotted against the CO_2 gas pressure. ○, this work; ●, Chiou et al. [8]; ▲, Wissinger and Paulaitis [9]; ■, Wang et al. [7]. Solid curve was calculated from Chow's model [25].*

References

[1] McHugh, M. and Krukonis, V., "Supercritical Fluid Extraction: Principles and Practice," Butterworths, Boston, 1986.
[2] Burgess, A. N. and Jackson, K., "The Removal of Carbon Tetrachloride from Chlorinated Polyisoprene Using Carbon Dioxide," *Journal of Applied Polymer Science*, Vol. 46, 1992, pp. 1395–1399.
[3] Berens, A. R., Huvard, G. S., Korsmeyer, R. W., and Kunig, F. W., "Application of Compressed Carbon Dioxide in the Incorporation of Additives into Polymers," *Journal of Applied Polymer Science*, Vol. 46, 1992, pp. 231–242.
[4] Koros, W. J., Coleman, M. R., and Walker, D. R. B., "Controlled Permeability Polymer Membranes," R. A. Huggins, Ed., *Annual Reviews of Material Science*, Vol. 22, 1992, pp. 47–90.
[5] Srinivasan, G. and Elliott, J. R., "Microcellular Materials via Polymerization in Supercritical Fluids," *Industrial Engineering Chemistry Research*, Vol. 31, 1992, pp. 1414–1417.
[6] Spalding, B. J., "Foamed Polymers Toughen up Their Act," *Chemical Week*, Vol. 30, 1988, p. 34.
[7] Wang, W.-C. V., Kramer, E. J., and Sachse, W. H., "Effects of High-Pressure CO_2 on the Glass Transition Temperature and Mechanical Properties of Polystyrene," *Journal of Polymer Science: Polymer Physics Edition*, Vol. 20, 1982, pp. 1371–1384.
[8] Chiou, J. S., Barlow, J. W., and Paul, D. R., "Plasticization of Glassy Polymers by CO_2," *Journal of Applied Polymer Science*, Vol. 30, 1985, pp. 2633–2642.
[9] Wissinger, R. G. and Paulaitis, M. E., "Glass Transition in Polymer/CO_2 Mixtures at Elevated Pressures," *Journal of Polymer Science: Part B: Polymer Physics*, Vol. 29, 1991, pp. 631–633.
[10] Hachisuka, H., Sato, T., Imai, T., et al., "Glass Transition Temperature of Glassy Polymers Plasticized by CO_2 Gas," *Polymer Journal*, Vol. 22, 1990, pp. 77–79.
[11] Assink, R. A., "Plasticization of Poly(dimethyl siloxane) by High-Pressure Gases as Studied by NMR Relaxation," *Journal of Polymer Science*, Vol. 12, 1974, pp. 2281–2290.
[12] Sanders, E. S., "Penetrant-Induced Plasticization and Gas Permeation in Glassy Polymers," *Journal of Membrane Science*, Vol. 37, 1988, pp. 63–80.
[13] Puleo, A. C., Paul, D. R., and Kelley, S. S., "The Effect of Degree of Acetylation on Gas Sorption and Transport Behavior in Cellulose Acetate," *Journal of Membrane Science*, Vol. 47, 1989, pp. 301–332.
[14] Wessling, M., Schoeman, S., Boomgaard, Th. van der, and Smolders, C. A., "Plasticization of Gas Separation Membranes," *Gas Separation and Purification*, Vol. 5, 1991, pp. 222–228.
[15] Sanders, E. S., Jordan, S. M., and Subramanian, R., "Penetrant-Plasticized Permeation in Polymethylmethacrylate," *Journal of Membrane Science*, Vol. 74, 1992, pp. 29–36.
[16] Kamiya, Y., Mizoguchi, K., Naito, Y., and Hirose, T., "Gas Sorption in Poly(vinyl benzoate)," *Journal of Polymer Science: Part B: Polymer Physics*, Vol. 24, 1986, pp. 535–547.
[17] Vieth, W. R., Dao, L. H., and Pedersen, H., "Non-equilibrium Microstructural and Transport Characteristics of Glassy Poly(ethylene terephthalate)," *Journal of Membrane Science*, Vol. 60, 1991, pp. 41–62.
[18] Sefcik, M. D., "Dilation and Plasticization of Polystyrene by Carbon Dioxide," *Journal of Polymer Science: Part B: Polymer Physics*, Vol. 24, 1986, pp. 957–971.
[19] Handa, Y. P., Capowski, S., and O'Neill, M., "Compressed-gas Induced Plasticization of Polymers," *Thermochimica Acta*, Vol. 226, 1993, pp. 177–185.
[20] Handa, Y. P., Hawkins, R. E., and Murray, J. J., "Calibration and Testing of a Tian-Calvet Heat-Flow Calorimeter," *Journal of Chemical Thermodynamics*, Vol. 16, 1984, pp. 623–632.
[21] Handa, Y. P., "Calorimetric Determinations of the Compositions, Enthalpies of Dissociation, and Heat Capacities in the Range 85 to 270 K for Clathrate Hydrates of Xenon and Krypton," *Journal of Chemical Thermodynamics*, Vol. 18, 1986, pp. 891–902.
[22] Sanders, E. S. and Koros, W. J., "Sorption of CO_2, C_2H_4, N_2O and Their Binary Mixtures in Poly(methyl methacrylate)," *Journal of Polymer Science: Polymer Physics Edition*, Vol. 24, 1986, pp. 175–188.
[23] Jordan, S. M., Koros, W. J., and Beasley, J. K., "Characterization of CO_2-Induced Conditioning of Polycarbonate Films Using Penetrants with Different Solubilities," *Journal of Membrane Science*, Vol. 43, 1989, pp. 103–120.
[24] Stern, S. A. and Kulkarni, S. S., "Solubility of Methane in Cellulose Acetate—Conditioning Effect of Carbon Dioxide," *Journal of Membrane Science*, Vol. 10, 1982, pp. 235–251.
[25] Chow, T. S., "Molecular Interpretation of the Glass Transition Temperature of Polymer-Diluent Systems," *Macromolecules*, Vol. 13, 1980, pp. 362–364.

H. G. Wiedemann,[1] G. Widmann,[1] and G. Bayer[2]

Glass Transition in Polymers: Comparison of Results from DSC, TMA, and TOA Measurements

REFERENCE: Wiedemann, H. G., Widmann, G., and Bayer, G., "**Glass Transition in Polymers: Comparison of Results from DSC, TMA, and TOA Measurements,**" *Assignment of the Glass Transition, ASTM STP 1249*, R. J. Seyler, Ed., American Society for Testing and Materials, Philadelphia, 1994, pp. 174–181.

ABSTRACT: The determination of the glass transition temperature, T_g, of polymers by differential scanning calorimetry (DSC) and thermomechanical analysis (TMA) is sometimes problematic and rather subjective. This was shown previously in the ICTA certificate (distributed by NBS as GM-754) for the certified reference material polystyrene (PS). The not very good reproducibility of the measured value of the onset is due to a variety of instrumental and experimental parameters. This is true also for the determination of the glass transition by TMA measurements. The main reasons are temperature gradients caused by the relatively high sample mass required for DSC and by the limited heat transfer in TMA, respectively. Our own experiments which were carried out with polystyrene and with [poly(ethyleneterephthalate)] (PET) proved that a combination of DSC with TOA (thermo-optical analysis or hot stage microscopy under polarized light) can solve some of these problems. TOA is a nonsubjective method since the changes in birefringence and light transmittance during the glass transition which are visible under the microscope are measured with a photocell. TOA allows T_g measurements of small samples (fraction of milligrams).

KEYWORDS: glass transition in polymers, simultaneous differential scanning calorimetry (DSC) analysis with thermo-optical analysis (TOA), thermomechanical analysis (TMA), hot stage, poly(ethyleneterephthalate) (PET), polystyrene (PS)

Polymeric materials may show varying degrees of anisotropy caused by structural features and due to stresses. The latter depend mainly on the thermomechanical history during the production, forming, and handling of these organic materials. The build-up and the relief of stresses, but also phase transformations due to the structural changes, melting, and decomposition of polymers can be determined by optical microscopy under polarized light through the use of hot stage. Results from such optical investigations can be correlated with differential scanning calorimetry (DSC) and thermomechanical analysis (TMA) measurements as will be shown in the following.

The glass transition is a characteristic phenomenon of all amorphous or partially amorphous matter. Below the glass transition temperature, the substance is in the solid, often brittle glassy state. Above this temperature, substances assume a liquid, viscous or rubbery state.

At the glass transition temperature, many physical properties change abruptly (Fig. 1):

---

[1]Ph.D. Senior scientist, and research engineer, respectively, Mettler-Toledo AG, 8606 Greifensee, Switzerland.
[2]Professor, Swiss Federal Institute of Technology, Institute of Non-metallic Materials, ETH-Zürich, CH-8092 Zurich, Switzerland.

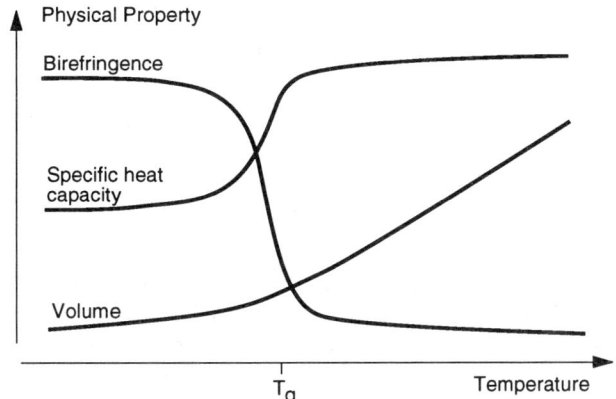

FIG. 1—*Physical property changes during glass transition.*

Young's and shear modulus, specific heat capacity, expansivity, dielectric permittivity, and birefringence. The last is caused by a certain anisotropy remaining from the production process (e.g., injection molding or extrusion).

Experimental

Instruments used include:

- Mettler FP84, hot stage with simultaneous DSC and thermo-optical analysis (TOA) controlled by the FP90 unit. To convert the light intensity into an electrical signal the photomonitor of the melting point (FP) system has been used (Fig. 2).
- TMA40 of the Mettler TA4000 system.
- Leitz Orthoplan microscope, with crossed polarizers, magnification 200×.

For our investigations we have chosen polystyrene (PS) and poly(ethyleneterephthalate) (PET) as examples. All measurements have been performed in still air.

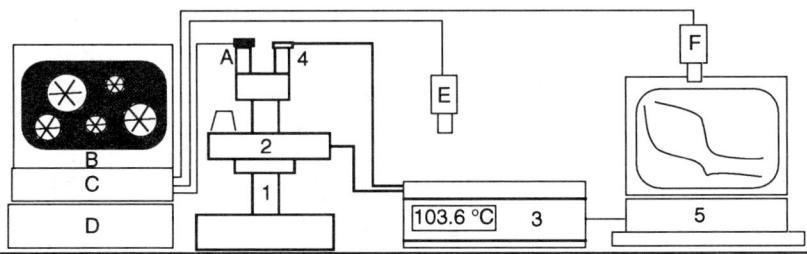

FIG. 2—*Instrumental setup for TOA-DSC measurements. 1: Microscope. 2: Hot stage with sample. 3: Control unit of hot stage. 4: Photocell. 5: Computer for curve storage and evaluations. For video films additional instrumentation is needed. A: Video camera. B: Video monitor. C: Mixer for the various cameras. D: Tape recorder. E: Video camera for the display of the control unit. F: Video camera for the computer screen showing the DSC-TOA curves.*

Measurement of the Glass Transition Temperature Using DSC [1–3]

At the glass transition temperature, there is an endothermic shift in the DSC curve corresponding to the increase in specific heat capacity (Fig. 1). According to ASTM D 3418, Test Method for Transition Temperatures of Polymers by Thermal Analysis, mainly the extrapolated onset temperature is used to characterize T_g. In addition, the midpoint and the extrapolated end temperature can be evaluated.

The evaluation is based on the following three auxiliary lines: extrapolated baseline before the transition, inflectional tangent through the inflection point during the transition, and the baseline extrapolated back after the transition.

Frequently, an endothermic peak occurs simultaneously with the change in specific heat capacity the first time a substance is heated above the glass transition temperature. It is caused by a relaxation phenomenon (enthalpy relaxation) and normally disappears in a second measurement after rapid cooling below T_g (thermal history dependence). That is why ASTM D 3418 suggests to carry out the measurement immediately after a preliminary thermal cycle to a temperature high enough to erase previous thermal history.

The measured glass transition temperature depends somewhat on the heating rate; a rate of 10 or 20°C/min is usually employed. To obtain a large enough signal, normally at least a 10 mg sample is used to obtain a good signal to noise ratio.

Measurement of the Glass Transition Temperature Using Dilatometry and TMA

As Fig. 1 illustrates, the slope of the volume curve increases at the glass transition (i.e., the expansivity increases abruptly). Two auxiliary lines are needed for the evaluation of the dilatometric curve: the extrapolated baselines before and after the transition. The sample is frequently deforming during the first heating, thus impairing the evaluation. Possible causes are relief of frozen stresses in the sample or penetration by a foreign body (dust particle). To prevent such problems, the sample is measured immediately after a thermal cycle to a temperature high enough to erase previous thermal history.

If the "softening point" is determined in the TMA mode (deformation under load), it is obvious that a large compressive stress begins to deform the sample somewhat below the glass transition temperature. In contrast to dilatometry, TMA is especially suitable for thin samples (< 0.1 mm). Figure 3 shows a flat sample of an injection-molded PS product that is measured between the TMA platform and a fused silica disk under a hemispherical probe.

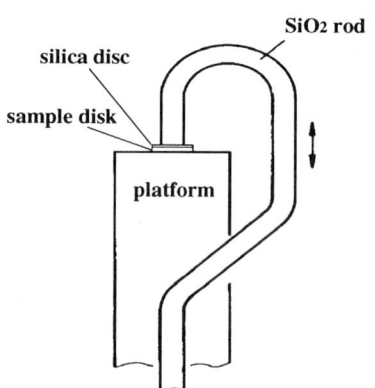

FIG. 3—*Sample holder for TMA measurements.*

Since the heat transfer in a TMA is slower than in a DSC, the heating rates usually are limited to approximately 10°C/min.

On first heating above the glass transition temperature, the frozen stresses relief causing dimensional changes of the polymer sample. Such dimensional changes are measured in a TMA provided the applied mechanical stress is negligible (Fig. 4, first run). This is achieved by a sample shaped as disk of, e.g., 5 mm diameter. To distribute the probe force F, the sample is covered by a fused silica disk. With such an arrangement the compressive stress σ_c is negligible

$$\sigma_c = F/A = 0.02 \text{ N} / 20 \text{ mm} = 0.001 \text{ N/mm}^2$$

where A is the surface area of the sample.

Another desired effect of the first heating is the formation of full contact of the sample with its support and cover-disk by smoothing out the original irregularities or pressing-in of foreign particles. A second measurement shows only a very small change in the slope of the TMA curve during the transition (Fig. 4, second run) due to the increasing expansivity. In the third measurement, by application of a high force of 0.5 N (or, 0.025 N/mm²) the sample is plastically deformed above the transition (Fig. 4, third run). The obtained orientation is frozen in after cooling down under the same stress.

A final measurement with a force of 0.02 N (Fig. 4, fourth run) shows an expansion above the transition, caused by the elastic recovery of the anisotropic sample. Principally all four curves allow the determination of the glass transition temperature from the intersection of the baseline before and a suitable tangent after the transition. But the different thermomechanical history leads to different results.

Usually in the first run the unknown thermomechanical history of the sample is reflected. A second measurement then is used to determine the glass transition temperature. These reproducible conditions give the highest accuracy.

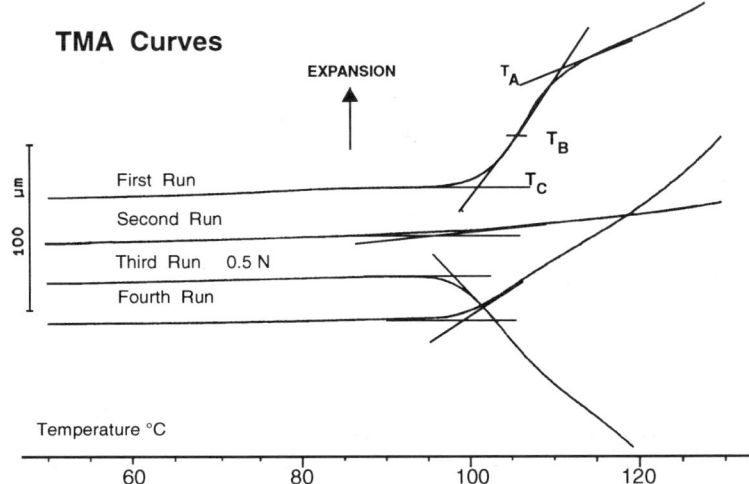

FIG. 4—*TMA curves of polystyrene: sample thickness 1.40 mm, rate 10°C/min, load 0.02/0.5 N. T_A onset: first run 102.0°C; second run 97.5°C; third run 98.0°; fourth run 99.5°C.*

178   ASSIGNMENT OF THE GLASS TRANSITION

Measurement of the Glass Transition Temperature with TOA

TOA is concerned with the visual observation of a sample subjected to a temperature program. Birefringent samples are studied between crossed polarizers and the light intensity is measured with a photocell to record the TOA curve (Fig. 2). Usually only the first heating curve is investigated since birefringence (or anisotropy) disappears more or less completely at the glass transition, and would reappear only when cooling the sample under stress. The latter can be used to induce a birefringent marker for subsequent analysis [4].

With TOA, even less distinct glass transitions of extremely small samples can be determined.

The combination of TOA with DSC is also important for optimum temperature calibration of the hot stage based on metal melting points. These are not visible in TOA; therefore, the obvious restriction of TOA are transparent samples.

Results and Discussion

Polystyrene

For the observation and photographic recording of the stresses in the PS sample microscopic investigations under polarized light were used. When the samples are heated on a hot stage, combined with a DSC sensor, the disappearance of stresses (shown by interference colors) can be followed. Also, the specific heat capacity of the samples changes during the heating process. The actual temperatures at which the interference colors change can be correlated to the changes in the heat flow (Fig. 5). T_A corresponds to the beginning of the relaxation (96.1°C), T_B to the general disappearance of stresses (100.9°C), and T_C to the attainment of the isotropic state (105.6°C).

A comparison of these characteristic temperatures T_A, T_B, and T_C is given in Table 1. All evaluations are based on the first heating curve at a rate of 10°C/min. It proves that the values obtained by the three different methods, TOA, TMA, and DSC, show good agreement.

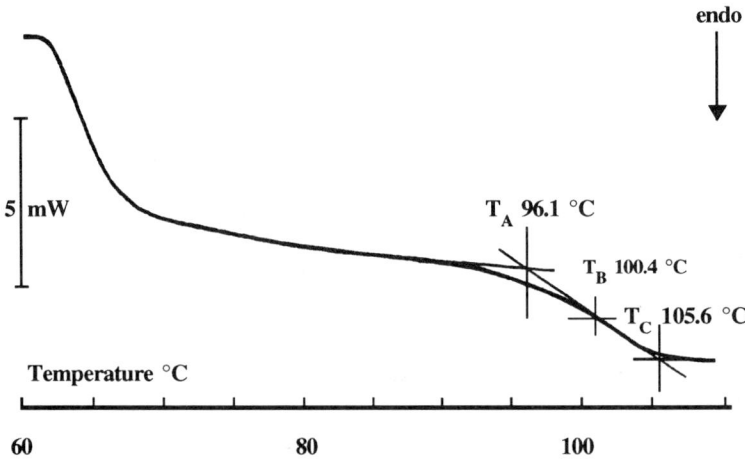

FIG. 5—*DSC curve of a polystyrene sample: sample weight 21.206 mg; rate 10°C.*

TABLE 1—*Glass transition of polystyrene.*

Method	T_A, °C	T_B, °C	T_C, °C	Sample weight, mg	Sample thickness, mm
1 DSC	93.4	98.1	102.7	12.1	—
TOA	95.4	99.8	102.5	—	0.5
2 DSC	94.3	98.4	102.5	12.5	—
TOA	95.9	98.0	100.0	—	0.5
3 DSC	100.7	101.9	103.1	25.9	—
TOA	98.9	101.4	103.9	—	1.0
4 DSC	92.8	94.5	97.5	24.3	—
TOA	98.9	101.1	104.4	—	0.5
5 TMA	102.1	104.8	109.5	—	1.40
6 TMA	98.1	100.8	103.5	—	0.37

Poly(ethyleneterephthalate)

During heating, the glass transition as well as the crystallization of PET was recorded by DSC (Fig. 6). Whereas the glass transition can be determined also with TOA, this is not possible for the crystallization [5].

Simultaneous TOA-DSC measurements of the glass transition were carried out under similar conditions as in the case of PS. The curves obtained by PET samples are shown in Fig. 7. A comparison and summary of the values obtained from the three different thermal methods TOA, DSC, and TMA, is given in Table 2. All evaluations are based on the first heating curve at a rate of 10°C/min. Within the range of sample thickness used in our measurements (0.2 to 0.5 mm) we could not observe any specific trend in DSC and TOA values.

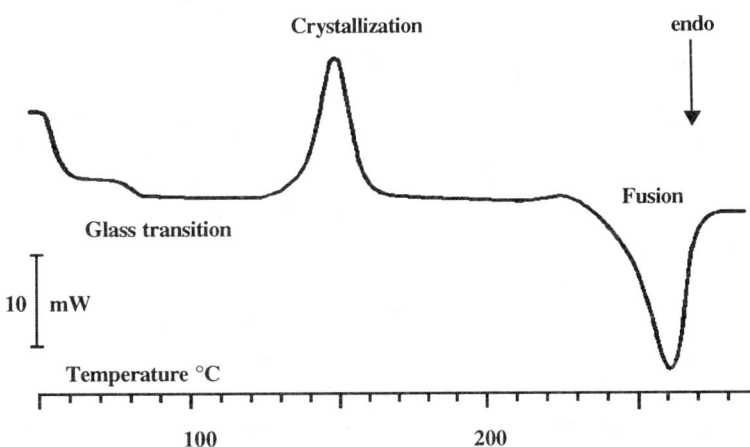

FIG. 6—*DSC curve of the glass transition, crystallization, and fusion of a PET sample quenched in liquid nitrogen. Sample weight 5.25 mg; rate 10°C/min.*

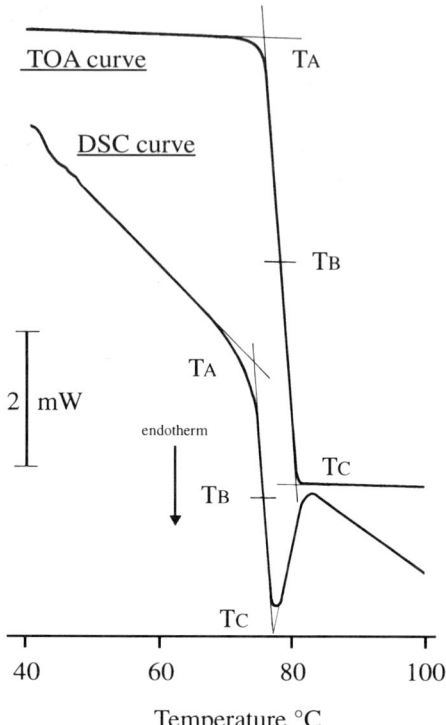

FIG. 7—*Glass transition of PET, simultaneous TOA-DSC curves. The ordinate of the TOA curve is in arbitrary units.*

TABLE 2—*Glass transition of poly(etheleneterephthalate).*

Method	T_A, °C	T_B, °C	T_C, °C	Sample weight, mg	thickness, mm
1 TOA	75.3	80.2	85.0	21.21	0.3
2 DSC	74.3	75.1	76.3	10.24	—
TOA	74.9	76.0	77.9	—	0.4
3 DSC	74.1	75.5	76.5	10.62	—
TOA	75.7	76.6	79.7	—	0.4
4 DSC	77.5	79.8	81.2	10.46	—
TOA	77.2	79.5	81.0	—	0.2
5 TMA	74.0	76.2	82.0	20	0.5

Conclusion

Results from thermal analytical measurements obtained on two polymers (PS, PET) proved that the TOA method offers some advantages in the determination of the glass transition compared with TMA and DSC. Especially with small samples, the signal to noise ratio is superior.

References

[1] Widmann, G., in *Thermochimica Acta*, Vol. 112, 1987, pp. 137–140.
[2] Wiedemann, H. G. and Bayer, G., in *Thermochimica Acta*, Vol. 169, 1990, p. 1.
[3] Wiedemann, H. G. and Bayer, G., in *Thermochimica Acta*, Vol. 85, 1985, p. 271.
[4] Wiedemann, H. G. and Bayer, G., in *Thermochimica Acta*, Vol. 92, 1985, p. 399.
[5] Kovacs, A. J. and Hobbs, S. Y., *Journal of Applied Polymer Science*, Vol. 16, 1972, p. 301.

Materials

Joseph Kincs,[1] *Jaephil Cho,*[1] *Donnie Bloyer,*[1]* *and Steve W. Martin*[1]†

Glass Transition and Heat Capacities of Inorganic Glasses: Diminishing Change in the Heat Capacity at T_g for xNa$_2$S + (1 − x)B$_2$S$_3$ Glasses

REFERENCE: Kincs, J., Cho, J., Bloyer, D., and Martin, S. W., "Glass Transition and Heat Capacities of Inorganic Glasses: Diminishing Change in the Heat Capacity at T_g for xNa$_2$S + (1 − x)B$_2$S$_3$ Glasses," *Assignment of the Glass Transition, ASTM STP 1249,* R. J. Seyler, Ed., American Society for Testing and Materials, Philadelphia, 1994, pp. 185–201.

ABSTRACT: The T_g's and heat capacity functions have been measured for a series of Na$_2$S + B$_2$S$_3$ glasses for the first time. Unlike the alkali borates, T_g decreases rapidly as Na$_2$S is added to B$_2$S$_3$. This effect, even in the presence of a rapidly increasing fraction of tetrahedrally coordinated borons, has been associated with the "over crosslinking" effect of the sulfide ion. Unlike the borate glasses where each added oxygen produces two tetrahedral borons, the conversion rate for the thioborates is between four and six. This behavior is suggested to result in the formation of local tightly-bonded molecular-like structures that exhibit less long-range network bonding than the alkali borate glasses. As a result, T_g decreases with added alkali in alkali thioborates rather than increases as in the alkali borate glasses.

The change in heat capacity at T_g, $\Delta C_p(T_g)$ has been carefully measured and is found to also decrease dramatically as alkali sulfide is added to the glass. Again this effect is opposite to the trends observed for the alkali borate glasses. The decreasing $\Delta C_p(T_g)$ occurs even in the presence of a decreasing T_g. We have tentatively associated the diminishing $\Delta C_p(T_g)$ values to the decreasing density of the configurational states above T_g. This is attributed to the high coordination number and site specificity caused by the added alkali sulfide. The glassy state heat capacities were analyzed and found to reach ~90% of the classical limiting DuLong-Petit value just below T_g for all glasses. This was used to suggest that the diminishing $\Delta C_p(T_g)$ values are associated with a unique behavior in the system to become a liquid with very little change in the density of configurational states.

KEYWORDS: glass transition, heat capacity, alkali borate, alkali thioborate, differential scanning calorimetry, glasses

The study of the thermal properties, the glass transition and heat capacity in particular, of the inorganic glasses has been an active field for many years. Silicate glasses [1–4] have received the bulk of attention, but borate [5–11], germanate [12], halide [13,14], phosphate [15,16], and more recently chalcogenide glasses have also been studied. This work has generated a large database of the composition, thermal history, and heating rate dependence of the glass transition and heat capacities of these glasses. In all of this work, differential

---
[1]Department of Materials Science and Engineering, Iowa State University of Science and Technology, Ames, IA 50011.
*Present address: GE Astrospace, Philadelphia, PA.
†To whom correspondence should be directed.

scanning calorimetry (DSC) has been the method of choice in determining the T_g and heat capacity of inorganic glasses [*18*], although differential thermal analysis (DTA) and dilatometry have been used as well.

One of the most common descriptions of the glass transition in terms of glass structure has been to relate T_g to the fraction of nonbridging (network breaking) and bridging bonds (network forming) in the glass. Hence, the dramatic decrease in T_g from ~1200°C for silica to ~500°C for a modified sodium silicate glass is associated with the increasing fraction of network degrading nonbridging oxygens in the glass. Likewise, for the alkali borate glasses where T_g *increases* some 200°C with a 30 mol% increase in the alkali oxide fraction, the increase in T_g is associated with an increasing fraction of tetrahedrally coordinated boron units. The crosslinking nature of the added oxygen is associated with this increase in T_g.

Little work has been done to date on inorganic chalcogenide glasses whose compositions are analogous to the oxide glasses. For example, almost no work has been done on the $M_2S + B_2S_3$ [*19–22*] or $M_2S + SiS_2$ [*23–26*] systems, where M = Na and Li. The reasons for this lack of effort on these seemingly simply constituted glasses arises from their reactivity with air and moisture, their poor glass-forming character in most cases, their reactivity with most crucible materials, the difficulty in handling them in pure form, and finally to the lack of commercially available inexpensive and high-purity starting materials.

However, these glasses still remain interesting for a number of reasons. First, they represent a class of little-studied glasses. In the work that has been reported, the structures and properties that have been determined are quite unique and in many cases quite different from that expected by extrapolating the structure and properties from the corresponding oxide glass family. Finally, some of the highly modified alkali sulfide and even selenide glass families have been found to exhibit very high ionic conductivities, especially for lithium, and in some cases the conductivity approaches $10^{-3}(\Omega cm)^{-1}$ at room temperature. This high conductivity has prompted many researchers to study the electrochemical properties of these glasses as potential solid electrolytes in new all-solid-state lithium batteries.

For these reasons, we have begun a thorough examination of the structures, properties, and the relationships between these two of the simply constituted alkali and alkaline earth chalcogenide glasses. Specifically, we have examined the glasses $M_2S + SiS_2$ [*26*], $M_2S + B_2S_3$, and to a lesser extent $M_2S + P_2S_5$ where M = Na and Li. In our studies, we have shown that the thiosilicate glasses are poor glass formers over very limited composition ranges, whereas the thioborates like the borates are almost impossible to crystallize for many compositions. For these reasons we have studied the thioborate glasses in more detail. We have already reported on the glass formation range [*27*], the T_g for limited compositions [*27*], the IR spectra of the glasses [*28*], compounds, and the density. In this study, we report detailed measurements of the glass transition over wide composition ranges and the heat capacity near the glass transition temperature for the low-alkali glasses. As will be described below, the sodium thioborate system forms glasses in a low-alkali sulfide region up to 33 mol%, and in a high-alkali sulfide region between 55 and 80 mol% Na_2S.

Experimental Methods

Preparation of the Glasses

As indicated above, one of the main reasons that the alkali thioborate glasses have received so little attention from the glass research community is that they are very difficult to prepare in pure form. This stems from the fact that not only are they hygroscopic and air-sensitive but they are very volatile, and almost no pure compounds are available commercially. For these reasons, we have expended significant effort to develop preparation schemes that enable us to obtain these glasses in very pure form. Much of the confusion that surrounds the structures and properties of these glasses can be attributed to varying degrees of contamination in the glasses that different researchers were studying.

The $x\text{Na}_2\text{S} + (1-x)\text{B}_2\text{S}_3$ glasses where $0 \leq x \leq 0.33$ and $0.55 \leq x \leq 0.75$ were prepared from high-purity glassy B_2S_3 ($\nu - \text{B}_2\text{S}_3$) prepared using the methods we have developed and reported on separately [32] and from Na_2S (Cerac, 99.9%, metals basis). The powders were ground finely in an agate mortar and pestle, loaded into a vitreous carbon crucible, and melted at 850°C in an inert electric atmosphere furnace. The melts were plate-quenched into a heated brass mold. Prior to the thermal measurements, the glass samples were not given any specific thermal treatments; this was done with the DSC. Total sulfur analyses of the B_2S_3 showed that it was 99% pure based on sulfur content [33]. The Na_2S was not further analyzed except that its XRD powder pattern was clean and representative of high-purity Na_2S. All operations were carried out inside a high-quality home built glove-box (which had <10 ppm of oxygen and water). Samples that were left out in the glove-box exhibited no visual signs of contamination even over many weeks.

Glass Transition Temperature Measurements

The glass transition temperatures were measured using a carefully calibrated Perkin-Elmer DSC-4 DSC interfaced to a PC which employed Laboratory MicroSystems thermal analysis software. The DSC was calibrated using indium, tin, lead, and zinc melting points with an accuracy of ±1°C. The effect of scanning rate on the temperature calibration was accounted for by running the indium standard at four different heating rates spanning the heating rate used in the experiments and fitting the scanning rate dependent melting point to a second order polynomial. The temperature axis was then corrected using the polynomial correction factor. In this way, the unavoidable effect of thermal lag in the instrument could be accounted for to ±1°C. All data reported in this paper were corrected in this manner. The heat flow axis of the DSC was calibrated by using the enthalpy of melting for each of the four metal standards. Again, a second order polynomial was fit to the enthalpies and then used to correct the heat flow axis measurements. In this way accuracies of ±2 to 5% could be achieved.

A uniform heating rate of 20°C was used in all T_g and heat capacity measurements. This rate was a compromise between signal sensitivity, steady-state heat flow conditions, and time constraints to perform the measurements. Heating rates slower and faster than this were used and gave equivalent results when the data were scaled to account for the effect of heating rate on T_g. All T_g's were taken after the sample had been heated for a second time through the transition region. The sample was first heated through T_g to $\sim T_g + 50°C$, cooled at the same rate to $\sim T_g - 50°C$ and then reheated through the transition region. Data collected in this way assure that all samples have the same thermal history and as such when the T_g is plotted against composition, only the effect of changing composition and structure are observed.

Heat Capacity Measurements

High-quality heat capacity measurements were made on each of the glasses using established methods [34–36]. To ensure the highest quality in the heat capacity data, the following procedures were required. First the temperature ranges over which the measurements were made had to be as small as reasonably possible. In this case, we kept the temperature range down to the minimum required to see a reasonable amount of the pre-transition and post-transition ranges. For most of the glasses this was 100 to 150°C.

Temperature intervals much larger than this resulted in widely varying baselines and a concomitant degradation in the reproducibility and accuracy of the heat capacity data. Second, we found it critical to give the DSC measuring head, standard reference specimens, specimen pans, and glass specimens all the same thermal history as described above for the T_g measurements. Hence, all instrument baselines, standard reference, and sample specimens were scanned three times over identical temperature ranges. Whenever it was necessary to change the temperature interval of the experiment (because of changing T_g's with composition), it was necessary to repeat this process. Only after paying such careful attention to the thermal history of the experiment was it possible to obtain high-quality heat capacity data. Standard specimens (typically 15 to 25 mg) of alumina (sapphire) were used as the reference heat capacity standard. Glass sample specimens were typically 10 to 20 mg, due to their slightly lower densities. Scans were obtained for all specimens that had been previously hermetically sealed into aluminum specimen pans. Masses were determined using a PE AD-2Z microbalance accurate to ±0.01 mg prior to measurements. The LMS software has a built-in heat capacity routine that calculates the heat capacity using the standard routine where the specimen signal minus the baseline signal is proportional to the heat capacity of the sample. The heat flow data were corrected for differences in the masses of the different specimen pans. This effect was found to be small but systematic, however. Using sapphire as a standard, accuracies of ±1% could be achieved with the heat capacity precision being slightly better than this between identical runs.

Results

Glass Transition Temperatures

The DSC scans made on the $x\mathrm{Na_2S} + (1-x)\mathrm{B_2S_3}$ glasses for $0 \leq x \leq 0.30$ are shown in Fig. 1 and for $0.55 \leq x \leq 0.80$ are shown in Fig. 2. As can be clearly seen, the signature of

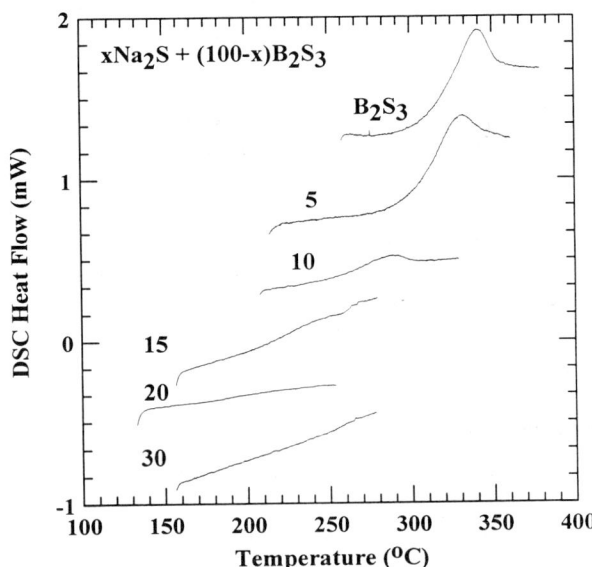

FIG. 1—*DSC scans for $x\mathrm{Na_2S} + (100 - x)\mathrm{B_2S_3}$ glasses with $0 \leq x \leq 30$. Heating and cooling rates were 20°C/min.*

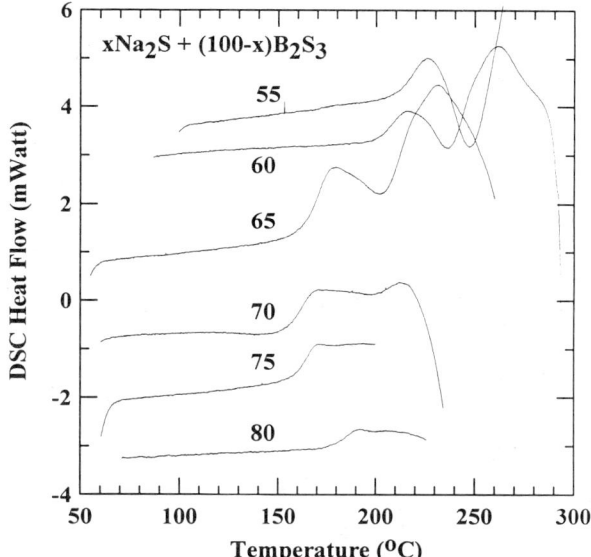

FIG. 2—*DSC scans for $xNa_2S + (100 - x)B_2S_3$ glasses with $55 \leq x \leq 80$. Heating and cooling rates were 20°C/min.*

T_g is quite strong for pure B_2S_3 but as Na_2S is added to the glass, the T_g signature decreases dramatically. Indeed, at $x = 0.20$ and 0.30, the T_g signature is practically absent. These samples were checked thoroughly for the effect of crystallization and all samples were found to be absent of crystallization. For the higher alkali glasses, a crystallization peak was observed after the T_g indicating that as expected, as the limit of the glass forming range is approached, the tendency to crystallize *above* T_g increases. All glasses were free of crystallinity below T_g though. The lack of a signature of a T_g in the DSC is direct measure of the glass to exhibit a diminishing change in the heat capacity at T_g. Hence, as Na_2S is added to B_2S_3, the effect to decrease the change in heat capacity at T_g to zero.

The other striking feature of the data given in Fig. 1 is that it shows that in complete contrast to that seen in the borate glasses, $Na_2O + B_2O_3$, T_g decreases as Na_2S is added to B_2S_3 rather than increases. Figure 3 shows that T_g decreases from ~300°C to ~170°C as the Na_2S fraction increases from 0 to 20 mol%. Contrary to this behavior, for the sodium borate glass series, T_g increases from ~260°C for B_2O_3 up to ~490°C with the addition of 30 mol% Na_2O [9–11]. This decrease suggests that the structural feature of the thioborate glasses that gives rise to a decreasing T_g must be quite different from those in the corresponding borate glasses. In the sections below, we will consider structural features beyond the short range order (SRO) and include effects in the intermediate range (IRO). It is believed that the T_g represents the temperature at which structural units in the glass well beyond the first coordination sphere become mobile on the timescale of a few hundred seconds [37–43]. If this is the case, it may well be that even for two glasses that might seem to have similar SROs due to their very similar chemistries such as the thioborates and the borates, they may well in fact have quite different IROs and as such exhibit very different T_g composition dependencies. It is interesting to note that the two glasses B_2S_3 and B_2O_3 which both have trigonal planar structures also have very similar T_g's of ~300°C.

After this initial rapid decrease in T_g, T_g then increases from its minimum value of ~170°C to ~270°C by the end of the low alkali glass forming limit of 30 mol%. Again, this

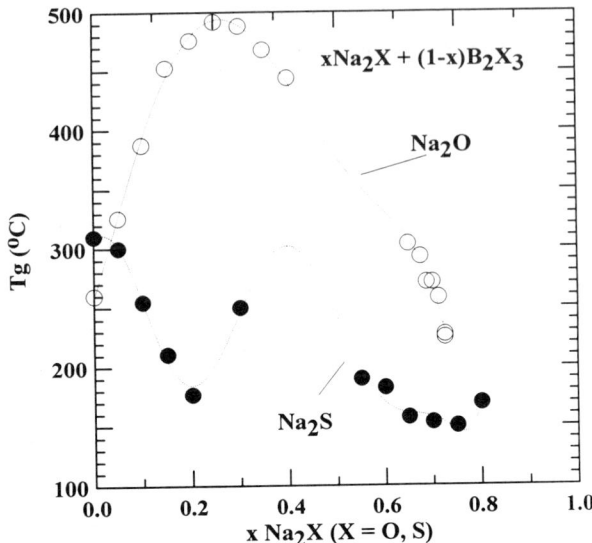

FIG. 3—*Composition dependence of T_g for $Na_2O + B_2O_3$ and $Na_2S + B_2S_3$ glasses. Data for the $Na_2O + B_2O_3$ glasses were taken from Ref 10.*

upswing is in complete contrast to the behavior in the oxide glasses. By this composition in the oxides, T_g has maximized and is now decreasing. This behavior suggests a very different structural model for these glasses.

Due to the large gap in the glass forming range between 30 and 55 mol% Na_2S, the compositional trend of T_g cannot be followed in this region. However, once glasses can be formed again at 55 mol%, T_g has now decreased substantially down to ~190°C. With further additions of Na_2S, T_g continues to decrease out to the limit of the glass forming range of 80 mol%. T_g does exhibit a modest increase at the very end and as we have seen in the density in this same region. We believe this to be associated with the densification of the structures a result of the structural depolymerization caused by the high fraction of alkali sulfide in the glass.

Heat Capacity Measurements

Separate DSC scans were made on carefully weighed samples to determine the heat capacities as discussed above. The heat capacities were determined for the low-alkali glasses, although work is in progress for the high-alkali glasses as well. Work is also in progress to determine the low temperature ($100 \leq T \leq 300$ K) and very low temperature ($1 \leq T \leq 100$ K) heat capacities of these glasses. This work will be reported on separately.

Figure 4 shows the heat capacity of the glasses examined so far. As indicated above, what was anomalous for these glasses is that the signature of T_g diminishes to zero as the low-alkali glass forming limit is approached. The data in Fig. 4 show this behavior quite dramatically, Hence, for pure B_2S_3, the change in heat capacity is quite large and easily observed and measured. As Na_2S is added, the change in heat capacity at T_g ($\Delta C_p(T_g)$) diminishes to nearly zero. The data in Fig. 4 are clearly striking. These data represent the very first observation of such a clearly observed diminishing $\Delta C_p(T_g)$ for any glasses studied to date. There have been reports of diminishing $\Delta C_p(T_g)$ values for some particular glasses

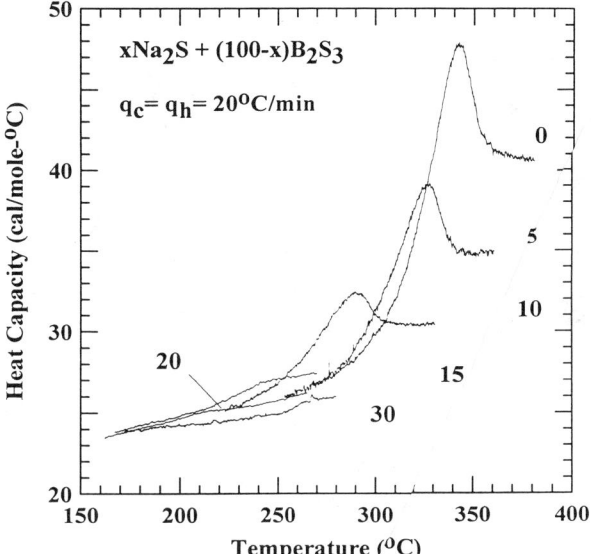

FIG. 4—*Heat capacity function for $xNa_2S + (100 - x) B_2S_3$ glasses in the transition range. Notice the dramatic decrease in the $\Delta C_p(T_g)$ as x approaches 20. All samples were ~20 mg in mass and were free of crystallization.*

such as for pure silica [*44*], but these are few and in the case of silica, the T_g is so high as to be out of the range of most sensitive DSCs and as such still remains an open question, although the $\Delta C_p(T_g)$ must be small or even insensitive DTA measurements could have observed a T_g signature.

Figure 5 shows the composition dependence of the $\Delta C_p(T_g)$. As can be seen, $\Delta C_p(T_g)$ decreases from ~56 J/mol-K for pure B_2S_3 to ~1.69 ± 0.5 J/mol-K for 20 mol% Na_2S. The decrease is quite dramatic. At this point, it is not clear whether this is the composition of the true minimum in $\Delta C_p(T_g)$ or not because of the large gap in glass-forming ability between 30 and 55 mol% Na_2S. The DSC scans in Fig. 2 do show, however, that the high-alkali glasses exhibit what might be considered as "normal" T_g DSC scans. Hence, by comparing the data in Fig. 1 with Fig. 2, it would be expected that the high-alkali glasses exhibit nonzero $\Delta C_p(T_g)$ values. We are currently measuring these values and will report on them later.

At this point, it is important to determine what is the source of the rapidly decreasing $\Delta C_p(T_g)$ values. Is it associated with the solid glassy state, where for some as yet not understood reason, the solid state heat capacity reaches values which are above the normal solid state DuLong-Petit value and as such approach liquid state values? Hence, is the decreasing $\Delta C_p(T_g)$ values associated with the rapidly increasing heat capacity of the solid? Or is it associated with the liquid state becoming more like the solid state in that the additional increments of heat capacity normally thought to arise from the extra degrees of rotational, translation, and libration degrees of freedom are insufficiently different from the vibrational modes in the solid glassy state? In this sense, the diminishing $\Delta C_p(T_g)$ values would be associated with an unusual structure and properties in the liquid state.

We will address these possibilities below by determining the extent to which the glassy heat capacity reaches the DuLong-Petit $3Rn$ value, where R is the ideal gas constant and n the number atoms in the vibrating unit, below T_g. If the effect of adding Na_2S is to cause the

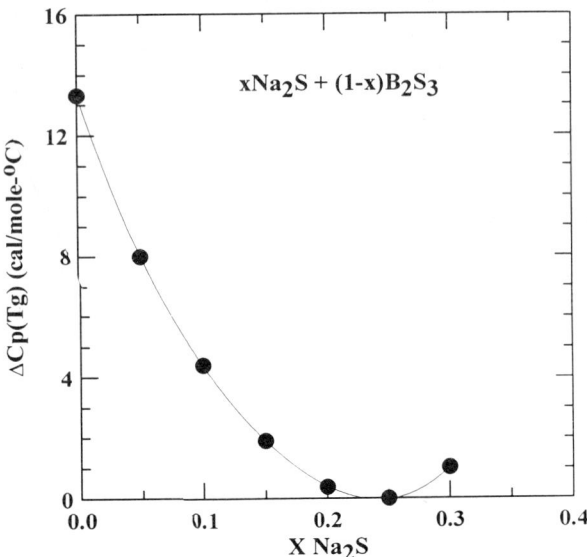

FIG. 5—*Composition dependence of the $\Delta C_p(T_g)$ for $xNa_2S + (1 - x)B_2S_3$ glasses.*

glass to override this value below T_g, then the unusual behavior can be attributed to the glassy state. Likewise, if the glasses all exhibit normal "sub-$3Rn$" values below T_g, then the diminishing $\Delta C_p(T_g)$ values must be attributed to some unique structure or properties, or both, of the liquid state.

Before we address the heat capacity values, however, we will spend some time trying to understand why T_g decreases with added alkali sulfide rather than increases as might have been expected based on the behaviors of the borate glasses.

Discussion

Composition Dependence of the Glass Transition Temperature

The composition dependence of the T_g for the sodium thioborate glasses in the low-alkali glass forming range has been shown above to be quite anomalous. It was expected from the outset that in similarity to the alkali borate that T_g would increase with alkali as the fraction of tetrahedral borons increased in the glass. Figure 1 shows either that tetrahedral borons do not form or that some other structural mechanism is operative in these glasses that is absent in the alkali borate glasses. Figure 3 shows the composition dependence of T_g for the sodium borate glasses over the full glass-forming range. Like the sodium thioborate glasses, these glasses also exhibit a low ($0 \leq x \leq 0.4$) and a high ($0.63 \leq x \leq 0.73$) alkali oxide fraction glass-forming range [9]. Unlike the thioborate glasses, however, T_g increases with alkali oxide fraction rather than decreasing.

For the alkali borate glasses, we and others have associated the increase in T_g with alkali oxide modifier content to the increasing fraction of tetrahedral by coordinated borons in the glass [9–11]. These tetrahedral borons crosslink two initially trigonal borons to form two tetrahedrally coordinated borons. The conversion rate in all the alkali borate glasses has been shown by Bray and others [45–50] to be the formation of two tetrahedrally coordinated borons for every additional oxide anion added to the borate glass. We have shown that this increase in T_g with tetrahedrally coordinated boron fraction to be a linear function [9]. When the fraction of tetrahedrally coordinated borons maximizes at $N_4 \sim 0.4$ at $xM_2O \sim 0.33$, the T_g also maximizes, thereafter T_g monotonically decreases out to the limit of the glass-forming range [9]. Figure 3 shows that the decrease is quite dramatic, especially at the very end of the high-alkali glass-forming range. We have associated this rapid fall off of T_g with alkali oxide in this range to the rapid depolymerization of the glass network [7]. The glass-forming ability of the liquids in this range also decreases rapidly to where at the end of the range, the glasses must be rapidly quenched between metal blocks into very thin sheets. The decrease of T_g for the high-alkali glasses is associated with the crossover in the structure from one where crosslinking tetrahedrally coordinated borons are the dominant structural motif for the low-alkali glasses to one where nonbridging oxygens are the dominant structural unit for the high-alkali glasses. All of these arguments are based upon SRO concepts of the structure of these glasses. Although the tetrahedrally coordinated boron and nonbridging oxygen units must manifest themselves structurally at longer ranges, the T_g of the glasses can be accurately modeled without invoking IRO structural arguments.

It would seem, therefore, that if the same were true of the sodium thioborate glasses then it would have to be concluded that no tetrahedrally coordinated borons form in these glasses, but rather nonbridging sulfurs form on the first addition of alkali sulfide. This would explain the rapid initial decrease in T_g, but not the increase at intermediate alkali sulfide modifier contents. Rather, our IR [27,28] and ^{11}B NMR [51] studies show conclusively that tetrahedral borons do form in these glasses. We have quantified the fraction of tetrahedral borons by comparing the area of the sharp to quadrapole broadened line, and Fig. 6 shows the composition dependence of the fraction of tetrahedral borons, N_4. These data are compared to those of the corresponding alkali borate glasses. What is striking is that for almost any given composition, N_4 is a factor of two to three times greater in the sulfide glasses than the oxide glasses. This indicates that not only is the SRO of these glasses similar, i.e., both

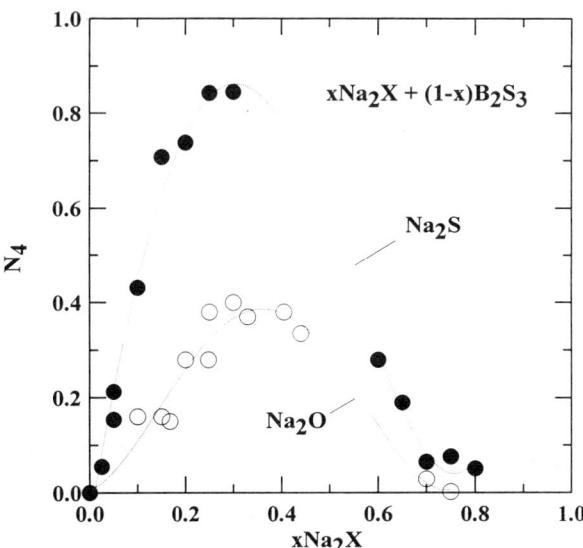

FIG. 6—*Composition dependence of the fraction of tetrahedral borons for $Na_2O + B_2O_3$ and $Na_2S + B_2S_3$ glasses. Data for the Na_2O glasses taken from Ref 45. Data for the Na_2S glasses taken from Ref 51.*

systems form tetrahedrally coordinated borons, but that there must be significant differences as well.

From these data, we have estimated that the conversion rate to tetrahedral borons from trigonal borons by the added sulfide ion must be on the order of four to six [51]. In this case rather than the sulfide ion being two-fold coordinated as the oxide ion is in the alkali borate glasses, the sulfide ion must be a highly coordinated species. This would indicate that the sulfide ion plays a dominate role in governing the properties of the liquid and glassy structures. Not only are the SROs of the two glasses likely to be different but the IROs are likely to be different as well. At present we do not have a complete understanding of the complicated structures of these alkali thioborate glasses, but a plausible structural model has been developed.

For example, if we assume the conversion rate is somewhere between four and six tetrahedrally coordinated borons for every sulfide ion added, this would suggest that four to six different tetrahedrally coordinated boron groups must be bonded to every sulfide ion added. One of the common alkali borate structures present in the low alkali borate glasses is the alkali diborate structure [45–50], shown in Fig. 7. Here the fraction of tetrahedral boron is 0.5. Now suppose since the sulfide ion has a larger ionic radius than the oxide ion (~1.8 Å versus 1.3 Å) and thereby form a large number of bonds to nearby boron centers. Hence, if a structure like the dithioborate were to form, it may be possible for the sulfur to achieve four-fold coordination by bonding to all of the borons as shown in Fig. 7. The test of this hypothesis would be to determine if the sodium dithioborate crystalline compound forms and if so does it form in the structure proposed in Fig. 7. Although this work is in progress [24], we have measured the IR and ^{11}B NMR spectra of a slowed cooled melt of the dithioborate composition and allowed it to thoroughly crystallize [51]. Both spectra show that *all* the borons are tetrahedrally coordinated in this polycrystal. It seems reasonable to conclude that structures like those shown in Fig. 7 may well be forming. Careful studies of the single crystal compounds will give a definitive answer, however.

We speculate that the rapid decrease in T_g for the sodium thioborate glasses arises from the way in which the tetrahedral borons form in these glasses. Given the high coordination of the added sulfide ion, it is reasonable to propose that this tends to create more "molecular-like" structures which, although locally quite extensively bonded together, this bonding

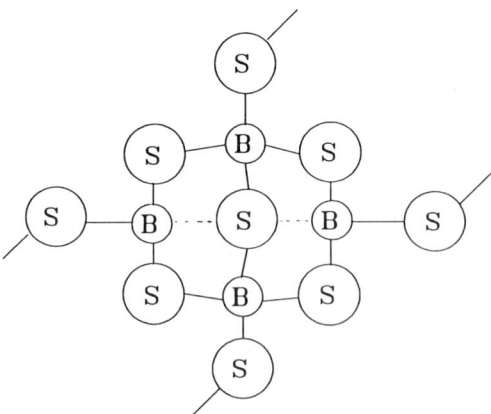

FIG. 7—*Proposed structure for $Na_2S : 2B_2S_3$ crystalline compound. This structure gives rise to a four coordinated sulfur which is supported by the ^{11}B NMR spectra.*

structure does not propagate out into the IRO of the glass. As a result this causes the overall network structure of the liquid to be degraded rather than enhanced.

What is completely not understood at this point is the dramatic upswing in T_g at the end of the low-alkali glass-forming range, if it is indeed a real effect. This increase could be associated simply with the fact that in order to prepare glasses of this composition, the liquid had to be rapidly quenched by blowing the liquid into a thin film. This much more rapid cooling rate will certainly increase T_g quite substantially. This effect does not, however, explain the increased T_g value for the $x = 0.30$ glass. For this composition, the glass could be quenched in the manner as for the other low-alkali glasses. Hence, the upswing in T_g is probably a real effect, but, perhaps due to the rapid quench rate used to make the $x = 0.33$ glass, is extenuated. Work is in progress to more fully characterize these glasses by collecting small fragments of the latter glass produced by plate-quenching.

Composition Dependence of the Heat Capacity in the Glass Transition Range

As Fig. 5 shows, the $\Delta C_p(T_g)$ decreases dramatically as the Na_2S fraction increases from 0 to 0.30. So far we have only been able to measure the low-alkali glasses, but the DSC scans of the high-alkali glasses suggests that the $\Delta C_p(T_g)$ values return to normal nonzero values.

Although all glasses exhibit a glass transition and hence a $\Delta C_p(T_g)$ value, albeit some, like those here, may be quite small, this area of research has been little studied. There are very few detailed studies of the heat capacity of glasses in the transition range [5,14,41,42]. We have found these measurements challenging at best to make and as a result this is likely to be the reason why such a dearth of data exists. For example, although the alkali borate glasses have been studied for some 30 to 50 years, there remains to date only one single report of the transition range heat capacity of these glasses. This report is now some 20 years old and the data were taken on what is now an obsolete DSC. The alkali silicate glasses have not been examined at all because their T_g values for the nonphase separated compositions are in the range of 500 to 1200°C and as such out of the range of older calorimeters. New calorimeters can make reliable heat capacity measurements up to 1100°C and this should prompt others to study these glasses.

Given the lack of good data that exist for inorganic glasses, there is concomitant lack of models that describe the composition dependence of the $\Delta C_p(T_g)$. One of the most widely cited studies is by Wunderlich et al. [52] who examined the $\Delta C_p(T_g)$ values for a number of organic polymers and compared them to the little data that were available for the inorganic glasses. In his work, he associated $\Delta C_p(T_g)$ with the number of independent units in the glass structure capable of diffusive motion. The hypothesis in this model is that the $\Delta C_p(T_g)$ is associated with independent structural units becoming thermally activated in limited diffusional motion and thereby providing the glass with the extra degrees of freedom that would account for the increased heat capacity. In this model, it would follow that glasses that exhibit a large number of possible structures or at least modes of motion at T_g would be those that exhibit large $\Delta C_p(T_g)$ values. In a qualitative fashion this is what is observed for a variety of inorganic glasses. Silica, for example, has a structurally rigid simple tetrahedral network with little diffusive motion possible for the units of the glass structure and exhibits almost no measurable $\Delta C_p(T_g)$ value. The fluoride glasses, on the other hand, with their complex compositions and variable coordination number are structurally quite complex. Correspondingly they exhibit large $\Delta C_p(T_g)$ values.

Contradicting this hypothesis are, quite interestingly, the alkali borate glasses. Although the data are of rather poor quality due to the lack of good instrumentation at the time, Uhlman et al. [5] showed that as the fraction of tetrahedrally coordinated borons increases in the glass which is associated with a corresponding increase in T_g and decrease in the average mobility of the glass structural units at any given temperature, the $\Delta C_p(T_g)$ values do not decrease as might be expected. Rather they remain near the value for pure B_2O_3 and even increase slightly. We are in the progress of repeating these measurements using a DSC with a much higher sensitivity over wider temperature ranges [53]. Our measurements to date, however, support the observation of Uhlmann et al. We also find that the $\Delta C_p(T_g)$ values do not change in the way that the sulfide glasses do. It would seem therefore that the alkali borate glasses are rather ordinary in their behavior where as the alkali thioborate glasses remain quite extraordinary.

For the alkali thioborate glasses where $\Delta C_p(T_g)$ decreases to zero, the hypothesis would be that the structure is becoming more rigid where individual structural units are less available for motion in the transition range. If the fraction of tetrahedral boron in these glasses increases in the ways that it is believed, this structural change would be expected to decrease $\Delta C_p(T_g)$ for the reasons described above. The decreasing T_g on the other hand, however, contradicts this hypothesis. With a decreasing T_g, the average structural rigidity of the network is actually a decreasing function of composition. The diminishing $\Delta C_p(T_g)$ with composition is therefore opposite to that expected from the T_g behavior. To study this problem further, we now look to the glassy state to see if there is some unusual behavior in the glassy heat capacity that gives rise to the vanishing $\Delta C_p(T_g)$ values.

To do this, we hypothesize that the T_g of a glass is analogous to the melting point (T_m) of a crystalline solid, i.e., the glass exhibits a T_g when the available thermal energy to the system is sufficient to cause rupture of the bonds defining the long range structure of the system. This analogy has been used very successfully in describing the mechanical properties of glass, for example [54]. This hypothesis suggests that the heat capacity of the glass should reach the classical DuLong-Petit value at T_g since it represents the thermal state of the system where the vibrational modes of motion in the system are fully saturated. Above this temperature, in order for the heat capacity to increase, the system must populate additional modes of motion beyond the vibrational modes. These would obviously be the transitional, rotational, and libration modes. It is these very modes that describe a viscous liquid and as such should account for T_g. Hence, in Table 1 we calculate the heat capacity below and above the glass transition region using the constructions shown on Fig. 1. The classical limiting DuLong-Petit value of $3Rn$ was calculated from the composition assuming that all atoms (ions) in the formula unit contributed to the heat capacity as a worst case assumption. The test is to determine whether or not the glass is in some way "over thermally" populated below T_g. By maximizing the number of modes available, we are maximizing the glassy heat capacity. If the glassy heat capacity is larger than this value, then the glassy state heat capacity is indeed quite unusual and hence may be responsible for the diminishing $\Delta C_p(T_g)$ values. On the other hand, if the glassy heat capacity falls short of the limiting classical value, the glassy state is not responsible for the diminishing $\Delta C_p(T_g)$ values.

As Table 1 shows, all of the glasses behave quite normally. All of the glassy state heat capacities approach 90 to 95% of the classical DuLong-Petit heat capacity below T_g. This would suggest that just as expected, the glass exhibits a T_g just when all or nearly all of the vibrational modes of motion in the glass are fully excited. In this sense, it means that we must look to the liquid state for the source of the diminishing $\Delta C_p(T_g)$ values. The conclusion must be that the liquid state structure is one which, on a thermal properties basis, is not significantly different from the glassy state structure. To investigate this hypothesis further, we return once again to the behavior of a few select crystalline materials.

If we take B_2O_3 as a representative sample, we find that at the melting point of 723 K, the C_p is 112.55 J/mol-K, versus the DuLong-Petit value of 124.73 J/mol-K, or 90.3%. Likewise

TABLE 1—T_g's and heat capacity values for $xNa_2S + (1 - x)B_2S_3$ glasses.

x	T_g (°C)	F.W. (g/mol)	n	$C_p(T_g-)$ (J/mol-K)	$C_p(T_g+)$ (J/mol-K)	$\Delta C_p(T_g)$ (J/mol-K)	$3Rn$ (J/mol-K)	$\dfrac{C_p(T_g-)}{3Rn}$ (%)
0.0	310.2	118	5	114.64	170.41	55.77	124.73	91.9
0.05	300	116	4.9	112.97	146.44	33.47	122.21	92.4
0.10	254.9	114	4.8	108.78	127.27	18.49	119.70	90.9
0.15	210.8	112	4.7	106.19	114.18	7.89	117.24	90.6
0.20	176.5	110	4.6	103.01	104.60	1.59	114.73	89.8
0.25		108	4.5	103.26	103.26	0	112.21	92.0
0.30	250	106	4.4	103.55	107.82	4.27	109.75	94.4

for SiO_2, we find that C_p at the melting point of 1883 K is 111.21 J/mol-K versus a $3Rn$ value of 99.75 J/mol-K, or 89.7% [55]. Other singly bonded oxide systems give similar values. In both cases, the change in heat capacity at the melting point is a modest value of ~12.6 J/mol-K. A possible explanation may well be that both of the systems are high-viscosity strongly bonded systems that exhibit very little SRO structural change upon entering the liquid state. There must be a large amount of IRO structural difference, however, to account for the liquid state structure. For the thioborate glasses, where the $\Delta C_p(T_g)$ decreases even in spite of a decreasing T_g, the reason must lie not in the rigidity of the glassy network but rather in the configurational similarity between the liquid and glassy states. For example, even though the isothermal viscosity is a decreasing function of the alkali sulfide content, the density of configurational states which are accessible just above T_g must not be any larger than those below T_g. If there were a large number of such states available then the heat capacity could access them and thereby produce a nonzero $\Delta C_p(T_g)$. We propose that the lack of an increasing number in the density of configurational states above T_g is linked to the unusual way in which the sulfide anions are incorporated into the glassy structure. The large number of tetrahedrally coordinated borons generated per each added sulfide suggests that the configurational "freedom" that the structure has available must be more severely restricted than were the sulfide anion not so heavily and preferentially coordinated. Indeed, in the halide glasses where the coordination numbers are high, but extremely variable, the $\Delta C_p(T_g)$ values are among some of the highest reported.

In Wunderlich's model [52] where the $\Delta C_p(T_g)$ is related to the number of independent "beads" which can contribute to the heat capacity, the number of beads must be decreasing in these glasses as alkali sulfide is added. If the structure is becoming more locally constrained by the high coordination number around the sulfide anion, this could account for a decreasing number of mobile beads in the system.

At this point, the ideas presented here are highly speculative and there are many unresolved problems with this model. For example, we have remeasured the heat capacities of a number of alkali borate glasses and find that $\Delta C_p(T_g)$ actually increases with T_g rather than decreases as might be expected from the arguments presented here. Even though the oxygen is still only two-fold coordinated in the alkali borate glasses when tetrahedral borons form, the effect is to still constrain the network as alkali is added. Another major unresolved problem is why the behaviors here are so different from the behaviors of so many other glasses. For example, there is almost a universal trend that if T_g decreases through a change in the composition, then the $\Delta C_p(T_g)$ increases. This is true for silicate, phosphate, halide, and other glasses. We are in the progress of analyzing these and other questions about these glasses.

Summary and Conclusions

The T_g's and heat capacity functions have been measured for a series of $Na_2S + B_2S_3$ glasses for the first time. Unlike the alkali borates, T_g decreases rapidly as Na_2S is added to B_2S_3. This effect, even in the presence of a rapidly increasing fraction of tetrahedral borons, has been associated with the "over crosslinking" effect of the sulfide ion. Unlike the borate glasses where each added oxygen produces two tetrahedral borons, the conversion rate for the alkali thioborates is between four to six. This behavior is suggested to result in the formation of local tightly-bonded molecular-like structures that exhibit less long-range network bonding than the alkali borate glasses. As a result, T_g decreases with added alkali, rather than increases.

The change in heat capacity at T_g has been carefully measured and is found to also decrease dramatically as alkali sulfide is added to the glass. Again this effect is opposite to the trends observed for the alkali borate glasses. The decreasing $\Delta C_p(T_g)$ occurs even in the presence of a decreasing T_g. We have tentatively associated the diminishing $\Delta C_p(T_g)$ values to the decreasing density of the configurational states above T_g. This is attributed to the high coordination and site specificity caused by the added alkali sulfide. The glassy state heat capacities were analyzed and found to reach ~90% of the classical limiting DuLong-Petit value just below T_g for all glasses. This was used to suggest that diminishing $\Delta C_p(T_g)$ values are associated with a unique behavior in the liquid state to become liquid with very little change in the density of configurational states.

Acknowledgments

This work was supported through NSF-DMR grants 87-01077 and 91-04460. The authors would like to thank F. Borsa and D. Torgeson of the Ames Laboratory and H. Patel, J. Sills, and J. Hudgens for their help in the preparation of the glasses.

References

[1] Eisenberg, A. and Takahashi, K., "Viscoelasticity of Silicate Polymers and Its Structural Implications," *Journal of Non-Crystalline Solids*, Vol. 3, 1970, pp. 279–293.

[2] Moynihan, C. T., Easteal, A. J., Tran, D. C., et al., "Heat Capacity and Structural Relaxation of Mixed Alkali Glasses," *Journal of the American Ceramic Society*, Vol. 59, No. 3-4, 1976, pp. 137–140.

[3] Hirao, K., Naohiro, S., and Masanaga, K., "Low Temperature Heat Capacity and Structure of Alkali Silicate Glasses," *Journal of the American Ceramic Society*, Vol. 62, No. 11–12, 1979, pp. 570–573.

[4] Shelby, J. E., "Property/Morphology Relations in Alkali Silicate Glasses," *Journal of the American Ceramic Society*, Vol. 66, No. 11, 1983, pp. 754–757.

[5] Uhlmann, D. R., Kolbeck, A. G., and De Witte, D. L., "Heat Capacities and Thermal Behavior of Alkali Borate Glasses," *Journal of Non-Crystalline Solids*, Vol. 5, 1971, pp. 426–443.

[6] Visser, T. J. and Stevels, J. M., "Rheological Properties of Boric Acid and Alkali Borate Glasses," *Journal of Non-Crystalline Solids*, Vol. 7, 1972, pp. 376–394.

[7] Ohita, Y., Shimada, M., and Koizumi, M., "Properties and Structure of Lithium Borate and Strontium Borate Glasses," *Journal of the American Ceramic Society*, Vol. 65, No. 11, 1982, pp. 572–574.

[8] Martin, S. W., Cooper, E. I., and Angell, C. A., "CO_2 Retention in High-Alkali Borate Glasses," *Journal of the American Ceramic Society*, Vol. 66, No. 9, 1983, c-153–154.

[9] Shelby, J. E., "Thermal Expansion of Alkali Borate Glasses," *Journal of the American Ceramic Society*, Vol. 66, No. 3, 1983, pp. 225–227.

[10] Martin, S. W. and Angell, C. A., "Glass Formation and Transition Temperatures in Sodium and Lithium Borate and Aluminoborate Glasses up to 72 mol % Alkali," *Journal of Non-Crystalline Solids*, Vol. 66, 1984, pp. 429–442.

[11] Affatigato, M., Feller, S. Khav, E. J., et al., "The Glass Transition Temperature of Lithium and Lithium-Sodium Borate Glasses over Wide Ranges of Composition," *Physics and Chemistry of Glasses*, Vol. 31, No. 1, 1990, pp. 19–24.

[12] Koritala, S., Farooqui, K., Affatigato, M., et al., "The Glass Transition Temperature of Lithium-Alkali Borates," *Journal of Non-Crystalline Solids*, Vol. 134, 1991, pp. 277–286.

[13] Mundy, J. N. and Lin, G. L., "Ionic Transport in Sodium Aluminogermanate Glasses," *Solid State Ionics*, Vol. 21, 1986, pp. 305–325.

[14] Iqal, T., Sharhriari, M. R., and Sgiel, G. H., "A New Glass Forming Ability Criterion for Multicomponent Glasses," *Materials Science Forum*, Vol. 67/68, 1992, pp. 225–232.

[15] Gavin, D. L. Chung. K. H., Bruce, A. J., et al., "Heat Capacities of Heavy Metal Fluoride Glasses," *Journal of the American Ceramic Society*, 1982, c-182–183.

[16] Martin, S. W. and Angell, C. A., "DC and AC Conductivity in Wide Composition Range Li_2O-P_2O_5 Glasses," *Journal of Non-Crystalline Solids*, Vol. 83, 1983, pp. 185–207.

[17] Hudgens, J. J. and Martin, S. W., "On the Glass Transition Temperature and Infrared Spectra of Low Alkali, Anhydrous Lithium Phosphate Glasses," *Journal of the American Ceramic Society*, Vol. 76, No. 7, 1993, pp. 1691–1696.

[18] Zhang, M. Mancini, S., Bresser, W., and Boolchand, P., "Glass Transition (T_g) Variation with

Average Coordination Number <m> in Network Glasses: Evidence of a Threshold Behavior in the Slope at the Rigidity Percolation Threshold (<m> = 2.4), *Journal of Non-Crystalline Solids*, in press.
[19] Naughton, J. L. and Mortimer, C. T., "Differential Scanning Calorimetry," in *IRS; Physical Chemistry Series 2*, Volume 10, Butterworths, London, 1975.
[20] Levasseur, A., Olazcuaga, R., Kbala, M., et al., "Synthesis and Properties of New Sulfide Glasses with High Ionic Conductivity," *Compte Rendus Acad. Paris*, Vol. 293, Series II, 1981, pp. 563–565.
[21] Susman, S., Boehm, L., Volin, K. J., and Delbecq, C. J., "A New Method for the Preparation of Fast-Ion Conducting Reactive Glasses," *Solid State Ionics*, Vol. 5, 1981, pp. 667–670.
[22] Wadas, H., Menetrier, M., Levasseur, A., and Haggenmuller, P., "Preparation and Ionic Conductivity of New B_2S_3-Li_2S-LiI Glasses," *Materials Research Bulletin*, Vol. 18, 1983, pp. 189–193.
[23] Menetrier, M., Hojjaji, A., Estournes, C., and Levasseur, A., "Ionic Conduction in the B_2S_3-Li_2S Glass System," *Solid State Ionics*, Vol. 48, 1991, pp. 325–330.
[24] Kennedy, J. H. and Yang, Y., "Glass-Formation and Structures in SiS_2-Li_2S-LiX (X = Br, I,)," *Journal of Solid State Chemistry*, Vol. 69, 1987, pp. 252–257.
[25] Tenhover, M., Boyer, R. D., and Henderson, R. S., "Magic Angle Spinning ^{29}Si Nuclear Magnetic Resonance of Si-Chalcogenide Glasses," *Solid State Comm.*, Vol. 65, No. 12, 1988, pp. 15-17–15-21.
[26] Pradel, A., Ribers, M., and Maurin, M., "^{7}Li NMR Study of the Li_2S-SiS_2 Glass System," *Solid State Ionics*, Vol. 28-30, 1988, pp. 762–765.
[27] Martin, S. W. and Sills, J., "^{29}Si and ^{27}Al MASS-NMR Studies of Li_2S + Al_2S_3 + SiS_2 Glasses," *Journal of Non-Crystalline Solids*, Vol. 135, 1991, pp. 171–181.
[28] Bloyer, D., Cho, J., and Martin, S. W., "Infrared Spectroscopy of Wide Composition Range xNa_2S + (1 − x)B_2S_3 Glasses," *Journal of the American Ceramic Society*, Vol. 76, No. 11, 1993, pp. 2753–2759.
[29] Martin, S. W. and Bloyer, D. R., "Preparation and Infrared Characterization of Thioborate Compounds and Polycrystals," *Journal of the American Ceramic Society*, Vol. 74, No. 5, 1991, pp. 1003–1010.
[30] Martin, S. W. and Polewik, T., "Density Measurements of Glasses in the Series xNa_2S + (1 − x)B_2S_3," *Journal of the American Ceramic Society*, Vol. 74, No. 6, 1991, pp. 1466–1468.
[31] Martin, S. W., Cho, J., and Polewik, T., "Wide Composition Range Studies of the Density of xNa_2S + (1 − x)B_2S_3 Glasses," to be submitted to *Journal of the American Ceramic Society*.
[32] Martin, S. W. and Bloyer, D. R., "Preparation of High-Purity Vitreous B_2S_3," *Journal of the American Ceramic Society*, Vol. 73, No. 11, 1990, pp. 3481–3485.
[33] Bloyer, D. R., "Preparation and Characterization of B_2S_3-based Glasses: Infrared and Scanning Calorimetry Studies of the Na_2S + B_2S_3 System," M. S. thesis, Iowa State University, 1989.
[34] Brown, M. E., *Introduction to Thermal Analysis Techniques and Applications*, Chapman and Hall, New York, 1988, ch. 4.
[35] Wunderlich, B., *Thermal Analysis*, Academic Press, Inc., New York, 1990, ch. 5.
[36] Suzuki, H. and Wunderlich, B., "The Measurement of High Quality Heat Capacity by Differential Scanning Calorimetry," *Journal of Thermal Analysis*, Vol. 29, 1984, pp. 1369–1377.
[37] Gibbs, J. H., "Nature of the Glass Transition and the Vitreous State," in *Modern Aspects of the Vitreous State*, J. D. Mackenzie, Ed., Butterworths, London, 1960, ch. 7.
[38] Gibbs, J. H. and DiMarzio, E. A., "Nature of the Glass Transition and the Glassy State," *Journal of Chemical Physics*, Vol. 28, No. 3, 1958, pp. 373–383.
[39] Narayanaswamy, O. S., "A Model of Structural Relaxation in Glass," *Journal of the American Ceramic Society* Vol. 54, No. 10, 1971, pp. 491–498.
[40] Angell, C. A. and Rao, K. J., "Configurational Excitations in Condensed Matter, and the 'Bond Lattice' Model for the Liquid-Glass Transition," *Journal of Chemical Physics*, Vol. 57, No. 1, 1972, pp. 470–480.
[41] DeBolt, M. A., Easteal, A. J., Macedo, P. B., and Moynihan, C. T., "Analysis of Structural Relaxation in Glass Using Rate Heating Data," *Journal of the American Ceramic Society*, Vol. 59, No. 1–2, 1976, pp. 16–21.
[42] Angell, C. A. and Sichina, W., "Thermodynamics of the Glass Transition: Empirical Aspects," *Annals of the New York Academy of Sciences*, Vol. 279, 1976, pp. 53–67.
[43] Moynihan, C. T., Macedo, P. B., Montrose, C. J., et al., "Structural Relaxation in Vitreous Materials," *Annals of New York Academy of Sciences*, pp. 15–35.
[44] Angell, C. A., "Oxide Glasses in Light of the 'Ideal Glass' Concept: II, Interpretations by

Reference of Simple Ionic Glass Behavior," *Journal of the American Ceramic Society*, Vol. 51, No. 3, 1968, pp. 125–134.
[45] Bray, P. J. and O'Keefe, J. G., "Nuclear Magnetic Resonance Investigations of the Structure of Alkali Borate Glasses," *Physics and Chemistry of Glasses*, Vol. 4, 1963, pp. 37–46.
[46] Kim, K. S. and Bray, P. J., "Nuclear Magnetic Resonance Studies of the Glass in the System $PbO-B_2O_3-SiO_2$," *Journal of Chemical Physics*, Vol. 64, 1976, pp. 4459–4465.
[47] Yun, Y. H. and Bray, P. J., "Nuclear Magnetic Resonance Studies of the Glasses in the System $Na_2O-B_2O_3-SiO_2$," *Journal of Non-Crystalline Solids*, Vol. 27, 1978, pp. 363–380.
[48] Bray, P. J., Leventhal, M., and Hooper, H. O., "Nuclear Magnetic Resonance Investigations of the Structure of Lead Borate Glasses," *Physics and Chemistry of Glasses*, Vol. 4, 1963, pp. 47–66.
[49] Greenblatt, S. and Bray, P. J., "A Discussion of the Fraction of Four-Coordinated Boron Atoms Present in Borate Glasses," *Physics and Chemistry of Glasses*, Vol. 8, 1967, pp. 213–217.
[50] Baugher, J. F. and Bray, P. J., "Magnetic Resonance Studies of Thallium Borate Glasses," *Physics and Chemistry of Glasses*, Vol. 10, 1969, pp. 77–88.
[51] Sills, J. A., Martin, S. W., Borsa, F., and Torgeson, D., "^{11}B NMR Studies of the Short Range Order in Wide Composition Range $Na_2S + B_2S_3$ Glasses," *Journal of Non-Crystalline Solids*, accepted for publication.
[52] Wunderlich, B., "Study of the Change in Specific Heat of Monomeric and Polymeric Glasses During the Glass Transition," *Journal of Phys. Chem.*, Vol. 64, 1960, pp. 1052–1056.
[53] Kincs, J. and Martin, S. W., "Heat Capacity and T_g Measurements of Sodium Borate Glasses," *Journal of Non-Crystalline Solids*, to be submitted.
[54] Hilton, A. R., Jones, C. E., and Bran, M., *Physics and Chemistry of Glasses*, Vol. 7, No. 4, 1966, p. 105.
[55] *CRC Handbook of Chemistry and Physics*, R. C. Weast, Ed., Vol. 58, CRC Press, 1977, D64–66.

Bruce Cassel[1] *and Alan T. Riga*[2]

Glass Transition of a Liquid Crystal Polymer

REFERENCE: Cassel, B. and Riga, A. T., "**Glass Transition of a Liquid Crystal Polymer,**" *Assignment of the Glass Transition, ASTM STP 1249,* R. J. Seyler, Ed., American Society for Testing and Materials, Philadelphia, 1994, pp. 202–213.

ABSTRACT: A commercially available liquid crystal polymer system having a very weak glass transition is investigated in order to illustrate the special problems of measurement and interpretation inherent in high-sensitivity T_g measurement. Differential scanning calorimetry, thermomechanical analysis, and dynamic mechanical analysis were employed over the temperature region from ambient to 175°C. A transition between 75 and 125°C with a midpoint around 100°C was detected by most techniques. The use of the maximum in the derivative of the measured signal as the primary assignment of the glass transition temperature is recommended. The results are discussed with reference to other literature on this polymer system.

KEYWORDS: glass transition, liquid crystal polymer, copolyester, mesophase, differential scanning calorimeter (DSC), thermomechanical analysis (TMA), dynamic mechanical analysis (DMA)

Background

The primary use of thermal analysis techniques is to qualitatively and quantitatively characterize properties of a polymer system as a function of temperature. Below about −170°C polymeric materials are rigid because at a molecular level adjacent polymer chains are frozen in a matrix with insufficient thermal energy for molecular rotational or translational motion to occur. With increasing temperature, the thermal energy results in an increasing degree of molecular motion, including rotation of polymer segments and side groups, and finally translation of adjacent polymer segments with respect to each other. This overall process is accompanied by increasing volume and heat capacity, and decreasing resistance to externally imposed stresses, either mechanical or electrical. Using thermal analysis techniques, the polymeric materials are heated (or cooled), and the changes in physical properties are recorded and analyzed as a function of temperature. These changes are usually concentrated in certain temperature regions, i.e., transition regions, which are interpreted in terms of morphological changes at a molecular level. For example, most polymers show transitions that can be identified as associated with the onsets of rotation of polymer subgroups or side groups, the glass transition, which corresponds to the onset of translation, normally in the amorphous phase, and the melting of the crystalline phase.

In contrast to semicrystalline polymers, where a fraction of the material is in the amorphous phase, a simple *crystalline* organic compound exists in a single crystalline state which allows no rotation or translational degrees of freedom. Above the crystalline melting point the crystal lattice breaks down, and the structure is amorphous and isotropic. A *liquid crystal* material exists in a crystalline state up to a temperature at which the material loses its long-range crystalline order while retaining some orientational and positional order up to the

[1]Principal scientist, Perkin-Elmer Corp., Norwalk, CT 06859.
[2]Senior scientist, Lubrizol Corp., Wickliffe, OH 44092.

isotropization point [1]. Thermal analysis has been used for some time to characterize the phase transitions of liquid crystalline materials [1,2].

Liquid crystal polymers (LCPs) are a class of polymers that form a rigid rod-like, or plate-like molecular structure which retains some degree of order above the crystalline melting point [3]. By annealing these materials above the melting point it is possible to develop a high degree of orientational order in the mesophase state. When cooled, this annealed polymer mesophase may solidify with a high degree of crystalline order and a low amorphous content, or under different thermal treatment it will freeze in the mesophase structure into a mesophase glassy state, again with a small amorphous content. As a liquid crystal polymer melt flows through an orifice, the stiff polymer chains orient in the direction of flow. This leads on cooling to a high degree of orientation which results in exceptional anisotropic strength in the direction of orientation. Other thermal properties that make such materials commercially interesting may include high thermal stability, low melt viscosity, and low coefficients of expansion.

The material investigated in this study is a commercially available random copolyester of 4-hydroxybenzoic acid (73%) and 2-hydroxy 6-naphthoic acid (27%), prepared and provided by the Hoechst Celanese Corporation. The material analyzed was provided as injection molded test bars that were cut using a water-lubricated precision saw. Specimens for mechanical analysis were cut parallel and perpendicular to the injection flow. Other samples of a nominally similar material were provided in pellet form, but thermal results on these were quite different (due presumably to their different thermal history or molecular weight distribution), and these results were not included in this paper.

Instrument Description

DSC

The calorimetric data were obtained with a Perkin-Elmer DSC 7 power compensation-type Differential Scanning Calorimeter (DSC), using the 7 Series/UNIX controller and software system. The DSC cooling system used was an ice water reservoir; the purge was nitrogen at 30 mL/min; the encapsulation was standard crimped aluminum pans, and sample size was approximately 20 mg. A scanning rate of 5°C/min was used in order to better provide a comparison to data taken by mechanical analysis. Temperature calibration followed ASTM E 967, Practice for Temperature Calibration of Differential Scanning Calorimeters and Differential Thermal Analyzers; the primary reference material for heat flow and temperature calibration was high-purity indium metal.

TMA and DMA

The dynamic mechanical data were obtained with a Perkin-Elmer DMA 7 Dynamic Mechanical Analyzer using a three-point bending geometry. Most samples were run using the 10-mm knife edge quartz measuring system in the "high-temperature" furnace. DMA sample dimensions were roughly 2 mm (height) by 2 mm by 15 mm. The DMA 7 was used in constant force mode for the thermomechanical analysis (TMA) data. Flexure analysis used the same mounting fixture and furnace as used with the DMA. For TMA expansion the same fixture was used with the sample (approximately 6 mm height) standing between the knife edges. The temperature calibration procedure for all mechanical data is given in ASTM E 1363, Test Method for Temperature Calibration of Thermomechanical Analyzers. Scanning rates were 3.0 or 4.0°C/min. Dynamic stress was on the order of 10^6 Pa. The frequency was 0.2, 1.0, or 5.0 Hz.

In addition to the standard two-point calibration for all calorimetric and mechanical data, an indium internal reference standard was used to correct the data. This method has been used previously to accurately compare the glass transition temperature of an epoxy secondary reference material by DSC, TMA expansion, TMA flexure, and DMA [4]. Figure 1 is a plot of typical DSC, DMA, and TMA data showing the LCP and indium reference transitions.

Results and Discussion

DSC

As a material is heated through the glass transition region the increased kinetic energy is sufficient to allow cooperative segmental motion of polymer chains. The new degrees of freedom associated with this translation result in a stepwise increase in specific heat. The heat flow output of the DSC is proportional to the sample weight, the specific heat of the sample, and the scanning rate. When this liquid crystal polymer was initially heated at 5°C/min from just above room temperature to 175°C there was some evidence of a weak glass transition having a delta C_p of 0.05 J/g°C (Fig. 2, bottom curve). Faster heating rates (25 and 50°C/min, data not shown) amplified the step in the heat flow signal in the 100°C region and indicated that this thermal event is a stepwise increase in specific heat.

In an attempt to sharpen the DSC heat flow signal in the 100°C transition region, a different thermal history was imposed. This technique, which is often successfully employed when analyzing highly crystalline thermoplastics, involves the following procedure. First, shock cool the specimen from above the crystalline melt to get as much material as possible into the amorphous state; then cool it slowly through the glass transition region to put the sample in a low enthalpy state [5]. When the material is subsequently heated at a moderate rate, the transition region is usually compressed into a short period of time, thus making the change more evident.

This technique was applied to analyzing this LCP specimen; however, when the sample is

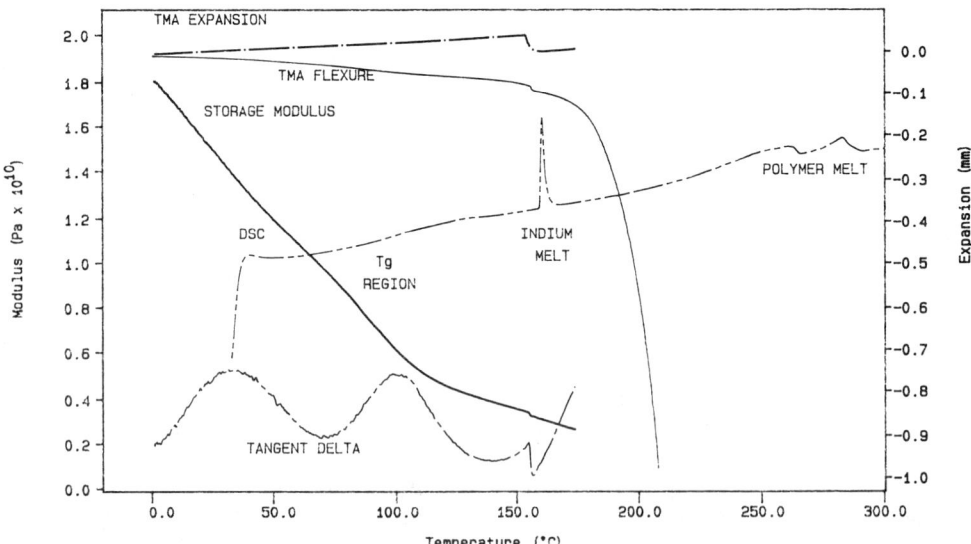

FIG. 1—*Thermal curves of a copolyester liquid crystal polymer, showing the use of indium as an internal reference standard.*

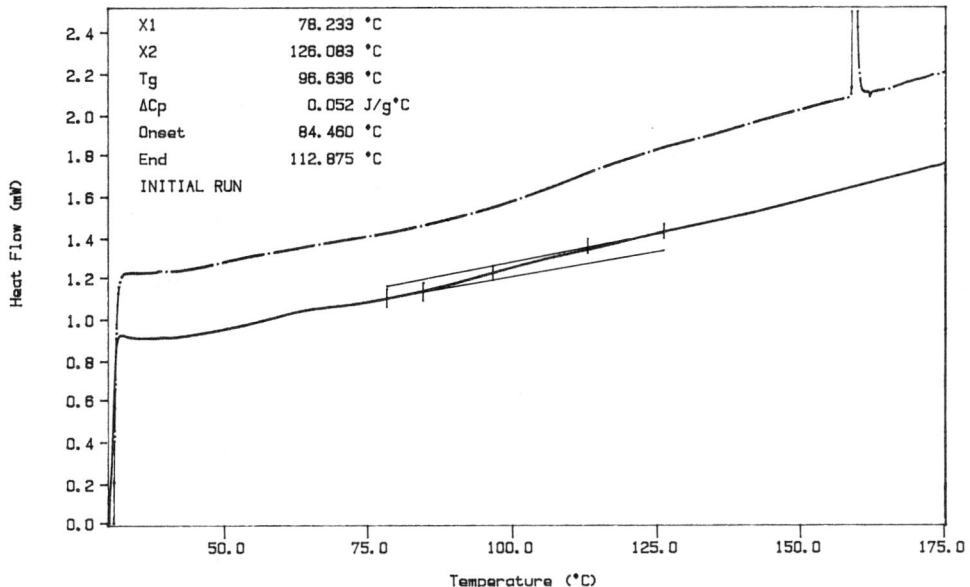

FIG. 2—*DSC of injection molded* (bottom curve), *and thermally cycled material, showing slight increase in heat capacity at the glass transition region.*

heated above the glass transition there is very little crystalline melting; that is, the melting endotherm is only around 1.4 J/g. Hence, in attempting this procedure the sample was heated with each attempt to an increasingly high temperature (300, 350, 400, and 450°C, in an inert atmosphere) in an attempt to achieve an amorphous liquid state prior to shock cooling. After each heat the material was cooled rapidly to 120°C, then cooled slowly from 120 to 85°C. When the sample was then analyzed by DSC it was found that with each thermal cycle the step in the specific heat became slightly more pronounced. This is interpreted as evidence that more material is being trapped in the amorphous phase, and that it is a transition in this phase that is being observed. The transition *temperature* also increased with this conditioning (Fig. 2, top curve), which suggests that the cycling treatment was causing some degradation, for example, transesterification which might be expected to effect the amorphous phase [6]. After the most extreme thermal treatment the specific heat step change appeared to be spread over several tens of degrees. Based on DSC data on similar systems Wunderlich has reported that the glass transition of the *mesophase* is extremely broad and extends all the way up to the apparent melting endotherm [7]. It is not clear that it is *this* T_g that is being detected at 100°C. The relative ineffectiveness of the thermal quenching treatment to produce more amorphous content is consistent with the possibility that melting of the mesophase to an anisotropic liquid occurs above the decomposition temperature [8].

TMA Expansion

As a polymeric material is heated through the T_g region the increased molecular mobility which gives rise to the change in mechanical and heat storage properties is also associated with changes in dimension. Hence, TMA is often used to determine T_g through monitoring changes in the rate of expansion. Expansion data were gathered on specimens cut parallel and perpendicular to the flow direction for this injection molded sample. Specimens were analyzed as received and after heating and cooling once over the T_g region.

Expansion data taken through the T_g on amorphous polymers show a distinct change in slope (i.e., change in coefficient of thermal expansion) in the T_g region. The intersection of the slopes above and below T_g have been shown to coincide with the midpoint of the heat capacity change as measured by DSC when all thermal history and thermal lag effects have been properly compensated for [*1*]. In the case of this liquid crystal polymer analyzed in the plane perpendicular to the orientation, there is a perceptible change in the coefficient of expansion through the T_g region; however, the break in the curve is not sharp enough to give confidence in the analysis. The smoothed derivative (Fig. 3, top curve) is also displayed to help identify the transition. A slight step change is seen in the derivative. On the initial heat there was some penetration of the expansion probe into the sample, and the profile for this penetration shows softening over two temperature ranges, the higher one centering around 95°C. No attempt was made to obtain improved penetration data since it would be expected to be similar to flexural data under equivalent stress.

When this material is similarly measured in the direction of the orientation the coefficient of expansion is negative over the entire temperature range. The negative expansion on heating can be seen to be partly reversible; that is, the sample expands on cooling and then again contracts on reheating. In order to demonstrate that instrumental effects were not responsible for this, a "no sample" baseline is also included (Fig. 4). No break can be seen in these curves in the 100°C region to indicate the presence of a glass transition. Apparently the small amorphous content plays little role in supporting the probe in this orientation, at this low level of stress.

TMA Flexure

Thermomechanical analysis has long been considered to be more sensitive than DSC for detecting weak glass transitions when the sample is mounted in a highly leveraged geometry. To explore this, the liquid crystal polymer sample was analyzed by TMA flexure analysis at two different force loadings. The primary deflection is well above 150°C; however, at the higher force level there is a small amount of flexure between 70 and 140°C. One problem with using TMA flexure to quantify T_g is that the onset and magnitude of deflection frequently depend upon the amount of stress applied to the sample. However, it is useful as a comparative technique, and it is one of the methods under development of ASTM Committee E-37 on Thermal Measurements. The first derivative signal is useful to help define the deflection region for the purpose of positioning the tangents that define the onset (Fig. 5). In the TMA analysis of this liquid crystal polymer system we can only say that there is a softening event—consistent with a weak T_g—in the same temperature region previously observed by DSC.

Storage Modulus

When a normal polymer is heated through the glass transition region there is a drop in the storage modulus, which characterizes the polymer stiffness and the ability of the material to elastically store mechanical energy. On a molecular scale this modulus drop through the T_g region is due to the onset of segmental translation in the amorphous phase. In this copolyester LCP the storage modulus can be seen to decrease continuously from well below room temperature to about 110°C, and then decrease at a lower rate (Fig. 6). The modulus then decreases smoothly at an ever increasing rate up to the 275°C endotherm.

One method of reporting the T_g which is under consideration in the ASTM Committee

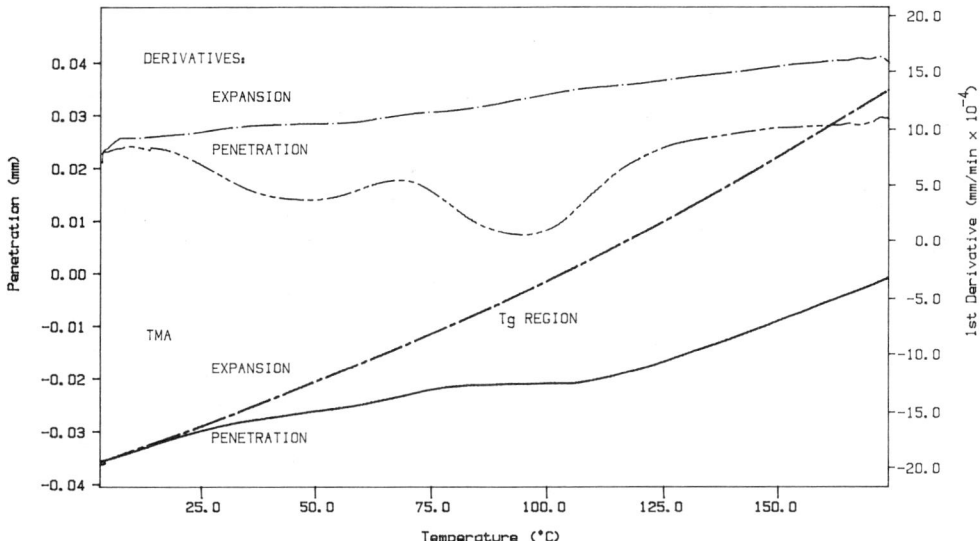

FIG. 3—*Initial and reheat data using the TMA with a 1-mm probe and a 10-mN load. The sample is aligned perpendicular to the orientation. The bottom two curves are the first heat ("penetration") and reheat ("expansion"). The top curves are the respective derivatives.*

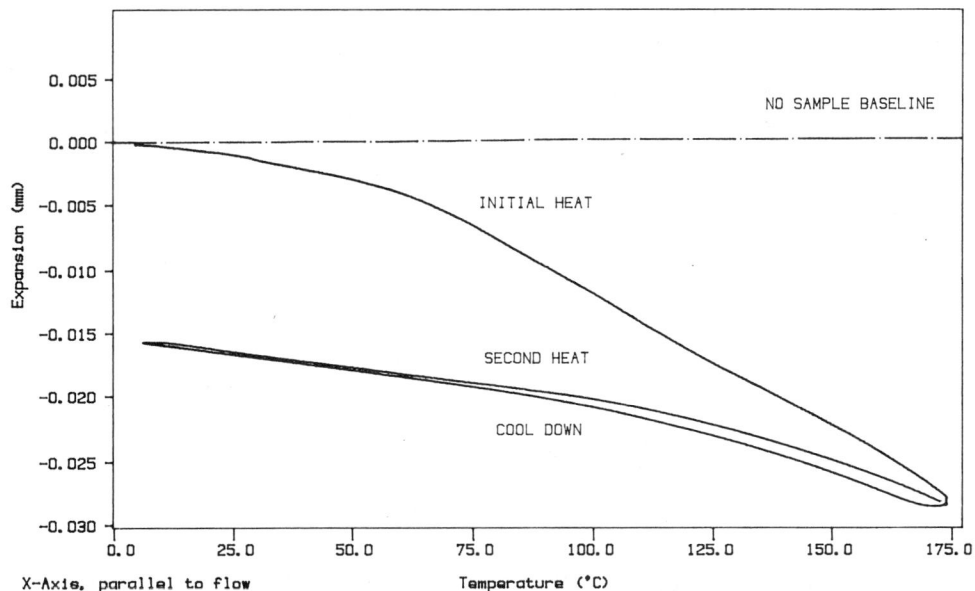

FIG. 4—*TMA expansion analysis (heat, cook, heat) in the direction of the orientation showing a negative coefficient of linear expansion.*

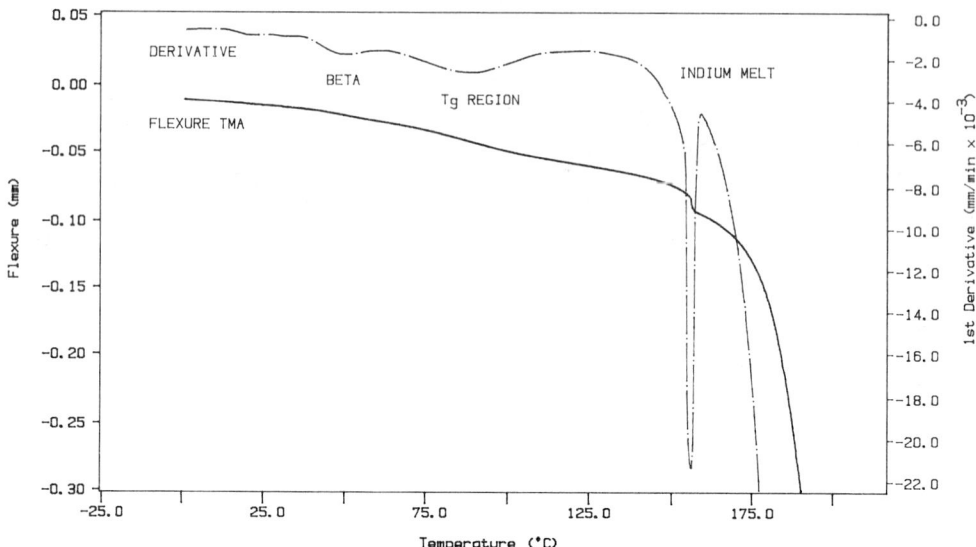

FIG. 5—*TMA flexure analysis (and derivative) using 1000 mN load, showing small deflections at 50°C and 93°C; and indium melting at 156.6°C.*

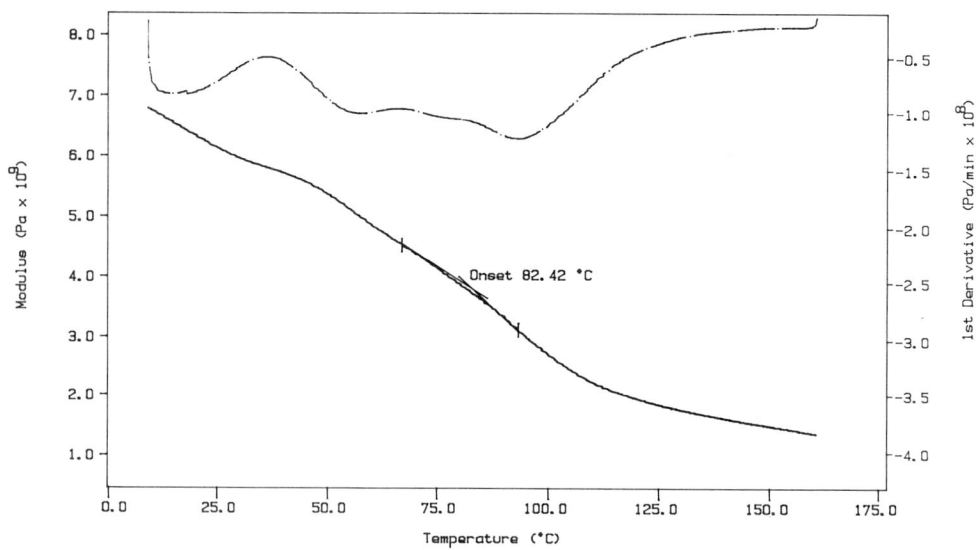

FIG. 6—*DMA storage modulus, initial heat, showing use of the derivative to select tangents for T_g.*

E-37 is that of determining the onset of the T_g modulus decrease. In order to objectively define the onset constructs for this calculation, the derivative of the storage modulus was used to select the positions of the onset tangents. While this technique is able to affix a number to the storage modulus onset, it does not appear to have identified a particularly noteworthy transition point (Fig. 6). The derivative of the modulus signal does indicate maxima that coincide with the derivative peaks observed with the TMA flexure. Experience with this sample suggests that for very weak transitions it would be better to report the derivative maximum of the storage modulus as the primary elastic parameter.

Loss Modulus

The measure of the ability of a material to dissipate mechanical energy into heat—the loss modulus—can also be obtained from DMA and analyzed for T_g. Figure 7 shows the tangent delta (the DMA primary output) and the loss modulus (a derived parameter) on the initial heat of this copolyester LCP. For this material the loss properties appear to isolate a discrete transition in the 100°C range. The frequency dependence of the DMA data is in the expected direction; namely, the data taken at a lower frequency is shifted to a lower temperature.

This loss transition and two lower temperature transitions have been studied for this polymer system [9–11]. By comparing the DMA and dielectric results of this polymer system and another having a different copolymer ratio, Blundell identified the two lower temperature transitions in terms of the onset of specific rotational degrees of freedom [9]. For example, the 50°C transition was attributed to the onset of rotation of the naphthalene group. It is likely that it is this transition event which is in evidence in the 50°C range in the above mechanical data. He identified the 110°C transition as the onset of cooperative motions along the chain, i.e., the glass transition. Trouten used DMA and dielectric analysis data taken over a range of frequencies to calculate activation energies for the three discrete transition regions [10]. The high activation energy of the 100°C transition is consistent with this as the glass transition. He developed a model wherein the liquid crystal mesophase aggregates are composed of stiff, interlocking polymer chains with a "sinuous" surface geometry. At low temperatures interchain interactions prevent matrix motion. Above the

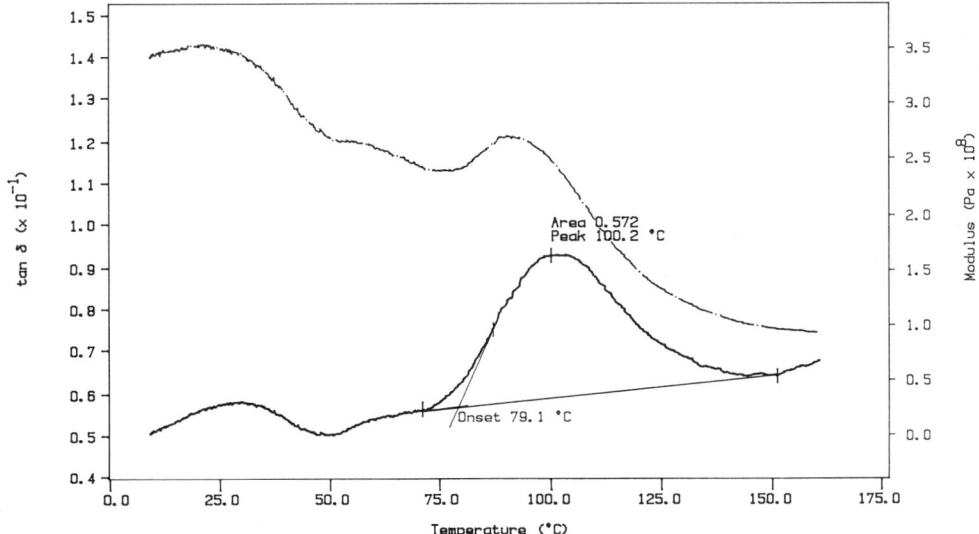

FIG. 7—*DMA loss modulus* (top curve) *and tangent delta, showing* T_g *characterization by peak analysis of the tangent delta curve.*

mesophase T_g, kinetic energy allows *lateral* slippage between adjacent chains but little slippage along the chain.

Effect of Anisotropy

One of the promoted attributes of liquid crystal polymer systems is their "self reinforcement" under flow [*12*]. Figure 8 shows the storage modulus and tangent delta of copolyester LCP, showing the effect of anisotropy produced in an injection molded bar. The results can be interpreted in the following way. In bending *across* the orientation a large fraction of the polymer backbones are being placed alternately in compression and tension from the DMA oscillation. In this geometry the storage modulus is high because of the strength of the molecular bonding, and the loss properties are low because the molecular bonds are acting as a spring to store the mechanical energy with minimal shear between adjacent chains. In the sample bending *along* the orientation the storage modulus is much lower because there are much fewer molecular backbones to support the stress, and the loss properties are higher because for a given amount of stress there is more lateral motion of adjacent chains.

Summary of Results

The results of the analysis are summarized in Table 1. These data indicate a glass transition region extending from approximately 75 to 125°C with a midpoint of 100°C. No statistical treatment was performed on the data because the primary source of uncertainty is the analyst's interpretation, i.e., the selection of constructs in performing the calculation. Another source of irreproducibility is the dependence of the material on previous thermal conditioning. Differences between various sections of the test bar (cut in the same direction) were apparent. Specimens treated to a 20°C/min cooldown from a previous run did not always reproduce. The figures shown are typical of the thermal curves. The use of the derivative signal was essential for selecting objective calculation constructs.

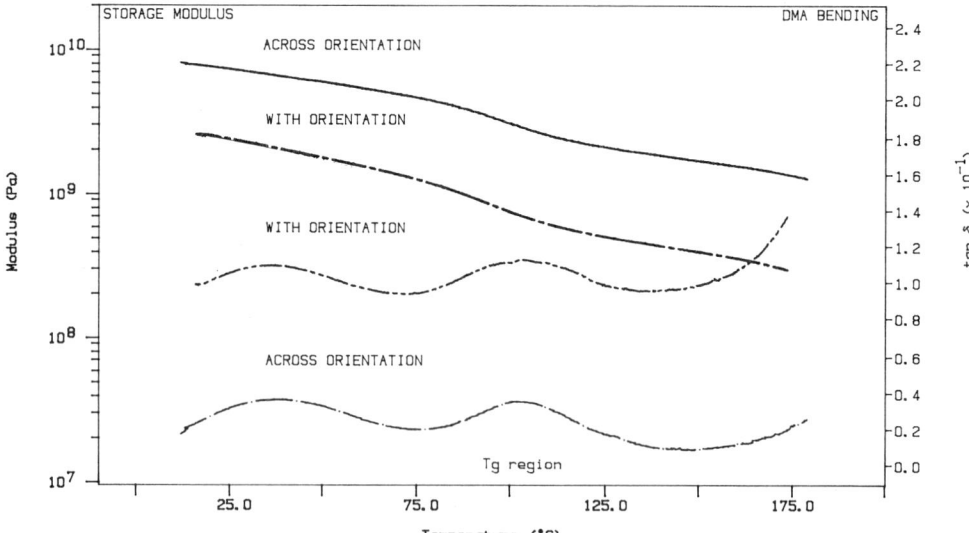

FIG. 8—*DMA storage modulus and tangent delta* (bottom two curves) *showing the effect of orientation on DMA data. Orientation has little effect on the transition temperatures.*

TABLE 1—*Comparison of Glass Transition Measurements.*

Method	Frequency, Hz	Onset, °C	Midpoint, °C	End, °C
DSC	—	76.2	97.5	113.3
TMA Expansion	—	81.2	95.4	—
TMA Penetration	—	82.8	100.4	114.1
TMA Flexure	—	77.2	93.1	105.6
DMA Storage Modulus	0.1	73.2	87	105.1
DMA Tan Delta	0.1	76	98	124
DMA Storage Modulus	1.0	87.3	93.3	108.5
DMA Tan Delta	1.0	75.3	102.2	129.1

Conclusion

A liquid crystal polymer system having a very weak glass transition was analyzed by DSC, TMA, and DMA. The conditions used for analysis were comparable scanning rates, and the data were corrected for thermal lag. The results tabulated indicate the range of values found by various techniques. The above sections describe considerations in assigning the transition temperature. Based on the difficulty of finding and identifying this glass transition it is suggested that for such weak transitions that the primary reported parameter be the midpoint of the transition based on the greatest rate of change of the measured parameter. The use of the derivative signal, which is now universally available on thermal analysis equipment, should be considered for incorporation in standard methods.

The particular polymer system selected for this study was chosen because its weak glass transition has been the subject of considerable interest. Unfortunately, for the intended purpose of this paper, to examine high sensitivity T_g techniques, it has considerable drawbacks. First, the thermal properties are dependent upon the processing history of the material, and on subsequent thermal and mechanical conditioning effects. Hence, by conditioning the specimen in the analysis region it is possible to shift the position of the apparent transitions. Second, the nature of the glass transition of a liquid crystal polymer mesophase may not be representative of other weak T_g's. Moreover, the interpretation of thermal curves may be less certain because of the equilibria between crystalline, mesomorphic, and amorphous phases.

Finally, the T_g results reported for this material are consistent with the DMA and DEA literature cited. This data support a T_g region extending over 50 to 70 degrees centigrade. An alternative treatment [7,11] based on thermodynamic projections proposes a much wider T_g range for the mesophase, one that extends up to the high temperature endotherm. It is hoped that further work will clarify this matter.

References

[1] Johnson, J. F. and Porter, R. S., *Liquid Crystals and Ordered Fluids*, Plenum Press, New York, Vol. 2, 1974.
[2] Brennan, W. P. and Gray, A. P. "Liquid Crystals: The Mesomorphic State," *Thermal Analysis Application Study*, Perkin-Elmer, Norwalk, CT, 1974.
[3] Wunderlich, B. and Grebowicz, J., *Thermotropic Mesophases and Mesophase Transitions of Linear, Flexible Macromolecules, Advances in Polymer Science 60/61*, 1984.
[4] Cassel, B. and Twombly, B., in *Materials Characterization by Thermomechanical Analysis, ASTM STP 1136*, A. T. Riga and C. M. Neag, Eds., American Society for Testing and Materials, Philadelphia 1991, pp. 108–119.

[5] Flynn, J. H., "Thermodynamic Properties from Differential Scanning Calorimetry by Calorimetric Methods," *Thermochimica Acta*, Vol. 8, 1974, p. 69.
[6] Economy, J., Storm, R. S., Matkovich, V. I., et al., *Journal of Polymer Science: Polymer Chemical Edition*, 1976, Vol. 14, p. 2207.
[7] Cao, M.-Y., Varma-Nair, M., and Wunderlich, B., "The Thermal Properties of Poly(oxy-1,4-benzoyl), Poly(oxy-2, 6-Naphthoyl), and Its Copolymers," *Polymers for Advanced Technologies*, John Wiley & Sons, Ltd., Vol. 1, 1990, pp. 151-170.
[8] Cao, M.-Y. and Wunderlich, B., "Phase Transitions in Mesophase Macromolecules. V. Transitions in Poly(oxy-1,4-Phenylene Carbonyl-co-oxy-2,6-Naphthaloyl," *Journal of Polymer Science: Polymer Physics Edition*, John Wiley & Sons, Inc., Vol. 23, 1985, pp. 521–535.
[9] Blundell, D. J. and Buckingham, K. A., "The b-loss Process in Liquid Crystal Polyesters Containing 2,6-naphthyl Groups," *Polymer*, Butterworth & Company, Ltd., Vol. 26, 1985, pp. 1623–1627.
[10] Alhaj-Mohammed, M. H., Davies, G. R., Abduljawad, S., and Ward, I. M., "The Dielectric Properties of Liquid-crystal Random Copolyesters of 4-Hydroxybenzoic Acid and 2-Hydroxy 6-Naphthoic Acid," *Journal of Polymer Science: Part B: Polymer Physics*, John Wiley & Sons, Inc., Vol. 26, 1988, pp. 1751–1760.
[11] Troughton, M. J., Davies, G. R., and Ward, I. M., "Dynamic Mechanical Properties of Random Copolyesters of 4-hydroxybenzoic Acid and 2-hydroxy 6-naphthoic Acid," *Polymer*, Vol. 30, 1989, p. 58.
[12] "Design Manual (VC-10)," *Vectra Liquid Crystal Polymer*, Hoechst Celanese, Engineering Plastics Division, NJ.

Discussion

B. Wunderlich[3] *(written discussion)*—This set of data represents a typical example for the variation in quality of DSC. In an earlier analysis of the same material (using also a Perkin-Elmer DSC) we could show quantitatively, by analyzing the heat capacities, that for the copolymer in question the glass transition range is about 200 K wide.[4] The different qualitative methods used in the present paper for the evaluation of T_g focus on different regions within the very broad glass transition. The results show that in this way it is not possible to assign a sharp "glass transition temperature." Each method that is not quantitatively linked to the degree of progress of the glass transition is sensitive to a different point within the region of softening, as can be seen from Fig. 13 of the first paper of this STP. Even the qualitative increase in heat flow in certain temperature ranges may look like a narrower glass transition region, but an analysis of ΔC_p would have revealed the partial nature of the measurement.

B. Cassel and A. T. Riga *(authors' closure)*—The quality heat capacity data and thermodynamic treatment of B. Wunderlich may support a higher and broader glass transition temperature for the LCP mesophase. However, the mechanical and dielectric data appear to support this paper's position. For example, there is no other T_g transition peak in the DMA tan delta signal between 100°C and 275°C (Fig. 9 added to support discussion.) If the high temperature modulus drop apparent in Fig. 9 is due to the mesophase T_g [7], then perhaps the 75 to 125°C transition is a T_g in the amorphous phase.

---

[3]University of Tennessee, Department of Chemistry, Knoxville, TN 37996-1600.
[4]Cao, M.-Y. et al., *Journal of Polymer Science*, Polymer Physics Ed., Vol. 23, 1985, p. 521; *Polymers for Advanced Technology*, Vol. 1, 1990, p. 151.

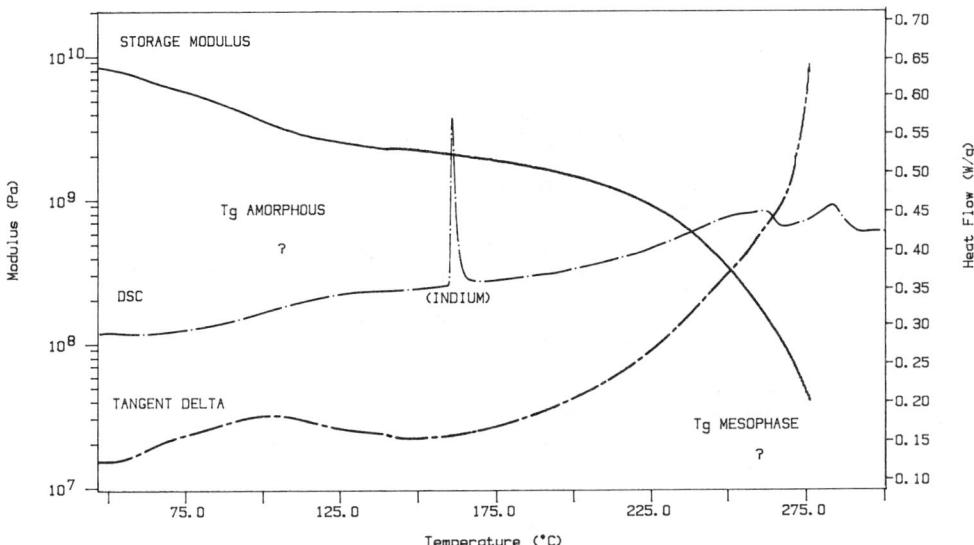

FIG. 9—*DMA storage modulus and tangent delta at 1 Hz and 1°C/min to 275°C, showing no evidence of transitions between 100 and 275°C.*

R. A. Weiss[1]

Glass Transition(s) of Ionomers

REFERENCE: Weiss, R. A., "**Glass Transition(s) of Ionomers**," *Assignment of the Glass Transition, ASTM STP 1249*, R. J. Seyler, Ed., American Society for Testing and Materials, Philadelphia, 1994, pp. 214–225.

ABSTRACT: Ionomers are predominantly nonpolar polymers that contain a small amount of bonded salt groups. Microphase separation of ion-rich microdomains occurs as a consequence of the thermodynamic incompatibility of the salt groups and the polymer matrix and associative interactions between salt groups. Associations of the salt groups usually increase the glass transition of the continuous matrix phase, presumably as a consequence of the inhibition of chain mobility that accompanies physical crosslinking. The central question raised in this paper is whether the dispersed ion-rich microphase exhibits a glass transition. Although no glass transition for the microphase is detected by calorimetry, a dynamic mechanical relaxation is commonly observed above the T_g of the matrix phase. This transition has some of the attributes of a glass transition, but it is not clear what is the actual relaxation process that is measured. This paper discusses the effect of the ionic groups on the matrix glass transition, the origin of the high-temperature dynamic mechanical transition, and the effects of the addition of plasticizers on the T_g of the matrix and the higher temperature mechanical relaxation.

KEYWORDS: ionomers, glass transition, calorimetry, mechanical relaxation, microphase separation, ionic domains

Ionomers are relatively nonpolar polymers, usually hydrocarbon or fluorocarbon polymers, that have been modified by the incorporation of a small to modest amount (usually less than 15 mol%) of bonded acid or neutralized acid groups [1–9]. Typical examples are poly(r-ethylene-co-r-methacrylic acid), poly(r-styrene-co-r-styrene sulfonic acid), and their various salts. The unique mechanical and rheological properties of these materials are due to a microphase-separated structure that results from the thermodynamic immiscibility of the salt groups and the nonpolar polymer backbone and from associative interactions between the ion-dipoles. Although the exact microstructure of ionomers remains an unresolved question, considerable evidence—primarily small angle X-ray scattering—indicates that the microphase separation consists of ion-rich domains with a characteristic dimension of 1 to 5 nm.

As one might expect, the glass transition of an ionomer is affected by the ionic interactions and, perhaps, by the microphase separation. With regard to the microphase separation, two pertinent questions are addressed by this paper: (1) How does microphase separation affect the glass transition temperature, T_g, of the continuous phase and (2) do the ionic microdomains exhibit a separate T_g? These points are discussed primarily with respect to experimental data for lightly sulfonated polystyrene ionomers.

[1] Associate Director, Institute of Materials Science, The University of Connecticut, Storrs, CT 06269-3136.

Experimental Details

Materials

Poly(r-styrene-co-r-styrene sulfonic acid), SPS, was prepared by sulfonating atactic polystyrene (Dow Chemical Co.; M_n = 103 000, M_w = 288 000) according to the procedure of Makowski et al. [10]. The sulfonating reagent, acetyl sulfate, was prepared by slowly adding sulfuric acid to a solution of acetic anhydride in 1,2-dichloroethane (DCE) at 0°C. The freshly prepared acetyl sulfate was then added to a well-stirred solution of the polymer in DCE at 50°C. After 1 h, the reaction was terminated by the addition of ethanol. The sulfonated polymer was isolated either by steam stripping the solvent or by precipitation in a large quantity of methanol, washed with methanol, dried at room temperature for 3 days, and dried under vacuum above T_g for another 3 days. The sulfonic acid content was determined by either titration of the polymer in a toluene/methanol mixed solvent (90/10 v/v) with methanolic NaOH or by elemental sulfur analysis.

The sulfonated polymers were fully converted to the sodium, zinc, or manganese salt by adding a 10% excess of sodium hydroxide, zinc acetate, or manganese acetate. The sulfonated polymer was first dissolved in a 90/10 toluene/methanol mixed solvent, and the base, dissolved in methanol or methanol with a minimal quantity of water, was added dropwise to the agitated solution. The solution was stirred for 30 min after all the neutralizing agent was added, and the ionomer was recovered, washed, and dried as described above. The sample notation used for the ionomers is $x.yM$-SPS where $x.y$ is the degree of sulfonation expressed as mole percent of metal sulfonate groups (i.e., $x.y$ sulfonate groups per 100 backbone styrene groups) and M denotes the cation (M = H, Na, Zn, and Mn for the free acid and the sodium, zinc, and manganese salts, respectively).

Unless specifically stated, the neat ionomer specimens were prepared by compression molding the materials into films at 200°C and at a sufficient pressure to accomplish adequate melt flow. Solution-cast 25 µm film specimens were also prepared by casting a 10% solution of the ionomer in a mixed solvent of THF/deionized water (90/10 v/v) onto glass. The solvent was evaporated in air and the film was dried under vacuum for 5 days at 60°C.

Thermal Analyses

Differential scanning calorimetry (DSC) measurements were made with a Perkin Elmer DSC-2 or a DSC-7 using a heating rate of 20°C/min.

Dynamic mechanical analyses (DMA) were performed with a Rheometrics System-4 mechanical spectrometer equipped with a 2000 g-cm transducer. Isochronal dynamic shear measurements were made from −150 to 250°C at five different frequencies between 0.03 and 3 Hz. Rectangular samples (10 × 63.5 × 0.7 mm) were used for measurements in the high-modulus region (>5 × 10^7 dyne/cm^2) and disk samples (25.4 mm diameter × 1 mm thick) were used for low-modulus (<5 × 10^7 dyne/cm^2) measurements. Isothermal dynamic shear data were obtained on thin-film specimens over the frequency range 0.003 to 3 Hz using a modified rectangular torsion fixture [11] with a 10 g-cm transducer. All measurements were made within the linear viscoelastic limit, which was determined from strain sweeps.

Moduli-frequency master curves, shift factors, and relaxation time distributions were calculated from the isothermal data. Vertical shift corrections were made by multiplying the moduli by T_g/T. Horizontal shifts involved an iterative scheme to determine the least-squares fit of a cubic equation to each shifted isothermal data set with respect to the accumulated master curve.

Small Angle X-ray Scattering (SAXS)

SAXS measurements were made at the National Center for Small Angle Scattering Research (NCSASR) at Oak Ridge National Laboratory using the 10-m camera. This instrument uses a rotating anode X-ray generator source (Cu K_α radiation; λ = 0.1542 nm), crystal monochromatization of the incident beam, pinhole collimation, and a two-dimensional (2-D) position-sensitive detector.

Results and Discussion

Effects of Ionic Groups on the Matrix Glass Transition Temperature

The now rather large body of literature on ionomers indicates that the incorporation of bonded salt groups to a polymer chain generally increases T_g. The usual explanation for this is that associative intermolecular interactions between ion-dipoles restrict local chain motions, thereby raising the temperature required for the segmental mobility associated with the glass transition. An alternative explanation is that the associations reduce *free volume*.

The magnitude of the increase can be quantified as dT_g/dc, where c is the ionic group concentration, and the value of dT_g/dc depends on a variety of molecular parameters, including (1) the type of ionic group attached to the matrix (e.g., cation versus anion, or in the latter case carboxylate versus sulfonate), (2) the choice of the counterion, (3) the placement of the ionic groups (e.g., random versus block), (4) polarity of the matrix, and (5) stiffness of the matrix [12–19]. Examples of some typical values are given in Table 1.

In nearly every report of the thermal behavior of ionomers, a single composition-dependent T_g was reported. Although those results suggest a single-phase material, a peak in SAXS experiments, often referred to as the ionic peak, indicates the existence of periodic electron density fluctuations involving the salt groups. This is believed to be due to ion-rich aggregates that also contain the associated backbone chain that is somewhat immobilized by the intermolecular associations of the ionic species. In the terminology first proposed by Eisenberg [20], these are denoted as ionic clusters. The SAXS data and the observation of a high-temperature relaxation in dynamic mechanical experiments provide evidence that the clusters represent a separate microphase. The actual structure of the clusters remains an unanswered question, but the Bragg spacing of the SAXS peak indicates a characteristic dimension for the electron density periodicity of 1 to 5 nm.

Like other phase-separated or microphase-separated materials, e.g., immiscible binary blends and block copolymers, one might expect the ionic clusters to exhibit a second T_g separate from that of the matrix phase. This is discussed in more detail in the next section of this paper. Galambos et al. [21] reported that the matrix glass transition temperature, $(T_g)_m$, is sensitive to the development of the microphase separation. For example, DSC thermal

TABLE 1—*Effect of ionic groups on the rate of increase of T_g.*

Ionomer	dT_g/dc, °C/mol ionic group	Reference
sulfonated polysulfone	1.2	[36]
poly(r-styrene-co-r-sodium methacrylate)	3.1	[37]
poly(r-styrene-co-r-sodium styrene sulfonate)	3.0	[38]
poly(r-ethylene-co-r-sodium methacrylate)	5.7	[39]
poly(r-butadiene-co-r-lithium methacrylate)	5.9	[40]
sulfonated EPDM	5.9	[41]

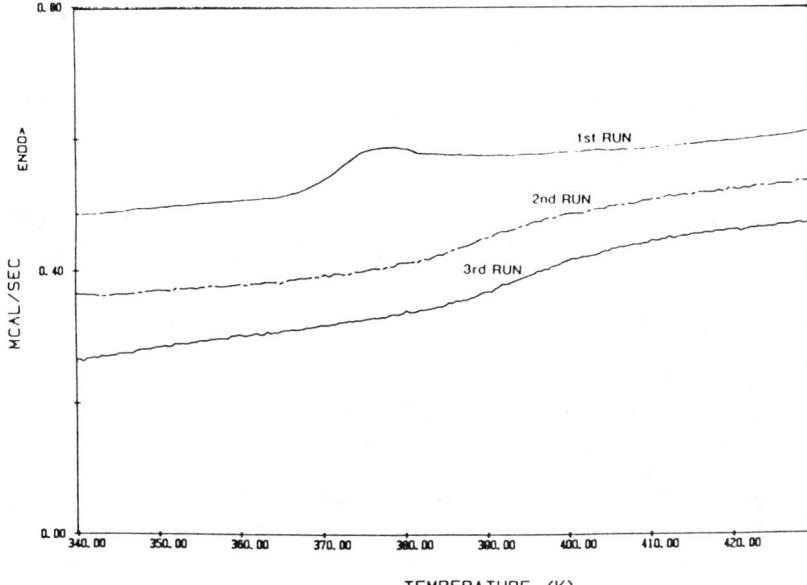

FIG. 1—*DSC thermal curves for 7.6Mn-SPS cast from THF/water. Curves are displaced vertically for clarity. (Reprinted with permission from Ref 21; copyright American Chemical Society.) (mcal/s = 4.184 × 10^3 W)*

curves for a 7.6Mn-SPS sample cast from THF/water (90/10) are shown in Fig. 1. Because of how that sample was isolated from solution, the SAXS pattern was devoid of the ionic peak, which suggested that the microphase separation had not occurred. SAXS measurements also showed, however, that when the sample was heated above T_g, the ionic peak in the SAXS pattern developed.

The as-cast 7.6Mn-SPS had a $(T_g)_m$ detected by DSC of ca. 97°C (Fig. 1), which increased to 116 and 123°C on the second and third heating scans, respectively. This result is contrary to expectations in that microphase separation is expected to deplete the hydrocarbon phase of ionic species. That would tend to decrease $(T_g)_m$ toward the value for polystyrene (PS), ~100°C. Instead, Fig. 1 indicates the nonmicrophase-separated material had a $(T_g)_m$ comparable to PS and microphase separation raised it. The reason for the low T_g for the as-cast sample is not known. One possibility was that some residual solvent, probably water, that solvated the ionic interactions may have remained in the sample, but infrared spectroscopy and thermogravimetric analysis provided no evidence of residual solvent. A T_g comparable to that of PS suggests that physical crosslinks arising from simple ion-dipole associations, i.e., ionic multiplets, following the nomenclature of Eisenberg [20], are either absent in that material or have no affect on the T_g. Those conclusions are surprising in that electron spin resonance studies [21] of the same sample indicated that ionic associations did exist and one would expect that intermolecular associations of any kind should increase T_g.

What is apparent, nevertheless, is that the incidence of microphase separation increased $(T_g)_m$. That may be a consequence of the crosslinking effect associated with the development of ionic clusters that are intimately connected to the matrix phase through the continuity of the polymer chain. As a result, the mobility of the matrix chains, which is responsible for $(T_g)_m$, is inherently coupled to that of the microdomains. Unlike block copolymers where the chains have only one or two couplings between phases or completely immiscible blends where there is no connectivity between phases, a single ionomer chain may pervade many microdomains. Therefore, the formation of the ionic microphase to some extent immobilizes the matrix chains and is responsible for the increase of $(T_g)_m$.

The alternative explanation, that the lower T_g of the as-cast material was due to residual water would be consistent with the report by Risen [22] that extremely small amounts of water can dramatically reduce the T_g of SPS ionomers. However, the T_g of a compression molded 7.6Mn-SPS film specimen that was soaked overnight in deionized water at room temperature changed only a few degrees with respect to a dried specimen. It appears, then, that water alone was not responsible for the very low T_g observed for the as-cast ionomer.

T_g of the Ionic Microphase

As stated above, no calorimetric evidence has been reported for a second, higher temperature glass transition for random ionomers. Weiss et al. [23] did observe by DSC a second T_g at ca. 257°C for a copolymer of styrene and sodium styrene sulfonate that was prepared by emulsion polymerization (Na-SPS)$_e$. In that case, however, they concluded that the sulfonated monomer distribution was blocky and, although no SAXS measurements were made of that polymer, one is tempted to assume that its microstructure was more analogous to that of a block copolymer than a random ionomer. The sulfonate concentration dependence of $(T_g)_m$ for the emulsion polymerized ionomers was also substantially different from that for random ionomers in that $dT_g/dc \rightarrow 0$ at sulfonate concentrations greater than 10 mol%. The chemical composition of Na-SPS and (Na-SPS)$_e$ were identical, and the differences in their glass transition temperatures was most likely a consequence of the placement of the ionic groups, i.e., random versus blocky.

A common observation in the DMA of ionomers is a mechanical relaxation at elevated temperatures, i.e., above $(T_g)_m$. This is shown in Fig. 2 for Na-SPS ionomers. The position of the high temperature peak generally increases with increasing concentration of the ionic species. Figure 3 gives the relaxation spectra for the ionomers in Fig. 2. These were calculated from G' versus frequency master curves, prepared by time-temperature superposition of isothermal DMA data, and application of Tschoegel's second approximation formula [24],

$$H(\tau) = G' \left[\left(\frac{d \log G'}{d \log \omega} \right) - \frac{1}{2} \left(\frac{d \log G'}{d \log \omega} \right)^2 - \frac{1}{4.606} \left(\frac{d^2 \log G'}{d(\log \omega)^2} \right) \right]_{1/\omega = \tau/\sqrt{2}}$$

Figure 3 shows that there are two distinct relaxation processes in the Na-SPS ionomers, one associated with the glass transition of the matrix polymer and one with a much longer relaxation time that results from the ion-rich microphase, i.e., the clusters. At a constant temperature, both relaxations decrease in frequency with increasing ionic concentration, which correspond to higher temperatures at a fixed frequency, c.f., Fig. 2. The relaxation time distribution broadened with increasing sulfonate content, which indicates that the ionic associations of the sodium sulfonate groups slowed down the diffusional relaxation processes. The peak due to the ionic groups shifted to lower frequency relative to the matrix glass transition and increased in intensity with increasing sulfonate concentration. It appeared that the relaxation of the ionic-rich regions became increasingly dominant with regard to the viscoelastic response of the ionomer as the level of functionalization increased.

The central question here is the origin of the low frequency (i.e., high temperature) mechanical transition. In analogy with the mechanical behavior of other microphase-separated systems, e.g., block copolymers, one is tempted to attribute the high temperature

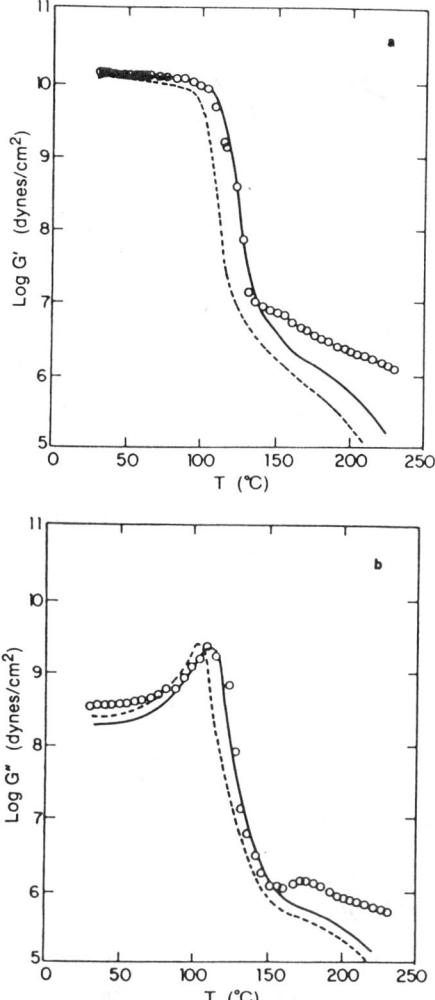

FIG. 2—G' and G'' versus temperature for (,,,) PS, (---) 1.82Na-SPS, (—) 3.44Na-SPS, and (○) 5.81Na-SPS; $f = 1$ Hz. (Reprinted with permission from Ref 35; copyright American Chemical Society.) (dyne/cm² = 0.1 Pa)

relaxation to a *glass transition* of the ionic domains and characterize it by a glass transition temperature, $(T_g)_i$ at a fixed frequency.

This interpretation of the high temperature relaxation in M-SPS is consistent with the model for the ionomer microstructure recently proposed by Eisenberg and coworkers [25,26], and it is also consistent with their conclusions for poly(r-styrene-co-r-alkali methacrylate) ionomers. In their model, the clusters are not distinct aggregates as assumed by previous cluster models such as in Refs 27 and 28. The basic entity in the cluster is the multiplet, and the intermultiplet spacing is responsible for the SAXS peak. A distribution of intermultiplet spacings yields a broad SAXS peak. The multiplets are too small to constitute a separate phase. The mobility of the polymer chains attached to the ionic groups participat-

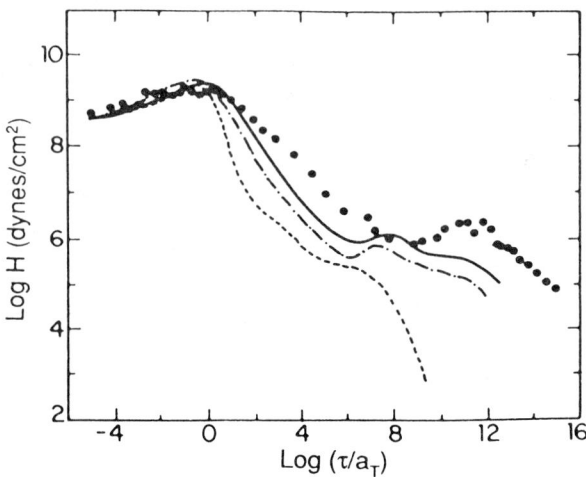

FIG. 3—*Relaxation time spectra for (---) PS, (-·-) 1.82Na-SPS, (—) 3.44Na-SPS, and (●) 5.81Na-SPS. The reference temperature was T_g. (Reprinted with permission from Ref 35; copyright American Chemical Society.) (dyne/cm^2 = 0.1 Pa)*

ing in the multiplet is inhibited by the ionic associations, which gives rise to a region of restricted mobility surrounding the multiplet. According to the Eisenberg model, a cluster is formed when the individual restricted mobility regions surrounding the multiplets overlap to form larger regions. Where their model deviates from prior cluster models is the relaxation of the concept of a distinct cluster with a characteristic geometry. The Eisenberg cluster is irregular and may have a thickness approaching the persistence length of the polymer.

The connection of the cluster concept to overlapping regions of restricted mobility allows a sort of percolation of the *microphase*. Eventually, as the ionic content of the ionomer is increased the clusters actually become the dominant phase. This interpretation is consistent with DMA results that indicate that the intensity (and area under the tanδ peak) of the cluster relaxation increases relative to that of the continuous phase glass transition with increasing ion concentration (Fig. 2). The increase of $(T_g)_i$ with increasing ion content of the ionomer would indicate a higher multiplet concentration within the clusters. The analogy here to other phase-separated systems is improved phase separation with increasing ionic concentration, or interphase area.

The Eisenberg model specifies $(T_g)_i$ as a glass transition in the classical sense. The argument for regions of restricted mobility larger than the 1 to 5 nm characteristic distance extracted from the SAXS ionic peak removes the objection that the clusters are too small to exhibit a glass transition. In addition, activation energies for the high temperature mechanical relaxation and the effects of plasticizers support that assignment. Hird and Eisenberg report an activation energy of 200-400 kJ/mol for the clusters in poly(r-styrene-co-r-alkali methacrylate) ionomers, which is the same order of magnitude as that for the primary glass transition in those materials, though lower by a factor of ca. 2.

A practical definition of the glass transition is that it represents the temperature or time for which the onset of segmental motion occurs. SAXS data obtained at elevated temperatures for *M*-SPS [29], where *M* = metal cation, support the idea that the high temperature mechanical relaxation observed in ionomers is indeed a glass transition for an ion-rich phase. For both the sodium and zinc salts, reorganization of the ion-phase structure was detected at a temperature consistent with $(T_g)_i$ measured by DMA.

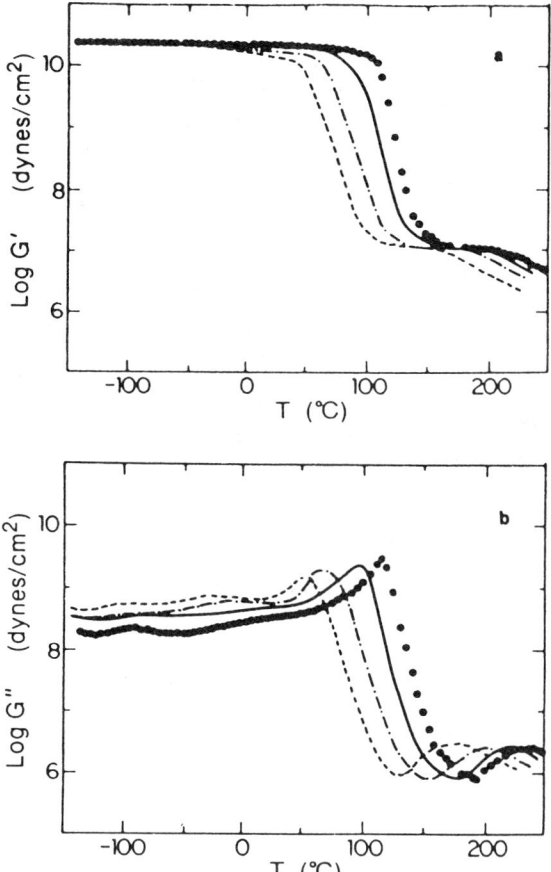

FIG. 4—G' and G'' versus temperature for DOP-plasticized 2.59Na-SPS: (●) 0%, (—) 5%, (-··-) 10%, and (---) 18% DOP; $f = 1$ Hz. (Reprinted with permission from Ref 31; copyright American Chemical Society.) (dyne/cm² = 0.1 Pa)

A number of studies has demonstrated that it is possible to selectively plasticize the two phases of an ionomer [30]. For M-SPS ionomers, SAXS and DMA studies showed that a relatively nonpolar diluent such as dioctyl phthalate, DOP, partitioned almost exclusively into the non-ionic phase of the ionomer and glycerol plasticized the clusters [31,32]. Figure 4 shows that the addition of DOP depresses both $(T_g)_m$ and $(T_g)_i$ for 2.59Na-SPS. The decrease of $(T_g)_m$ was due to plasticization of the non-ionic phase by DOP. SAXS for the same system, however, indicated that no DOP was in the clusters, and it is believed that the origin of the decrease of $(T_g)_i$ was a *thermal stress* effect similar to that observed in block copolymers [33,34]. Above $(T_g)_m$, motion of the chain segments in the non-ionic phase can create stresses at the boundary between phases and these can influence the relaxation behavior in the microphase—in this case, the clusters. That DOP only plasticized the non-ionic phase is more clearly shown by the relaxation spectra in Fig. 5. In this case, the relaxation times were normalized with respect to $(T_g)_m$, and the effect of DOP addition is seen to be an increase in the relative relaxation times of the clusters viz à viz those for the matrix glass transition.

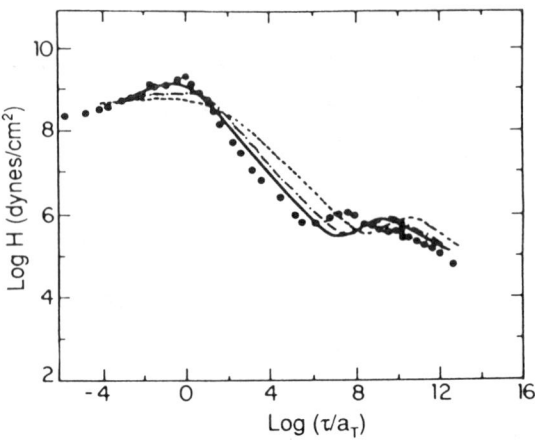

FIG. 5—*Relaxation time spectra for DOP-plasticized 2.59Na-SPS:* (●) *0%,* (—) *5%,* (-··-) *10%, and* (—) *18% DOP. (Reprinted with permission from Ref 31; copyright American Chemical Society.) (dyne/cm^2 = 0.1 Pa)*

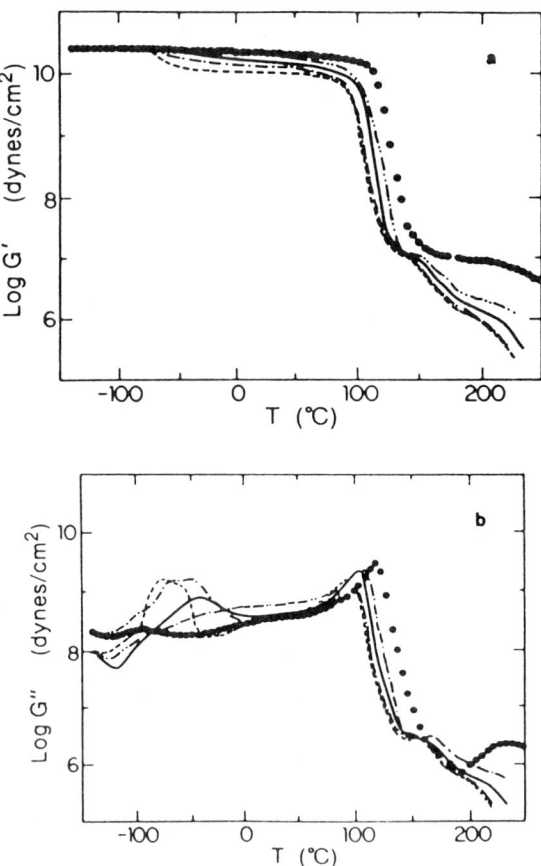

FIG. 6—*G' and G" versus temperature for glycerol-plasticized 2.59Na-SPS:* (●) *0%,* (-···-) *3%,* (—) *8%,* (-··-) *12%, and* (--) *23% glycerol. (Reprinted with permission from Ref 31; copyright American Chemical Society.) (dyne/cm^2 = 0.1 Pa)*

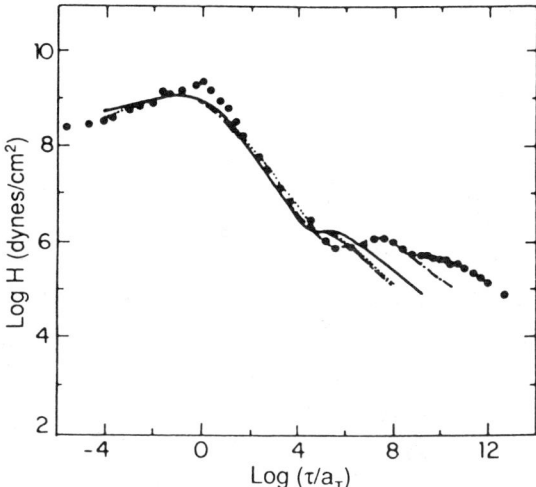

FIG. 7—*Relaxation time spectra for glycerol-plasticized 2.59Na-SPS: (●) 0%, (-···-) 3%, (—) 8%, (-··-) 12%, and (--) 23% glycerol. (Reprinted with permission from Ref 31; copyright American Chemical Society.) (dyne/cm² = 0.1 Pa)*

Polar plasticizers preferentially plasticize the ion-rich microphase. The addition of glycerol to 2.59Na-SPS resulted in a small decrease of $(T_g)_m$ (Fig. 6), but a substantial decrease of $(T_g)_i$. The effect on $(T_g)_m$ was probably due to limited solubility of glycerol in the hydrocarbon-rich phase as a consequence of the presence of some ion-pairs or multiplets in that phase. The major effect of the addition of glycerol, however, was depression of the relaxation times of the clusters, shown relative to those for the matrix glass transition in Fig. 7. This result appears to be consistent with the assignment of two glass transitions in ionomers.

One unanswered question that remains is why a higher temperature T_g is not resolved for ionomers by DSC measurements. This may be a consequence of a very broad transition, as might be suggested by the broad SAXS peak, which indicates a broad distribution of intermultiplet distances (using the Eisenberg interpretation [25]). That may, in turn, be responsible for the activation of segmental motion over a wide temperature range. If so, a T_g should be manifest in a DSC thermal scan as a gradual change in slope. We are not aware of any reports of this, or, if having been observed, the phenomenon was misinterpreted as drift in the thermal curve baseline. Based on the DMA data, that transition is expected to be ~200°C for M-SPS.

Conclusions

Ionomers are microphase-separated materials composed of a relatively non-ionic phase that may contain a small fraction of ion-pairs or associated ion-pairs and a ion-rich phase. Although the exact features of the microstructure of ionomers remain elusive, considerable evidence, primarily scattering and viscoelastic data, exists for microphase separation. As with other phase-separated polymer systems, each phase of an ionomer exhibits its own glass transition, though the evidence for a T_g for the ion-rich microphase is not unambiguous. On one hand, a mechanical relaxation is observed at elevated temperatures, and it corresponds to the onset of motion in the ion-rich phase. The high temperature mechanical relaxation also responds to plasticizers specific to the ion-rich phase in a manner expected for a glass transition. On the other hand, calorimetry experiments do not resolve a separate glass transition for the ion-rich phase.

References

[1] Otocka, E. P., *Journal of Macromolecular. Science: Macromolecular Chemistry*, Vol. 5, 1971, p. 275.
[2] *Ionic Polymers*, L. Holliday, Ed., Applied Science Publ., London, 1975.
[3] Eisenberg, A. and King, M., *Ion-Containing Polymers*, Academic Press, New York, 1977.
[4] *Developments in Ionic Polymers—I*, A. D. Wilson and H. J. Proser, Eds., Applied Science Publ., London, 1983.
[5] *Ions in Polymers*, A. Eisenberg, Ed., Advances in Chemistry Series, Vol. 187, American Chemical Society, Washington, DC, 1980.
[6] MacKnight, W. J. and Earnest, T. R., *Journal of Polymer Science, Macromolecular Reviews*, Vol. 16, 1981, p. 41.
[7] *Coulombic Interactions in Macromolecular Systems*, A. Eisenberg and F. E. Bailey, Eds., Symposium Series, Vol. 302, American Chemical Society, Washington, DC, 1986.
[8] Fitzgerald, J. J. and Weiss, R. A., *Macromolecular Science: Reviews in Macromolecular Chemistry and Physics*, Vol. C28, 1988, p. 99.
[9] *Multiphase Polymers: Blends and Ionomers*, L. A. Utracki and R. A. Weiss, Eds., Symposium Series, Vol. 395, American Chemical Society, Washington, DC, 1989.
[10] Makowski, H. S., Lundberg, R. D., and Singlhal, G. H., U.S. Patent 3,870,841, 1975.
[11] Foss, P. H., Ph.D. dissertation, University of Connecticut, 1985.
[12] Navratil, M. and Eisenberg, A., *Macromolecules*, Vol. 7, 1974, p. 84.
[13] Brockman, N. L. and Eisenberg, A., *Macromolecules*, Vol. 23, 1985, p. 1145.
[14] Lundberg, R. D. and Makowski, H. S., in *Ions in Polymers*, A. Eisenberg, Ed., Advances in Chemistry Series, Vol. 187, American Chemical Society, Washington, DC, 1980, p. 21.
[15] Gauthier, M. and Eisenberg, A., *Macromolecules*, Vol. 23, 1990, p. 2066.
[16] Weiss, R. A., Lundberg, R. D., and Turner, S. R., *Journal of Polymer Science, Polymer Chemistry*, Vol. 23, 1985, p. 549.
[17] Weiss, R. A., Agarwal, P. K., and Lundberg, R. D., *Journal of Applied Polymer Science*, Vol. 29, 1984, p. 2719.
[18] Wollmann, D., Gauthier, S., and Eisenberg, A., *Polymer Engineering and Science*, Vol. 26, 1986, p. 1481.
[19] Gauthier, S., Duchesne, D., and Eisenberg, A., *Macromolecules*, Vol. 20, 1987, p. 753.
[20] Eisenberg, A., *Macromolecules*, Vol. 3, 1980, p. 147.
[21] Galambos, A. F., Stockton, W. B., Koberstein, J. T., et al., *Macromolecules*, Vol. 20, 1987, p. 3091.
[22] Risen, W. M., paper presented at the 193rd Meeting of the American Chemical Society, Denver, April 1987.
[23] Weiss, R. A., Lundberg, R. D., and Turner, S. R., *Journal of Polymer Science, Polymer Chemistry*, Vol. 23, 1985, p. 549.
[24] Ferry, J. D., *Viscoelastic Properties of Polymer*, 3rd ed., Wiley-Interscience, New York, 1970.
[25] Eisenberg, A., Hird, B., and Moore, R. B., *Macromolecules*, Vol. 23, 1990, p. 4098.
[26] Hird, B. and Eisenberg, A., *Journal of Polymer Science, Polymer Physics*, Vol. 28, 1990, p. 1665.
[27] MacKnight, W. J., Taggart, W. P., and Stein, R. S., *Journal of Polymer Science, Polymer Symposia*, Vol. 45, 1974, p. 113.
[28] Yarusso, D. J. and Cooper, S. L., *Macromolecules*, Vol. 16, 1983, p. 1871.
[29] Weiss, R. A. and Lefelar, J. A., *Polymer*, Vol. 27, 1986, p. 3.
[30] Bazuin, C. G., in *Multiphase Polymers: Blends and Ionomers*, L. A. Utracki and R. A. Weiss, Eds., Symposium Series, Vol. 395, American Chemical Society, Washington, DC, 1989, ch. 21 and refs. within.
[31] Weiss, R. A., Fitzgerald, J. J., and Kim, D., *Macromolecules*, Vol. 24, 1991, p. 1064.
[32] Fitzgerald, J. J. and Weiss, R. A., *Journal of Polymer Science, Polymer Physics*, Vol. 28, 1990, p. 1719.
[33] Morese-Segvela, B., St. Jacques, M., Renaud, J. M., and Prud'homme, R., *Macromolecules*, Vol. 13, 1980, p. 100.
[34] Granger, A. T., Wang, B., Krause, S., and Fetters, L. J., in *Multicomponent Polymer Materials*, Advances in Chemistry Series, Vol. 211, American Chemical Society, Washington, DC, 1986, p. 127.

[35] Weiss, R. A., Fitzgerald, J. J., and Kim, D., *Macromolecules,* Vol. 24, 1991, p. 1071.
[36] Noshay, A. and Robeson, L. M., *Journal of Applied Polymer Science,* Vol. 20, 1976, p. 1885.
[37] Eisenberg, A. and Navratil, M., *Macromolecules,* Vol. 7, 1974, p. 90.
[38] Kim, D. S., Ph.D. dissertation, University of Connecticut, 1987.
[39] Otocka, E. P. and Kwei, T. K., *Macromolecules,* Vol. 2, 1968, p. 110.
[40] Otocka, E. P. and Eirich, F. R., *Journal of Polymer Science,* Part A-2, Vol. 6, 1968, p. 921.
[41] Agarwal, P. K., Makowski, H. S., and Lundberg, R. D., *Macromolecules,* Vol. 13, 1980, p. 1679.

A. K. Sircar[1] *and R. P. Chartoff*[1]

Measurement of the Glass Transition Temperature of Elastomer Systems

REFERENCE: Sircar, A. K. and Chartoff, R. P., "**Measurement of the Glass Transition Temperature of Elastomer Systems,**" *Assignment of the Glass Transition, ASTM STP 1249,* R. J. Seyler, Ed., American Society for Testing and Materials, Philadelphia, 1994, pp. 226–238.

ABSTRACT: The glass transition temperature of polymers depends on both the experimental procedures and the experimental techniques used for its evaluation. However, the value of the published data is often diminished because these details are omitted. Examples are cited to show how the magnitude of differential scanning calorimetry (DSC) T_g depends on the T_g location and different calibration methods. Some comments are also made about thermomechanical analysis (TMA) and dynamic mechanical analysis (DMA) methods, although these are treated only briefly.

The factors that are unique to elastomer glass transition temperatures are: (1) the difficulties of calibration of the instruments at subambient temperature, (2) increase of T_g by the extent of vulcanization and filler loading, (3) differences due to microstructure and microphase separation, and (4) the availability of different grades of the same elastomer with different composition or added components, which alter their T_g. Thus, the literature value of T_g should include not only the calibration and procedure for the experiment, but also the exact name and description of the elastomer with the number index, the recipe used, and the cure conditions.

KEYWORDS: glass transition temperature, elastomers, differential scanning calorimetry (DSC), thermomechanical analysis (TMA), dynamic mechanical analysis (DMA), vulcanization, cure

The glass transition temperature (T_g) is a very important parameter for elastomer characterization. Glass transition temperature defines the low temperature limit before the elastomer stiffens and loses its elastomeric properties. Many of the physical properties of elastomers such as rebound and hysteresis are related to T_g. T_g values have been used as a primary basis or key component for a remarkable range of polymer studies including product uniformity, quality control in elastomer manufacturing processes, elastomer identification, copolymer composition, plasticizer efficiency, microstructure of elastomers, and degree of cure. It is also used for blend compatibility, phase composition, and super molecular structure in block copolymers and ionomers constituting the thermoplastic elastomers.

The object of this presentation is to discuss some difficulties in assigning the glass transition temperature of elastomers. Some of these factors are common to all polymers, others are unique to elastomeric systems. The primary technique included for discussion is mainly differential scanning calorimetry (DSC), with only brief mention to thermomechanical analysis (TMA) and dynamic mechanical analysis (DMA).

[1]Research polymer scientist and professor of materials engineering, respectively, Center for Basic and Applied Polymer Research, University of Dayton, Dayton, OH 45469-0130.

Discussion

It is well known that T_g is dependent on both the experimental procedure used as well as the details of experimental technique. Very often the T_g value is treated in the literature as if it is a constant. The value of published data is often diminished because of the inadvertent omission of experimental details such as experimental techniques (DTA, DSC, TMA, or DMA), method of calibration, heating rate, exact location of T_g on the curve (onset, extrapolation, or midpoint), etc. Loadman [1] notes that only two of the publications [2,3] that quoted the T_g values for uncured natural rubber (NR) gave relatively complete details of the above parameters.

DSC Experiments

Location of T_g—Figure 1 shows a representative DSC curve [1] of heat flow versus temperature in the T_g region of an elastomer. All the temperatures indicated (T_0, $T_{0.5}$ and T_e or T_{e0}) have been cited by various authors as T_g with no clear-cut definition. T_0, the "onset" temperature is defined as the temperature at which the first departure from the baseline on the low temperature side is observed. T_0 is not reproducible, is too subjective and is often difficult to determine because of the baseline slope. T_e or T_{e0} (extrapolated onset) is the temperature at the cross-section of the extrapolated baseline and tangent at the maximum slope. It is generally reproducible and is the most often quoted value in earlier publications. Lombardi [4] recommends the T_e method. In the case of thermoplastics, T_e may define the limit of useful properties and is technically significant. However, this is not true for elastomers since its useful properties lie above the glass transition range. The temperature at which an elastomer loses elasticity (brittleness temperature) is determined by progressive cooling of the sample and may be closer to T_f (Fig. 1).

$T_{0.5}$, the endothermic shift temperature at half-height is used in many recent publications [5]. The auto-analysis feature available in all modern thermal analysis equipments makes this determination easy and reproducible. It has better reproducibility than the extrapolation

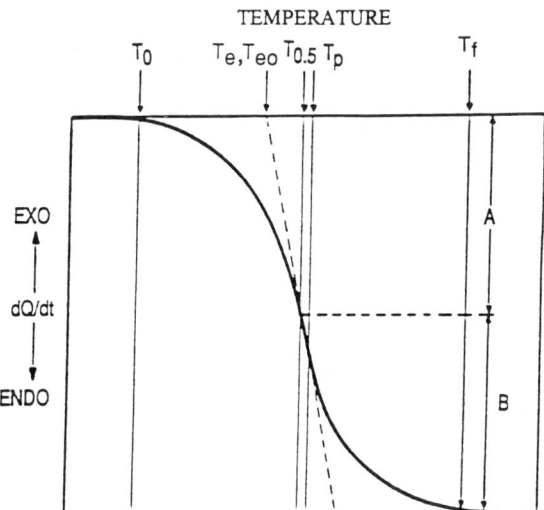

FIG. 1—*Representative DSC curve of heat flow versus temperature in the T_g region of an elastomer* [1].

method and locates a temperature where the transition has reached about the maximum rate. ASTM Test Methods for Determining Glass Transition Temperature by Differential Scanning Calorimetry or Differential Thermal Analysis (E 1356) and Transition Temperature of Polymers by Thermal Analysis (D 3418) recommend both T_e and $T_{0.5}$.

Apart from the above, other methods have been suggested for T_g location. Wunderlich [6] recommends T_g measurements on the cooling side of the T_g curve since the sample is in thermal equilibrium at the start of the measurement and enthalpy relaxation is avoided. For the reasons stated above, this method would be the more relevant for elastomer studies. However, instrumental drawbacks, such as precise control of the cooling rates in some DSC models, preclude general use of cooling curves to determine T_g.

The derivative DSC (DDSC) curve shown in Fig. 2 for NR (SMR-5) offers another alternative for locating T_g. Such curves exhibit a peak corresponding to the maximum slope of the endothermic shift accompanying T_g. The rate of change of slope is zero at the peak. This peak temperature (T_p, Fig. 1) is close to the midpoint value of the T_g endotherm. Landi [7] reported DDSC curves for a series of NBR elastomers, varying in composition; DDSC curves may also be helpful in locating secondary transitions or the presence of multiple transitions as shown in Figs. 3a and 3b for NBR with 33% and 20% acrylonitrile, respectively. However, in certain elastomer compositions (high loading) the T_g transition is not very sharp and, consequently, locating the DDSC peak is rather difficult. Such broad T_g transitions are illustrated in Fig. 4 for DTA curves of styrene-butadiene rubber (SBR) and butadiene rubber (BR) blends, with 50 phr carbon black in the recipe.

The T_g data for cured and uncured common elastomers, as determined by extrapolated onset, half-height, and DDSC peaks are presented in Table 1. As would be expected different T_g values are obtained for different locations. The extrapolated onset has a lower value than either half-height or DDSC peak which are about the same in most cases.

Derivative curves also have been found useful for locating T_g by TMA [8]. This is illustrated in Fig. 5 for chloroprene rubber (CR) vulcanizates evaluated by Brazier [8] over a temperature range from -80 to $+20°C$ at a heating rate of $5°C$ min^{-1}. The derivative signal has been used here for both penetration and tension measurements.

Glass Transition Temperature of Natural Rubber—A substantial amount of work has been carried out recently to decide on a standard procedure to determine the T_g of NR and its

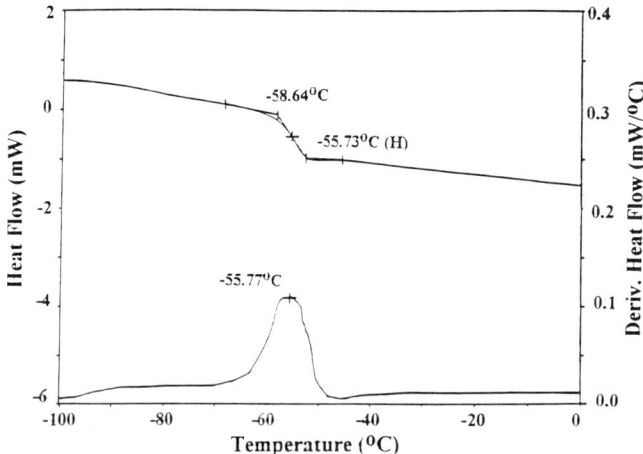

FIG. 2—*DSC and DDSC curve for NR, 20°C/min.*

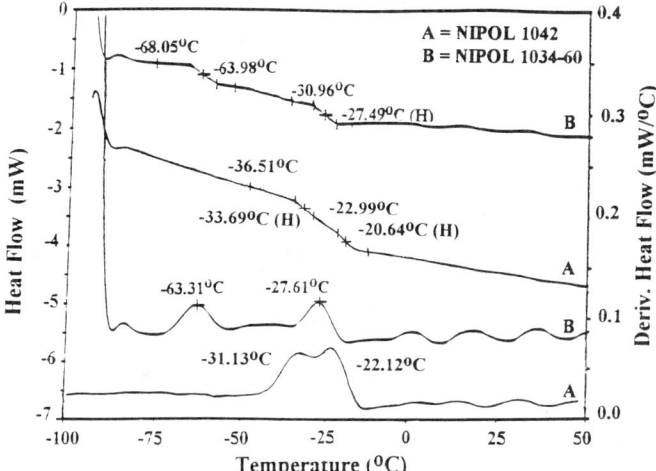

FIG. 3—(a) *DSC and DDSC curve for NBR with 21% ACN;* (b) *DSC and DDSC curves for NBR with 33% ACN at 10°C/min.*

analogues by DSC. As is well known, T_g varies with heating and cooling rates. Also, T_g determined by the same type of equipment supplied by different manufacturers is often different (machine path error). An interesting study that scrupulously addressed all possible errors was reported by Loadman [1]. Loadman [1] observed that heating rate dependence of the transition for the polymer and the calibrant (cyclohexane) may be different. This was attributed to the difference in molar mass between the two. The polymer chains have a high molar mass which causes them to relax more slowly. Loadman calibrated the Perkin-Elmer DSC 2 equipped with subambient accessories but with deliberate mismatch of the sample

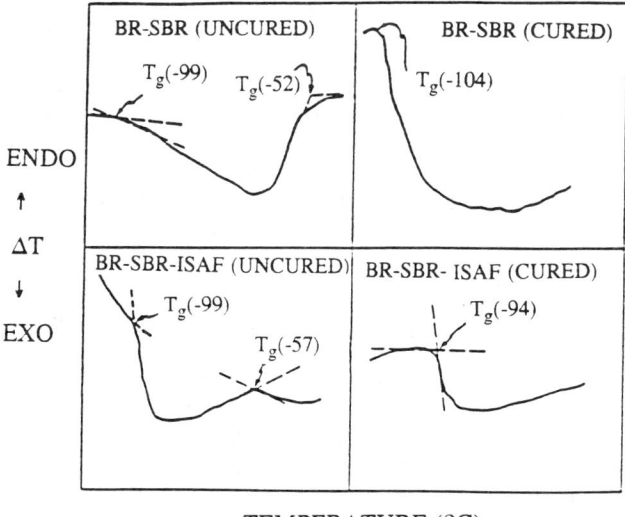

FIG. 4—*DTA T_g determination of SBR-BR blends* [18].

TABLE 1—*DSC glass transition temperature of common elastomers.*[a]

Elastomer	Onset	Uncured T_g Half-Height	DDSC Peak	Onset	Cured T_g Half-Height	DDSC Peak
NR (SMR 5)	−61	−60	−59	−58	−56	−56
IR (Natsyn 2200)	−61	−58	−58	−60	−58	−58
SBR 1500	−52	−49	−50	−47	−43	−44
BR (Budene 1207)	−103	−98	−99	−100	−96	−97
NBR (Nipol 1042)	−34	−24	−27	−27	−19	−22
			−20			−14
IIR (Butyl 077)	−63	−59	−59	−63	−58	−59
CR (Neoprene GN)	−38	−36	−36	−36	−33	−33
EPDM (Nordel 1040)	−65	−62	−62	−63	−60	−60

[a] Unpublished work by Sircar.
Heating rate, 20°C min^{-1}; annealed at 100°C 10 min in N$_2$; quench-cooled to −120°C, held for 10 min before ramping.

and reference cell covers (to induce an instrument effect). He then used mercury and cyclohexane for temperature calibration at different rates of heating of the calibrant (cyclohexane) and extrapolated the data to zero rate using the same empirical relationship (correlating rate of heating and T_g) as for NR. The NR sample (8 mg to 10 mg in thin slices, sealed in high-pressure aluminum capsules) was annealed to 370 K, held at this temperature for one minute (to eliminate any error due to prehistory), and quenched at a very high rate (320°/min) to some 30 K below the anticipated T_g. The NR run at various heating rates (2.5 K/min to 80 K/min using the mismatch and helium as purge gas) was started as soon as the recorder indicated a stable baseline. The effect of sample size, within limits (not specified) was insignificant. The data were extrapolated to zero rate by plotting T_e versus (rate)$^{1/2}$ [9,10] and (ΔT_e^{-1}) versus log rate [11], the latter being preferred because it yields a linear relationship. The reference holder contained 10 mg of aluminum to minimize the baseline slope and the liquid nitrogen reservoir was kept at least half full at all times. Following the above procedure the T_e value at zero heating rate for NR was 200.5 ± 0.5 K, which the author recommended as the standard value for NR under the specified conditions. Purification of the rubber by reprecipitation has no effect [3] on the shape or value of T_g.

Burfield and Lim [3] did not observe the variable temperature dependence of NR and the calibrant and recommended a simpler procedure based on an optimized fixed scan rate of both the calibrant and the polymer. They recommended an instrument calibration against mercury (−38.5°C), water (0°C), and indium (156.6°C) and for crucial measurements a check against cyclohexane (−83°C, 6.5°C), before and after running the sample. These recommendations were based on the following observations:

1. Extrapolation procedures are much more time consuming.

2. Any errors due to thermal lag are largely compensated if both the sample and the standard are run at the same scan rate.

3. The extrapolation procedure requires measurements at low scan rates. The reproducibility at low scan rate is poorer. This leads to lower precision of the determined T_g than that determined by repetitive measurements at the optimized fixed scan rate (20 K/min for NR). Support from earlier literature [12] was cited for this observation.

4. Extrapolations are rather unsatisfactory for trans-1,4-polyisoprenes (TPI); guttapercha consistently gives scattered data points and a positive slope for the T_g onset values while balata generally shows an increase in T_g after extrapolation to zero rate. This odd behavior

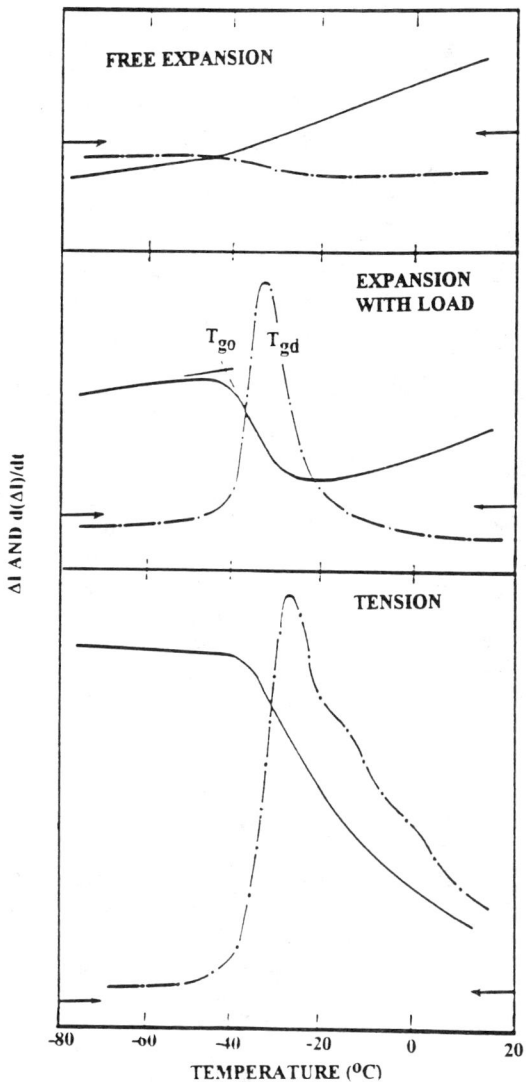

FIG. 5—*TMA-free expansion, indentation and tension thermograms: Neoprene vulcanizate 5°C/min., 50 g load in indentation and tension* ___Δl *versus temperature;* ___.___. $d(\Delta l)\ dt$ *versus temperature* [8].

was explained as due to the development of crystallization during the experiment after quench-cooling.

5. Different extrapolation procedures lead to somewhat different values as exemplified by the re-evaluation of published data [2] for high cis 1,4-polyisoprene in Table 2 by Burfield and Lim [3].

The method followed by Burfield and Lim [3] was as follows: Elastomer samples (~10 mg) were encapsulated in the standard aluminum pans, annealed at 400 K for 5 min, and

TABLE 2—*Glass transition measurements for various amorphous cis/trans-polyisoprenes[a].*

Sample	T_g,[b] K		ΔC_p, Cal g^{-1} K^{-1}
	Onset	Midpoint	
Cis			
Natural rubber	206.3 ± 0.1	208.2 ± 0.2	0.104 ± 0.007
Cis polyisoprene	207.8 ± 0.2	209.8 ± 0.3	0.110 ± 0.004
Trans			
Balata	201.8	203.7	0.101
Gutta-Percha	203.0 ± 0.1	205.0 ± 0.1	0.107 ± 0.003
Trans polyisoprene	201.6	203.4	0.103

[a]From Ref *3*.
[b]Mean of five values except for balata and trans polyisoprene.

quench-cooled to 50 K below the anticipated T_g. The samples were then heated at a specified rate (10 to 20 K/min) to determine T_g. The samples which undergo rapid crystallization (such as balata and synthetic TPI) were dropped directly into liquid nitrogen after high-temperature annealing and inserted into the DSC cell at 150 K—well below the T_g. While cooling in liquid nitrogen and transferring the cooled sample to the DSC cell extreme caution must be exercised to prevent or minimize moisture absorption. A suggestion by Tye [*13*] to evacuate the DSC cell and backfill it with dry gas seems appropriate, but does not solve the absorption of moisture while transferring the cooled sample to the DSC cell. The T_g values reported by Burfield and Lim [*3*] at a scan rate of 20 K for various amorphous cis/trans-polyisoprenes are shown in Table 2. Note the higher midpoint T_g (ca. 2 K). The NR T_g is marginally lower than the synthetic cis analogue. This was ascribed to a trace amount of 3,4-isomer in IR which reportedly increases the T_g of polyisoprenes [*2,9*]. However, a recent NMR examination revealed a 3,4-branch content of only 0.3 mol percent in the cis-polyisoprene used. This amount is insufficient to explain the discrepancy in T_g values. Later work with deproteinized and/or acetone-extracted NR revealed a plasticizing effect of the acetone extractable materials which lowers the T_g of NR. Thus, NR samples derived from different clones, which vary in the amount of extractable nonrubbers, should show slightly different T_g values. However, the amount of variation should be small and should not detract from the proposal of Loadman to use NR as a low-temperature DSC reference standard for polymer T_g studies. As may be observed in Table 2, the T_g for NR by Loadman (T_e) (200.5 ± 0.5 K) and Burfield and Lim (T_e) (206.3 ± 0.1) differ by approximately 6 K, mainly due to different experimental procedures. Also, T_g variation due to tailor-made microstructural variation is common to many elastomers [*2,9,10*].

The above discussion illustrates the complexity of assignment of exact T_g values for elastomers. The authors would recommend the method of Burfield and Lim, mainly for its simplicity and economy of time and labor.

It may be added that much of the published T_g data show comparative values brought about by composition or structural changes under the same conditions of measurement. These are completely valid. Also, for routine quality control measurements where the experimental values are judged against a standard, a rigorous procedure is neither necessary nor justified for the high cost it would incur. A higher heating rate, in addition to saving time, leads to a concomitant increase in the magnitude of baseline shift improving resolution. However, published literature should always include the experimental details, including the heating or cooling rate, temperature, and method of calibration.

Factors Unique to Elastomer T_g

Apart from the factors that are common to both thermoplastics and elastomers, such as heating or cooling rate, annealing procedure, effect of fillers and plasticizers, polymer crystallinity or microstructure, there are several other factors that are unique to elastomers themselves. These relate to the subambient T_g and special conformation of the molecular chains which need to be crosslinked with suitable curatives and reinforced with fillers in order to attain the level of properties suitable for different uses. All these treatments have their effects on T_g, which have to be taken into consideration. These are discussed below.

Subambient Operation—As discussed above, it is desirable to calibrate the instrument at the same temperature range as the T_g of the elastomers, which range from slightly below room temperature (high nitrile NBR, HNBR, CSM) to about $-120°C$ for silicone rubber. Mercury ($-38.88°C$), water ($0°C$), and cyclohexane ($-83°C$, $6.55°C$) have been suggested in the literature [*1,3*] for calibration. Other calibrants that may be used are: hexane ($-95°C$), octane ($-56.8°C$), decane ($-29.7°C$), dodecane ($-9.6°C$), and tetradecane ($5.9°C$). It is important that the calibrants be as pure as possible, since the melting point is strongly dependent on percent purity.

Moisture condensation is a common problem in subambient operation of thermal instruments. The instrument should not be opened at subambient temperatures. This is easy to follow for DSC and TMA experiments except for the special procedure described above for balata, but can not be avoided for some DMA experiments where the final adjustments of the sample are done in the glassy state. In some DMA instruments the thermocouple has to be shielded for subambient operation. This introduces a temperature lag. Condensation of moisture on the thermocouple shield further aggravates the problem.

In TMA experiments with an expansion probe it was observed in our laboratory that placing a load on the pan at room temperature, as recommended in ASTM E 831, Test Method for Linear Thermal Expansion of Solid Materials by Thermomechanical Analysis gives curves that are not reproducible. This was ascribed to the high nerve in the elastomer sample which is compressed by the small weight (1 g) placed on the pan. An alternative procedure is to place the static weight when cooled below T_g and measure sample dimension in the glassy state. This procedure gives much more reproducible coefficient of expansion data, relatively free from spurious spikes, as shown in Table 3. T_g values are also quite different in some cases. Only a minimum static weight necessary to keep the probe in contact with the sample, should be used in order to avoid compressing the sample in contrast to the 3 g recommended in ASTM E 831. It may be added that not all elastomer samples are amenable to TMA study. For example, a flat surface necessary to allow intimate contact of the sample and the probe cannot be cut from a gum elastomer with high nerve. Maurer [*14*] has recommended hot pressing, which works for low nerve IIR gum samples. Silica powder also has been used to level the surface of gum elastomers [*15*]. Maurer [*16,17*] published excellent reviews of the basic TMA technique that also includes the procedural and elastomeric variations that can influence the results. Another informative review is included in "Applications of Thermal Analytical Procedures in the Study of Elastomers and Elastomer Systems" by Brazier [*8*].

Effect of Curing—A unique feature of the elastomers is that they must be cured in order to have useful properties. Curing increases the glass transition temperature of elastomers (Table 1). However, the effect is not as pronounced as in some thermosets where a large increase of T_g is more common. T_g increases with carbon black loading, but the increase is limited, in most cases, to about 2 to 6°C as is evident in Tables 1 and 4. Carbon black introduces a dependence of modulus on strain in DMA curves, which will not be included in this discussion. Taking the extrapolated onset values for T_g in Table 1 the difference in ΔT_g

TABLE 3—*Comparison of TMA expansion data.*[a]

	Load at Run Temp.				Load Below T_g		
	α_g (μm/m°C)	T_g	α_R (μm/m°C)	Std. Dev. (α_g)	T_g	α_R (μm/m°C)	Std. Dev. (α_g)
NR (SMR-5)	75.8	−68.7	191	10.09	−42.8	226	3.23
	56.3	−60.4	199		−43.0	228	
	52.9	−60.9	204		−45.2	233	
SBR 1500	65.4	−30.3	227	7.50	−29.7	245	1.84
	80.4	−29.0	223		−34.6	254	
					−31.0	266	
IR (Natsyn 2200)	62.7	−45.0	138	8.60	−47.8	243	11.25
	42.6	−49.4	110		−57.0	284	
	58.1	−39.8	149		−49.4	227	
BR (Budene 1207)	39.7	−92.8	115	18.54	−91.0	118	1.90
	85.4	−90.0	201		−91.0	116	
	82.8	−84.5	220				
	60.6	−100.3	224				
NBR (Nipol 1042)	63.8	−11.2	448	7.90	−10.3	327	9.78
	55.5	−8.6	345		−10.2	358	
	74.8	−16.7	933		−7.6	287	
IIR (Butyl 077)	47.6	−35.5	183	5.81	−44.0	232	3.59
	34.4	−32.6	209		−39.3	218	
	45.6	−33.4	271		−41.3	230	
CR (Neoprene GN)	16.5	−25.3	263	32.64	−25.9	323	1.60
	54.1	−27.7	386		−23.6	283	
	96.4	−30.9	536				
EPDM (Nordel 1040)	117.0	−47.5	456	2.75	−38.5	225	0.45
	55.2	−41.6	267		−41.6	210	
	49.7	−47.5	261		−42.6	203	

[a]Subscripts g and R after α refer to coefficient of expansion at glassy and rubbery states, respectively.

	α_g (μm/m°C)		
	40.6		
	47.4		
	47.5		
	55.9		
	53.9		
	58.4		
	32.4		
	48.7		
	21.3		
	50.6		
	46.8		
	52.1		
	33.2		
	29.9		
	26.2		
	32.4		
	34.7		
	46.4		
	49.6		
	42.2		
	48.9		
	49.8		

TABLE 4—*Glass transition temperature (T_g) of common elastomers by DSC.*[a]

Elastomer	T_g, °C (Raw Rubber)[b]	T_g, °C (Vulc)[b] 50 phr, N-347
NR	−60	−56
IR	−57	−54
CIS-BR	−105	−95
SBR-1500	−49	−47
NBR (40% ACN)	−26	−20
CR (W TYPE)	−40	−34
CIIR	−61	−56
EPDM (NORDEL 1040)	−58	−52
CSM (HYPALON 40)	−16.5	—

[a]From Ref *18*. T_g determined at the extrapolated onset.
[b]Heating rate, 30°C min; annealed at 100°C, 5 min, quenched at −120°C.

for the cured and uncured state ranges from 0°C for IIR to around 8°C for NBR. The difference is somewhat higher in Table 4 [*18*] where raw and vulcanized elastomers are compared at 30°C/min instead of 20°C/min in Table 1. Also, the vulcanized samples in Table 4 contain carbon black; however, it is difficult to delineate the carbon black effect from heating rate effect, both of which will increase T_g. Burfield and Lim [*3*] report that typical sulfur cures, using either a conventional (S, 2.5; accelerator, 0.6 parts per hundred parts of rubber (phr)) or an efficient vulcanization system (S, 0.5; accelerator, 2.5 phr) increase T_g of NR by 3°C. A peroxide cure causes an increase of almost 1°C for each part of peroxide used per hundred parts of rubber.

Compatibility Considerations—Thermal analysis (DSC, DMA) is the most popular method to ascertain the degree of compatibility of polymers. An interesting case is provided by SBR-BR blends which are used for tread rubber compounds in passenger-car tires, combining the good hysteresis properties of BR with the abrasion resistance of SBR.

BR-SBR blends are mostly mentioned as compatible systems. However, a distinction may be made between the unvulcanized and vulcanized compounds. Callan et al. [*19*] studied a wide range of both cured and uncured blends and the effect of fillers on blending by using DTA. They employed a special electron microscopic technique whereby the contrast was greatly enhanced and discrete zones were observed for gum blends of 50:50, SBR-BR. Ziegler and butyl lithium polymers gave similar results, except that the latter gave higher T_g values reflecting the higher vinyl content in these elastomers. In all uncured BR-SBR blends, with or without carbon black, two T_gs were observed, whereas a single T_g was observed for the peroxide cured blends.

Callan et al. [*19*] suggest that uncured BR-SBR blends are microheterogeneous with very small domain sizes which are capable of covulcanization and, thus, give a single T_g when cured. This leads to a broadening of thermal response, giving a diffuse T_g interval that is difficult to detect. In more heterogeneous systems (NR-BR), the separate domains are sufficiently large that the response is less sensitive to the restriction imposed by crosslinking.

A different situation is presented by NBR copolymers. Collins et al. [*20*] reported that all commercial noncrosslinked NBR copolymers having less than 35% nitrile exhibit two glass transitions (Table 5) [*21*]. This was attributed to synthesis of copolymers richer in nitrile content in the earlier part of the reaction sequence, rather than those in the later part, which are deficient in nitrile content, in effect producing two separate phases in the same polymer.

Presumably, as the total nitrile content in the copolymer increases the product tends to

TABLE 5—*Glass transition temperatures of various NBR copolymers (rate of heating 10°C/min).*[a]

Trade Name	Bound ACN, %	Elastomer	
		$T_{g1}{}^d$	$T_{g2}{}^d$
PAN	100	150.4	
Nipol 1000X132	50.5	−12.2	
Nipol 100X88	43.9	−17.6	
Nipol 1041LG	40.9	−20.6	
Nipol 1042	33.0		−33
Nipol 1453 HM	28.3	−26.7	−52.4
Nipol 1034-60	20.3	−30.7	−68.3
Nipol 1432[b]	33.0	−24.3	−41.8
Nipol 1020[c]	44.0	−22.8	

[a] Unpublished work by Sircar.
[b] Contains >20% VYHH (Vinyl acetate-Vinyl chloride copolymers).
[c] Hydrogenated NBR, unsaturation 10%.
[d] T_g reported are T_{e0} values.

become more homogeneous as may also be inferred from the difference of two T_gs (Fig. 6, T_p values from DDSC). (T_{g1}-T_{g2}) is approximately 35°C and 9°C for NBRs with 21% and 33% acrylonitrile, respectively (Fig. 3). The value of ΔT_g decreasing with increasing acrylonitrile (ACN) concentration is consistent with the explanation with regard to the distribution of ACN between the phases, as described above. When Nipol 1042 (33% ACN) is vulcanized the DDSC peaks shown in Fig. 7 coalesce together with some inflections. The T_gs determined from the inflection points indicate a simultaneous increase of both T_gs with a (T_{g1}-T_{g2}) of 8°C, which is about the same as the uncured compound, but lower than the raw elastomer (9°C, Fig. 3). Interestingly, two DDSC peaks indicate separate phases in the cured

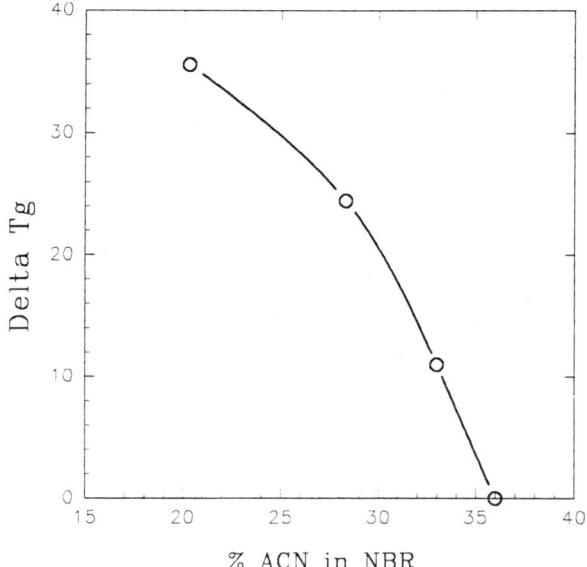

FIG. 6—ΔT_g *as a function of percent ACN in NBR copolymers.*

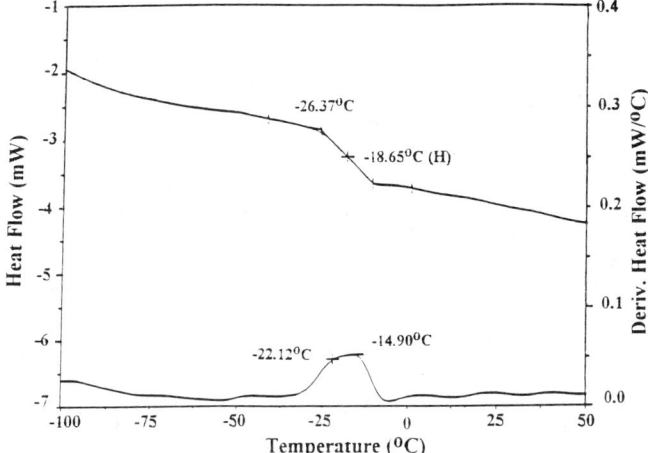

FIG. 7—DSC and DDSC curve of vulcanized NBR with 33% ACN.

compound and allow assignment of T_gs of the individual components, whereas the extrapolated onset and half-height indicate only one T_g. The usefulness of DDSC in quantifying the separate phases in blends has been reported [21].

Elastomer Nomenclature—Literature data often specify the elastomer used without giving complete information. For example, within SBR type rubbers (23.5% styrene), there are the 1000 series (hot polymerization type), 1500 series (cold polymerization type), 1700 series (oil-extended, cold type), 1600 and 1800 series (oil and carbon black extended), and there are SBRs with higher and lower percentages of bound styrene. All of these have different T_gs. Even within the same group T_g would be slightly different with the amount and type of oil and carbon black loading. It is, therefore, necessary to identify the polymer used by its full name (e.g., SBR 1712, not just SBR). The same is true for NBR rubber. Its nomenclature should indicate the percent bound acrylonitrile, since T_g increases with bound acrylonitrile. Following the same logic, when references are made from other literature values it is necessary to know the recipe and the cure conditions used in order to make a complete judgment of the relative values.

Conclusions

For the sake of consistency in literature data the location for specifying the T_g value on the DSC curve and the experimental conditions should be fully reported. The calibrant used should have a transition in the temperature range close to that of the polymer. For annealing, the method of Burfield and Lim [3] is recommended.

TMA expansion experiments should use a minimum load on the pan (1 g). The load should be applied only after the elastomer is cooled below T_g.

Temperature calibration in DMA should be with subambient liquids and calibration should be carried out exactly under the same conditions as the sample run. Moisture condensation should be kept to a minimum.

The description of the sample should include the exact name of the elastomer with the number index, the recipe used, and the cure conditions.

It is hoped there will be much more order in the literature data if the above recommendations are implemented.

Acknowledgments

The authors would like to thank M. Galaska and S. Rodrigues for their assistance in generating experimental data and J. Miller for secretarial help.

References

[1] Loadman, M. J. R., *Journal of Thermal Analysis*, Vol. 30, 1985, pp. 929–941.
[2] Kow, C., Morton, M., Fetters, L. J., and Hadjichristidis, N., *Rubber Chemistry and Technology*, Vol. 55, 1982, pp. 245–252.
[3] Burfield, D. R. and Lim, K. L., *Macromolecules*, Vol. 16, 1983, pp. 1170–1175.
[4] Lombardi, G. E., *For Better Thermal Analysis*, Vol. II, ICTA, Rome, 1980.
[5] Roychowdhury, N., Chaki, T. K., and Bhowmick, A. K., *Thermochimica Acta*, Vol. 176, 1991, pp. 149–161.
[6] Wunderlich, B., in *Thermal Characterization of Polymeric Materials*, E. A. Turi, Ed., Academic Press, New York, 1981, pp. 170–172.
[7] Landi, V. R., *Rubber Chemistry and Technology*, Vol. 45, 1972, pp. 222–240.
[8] Brazier, D. W. and Nickel, G. H., *Thermochimica Acta*, Vol. 26, 1978, pp. 399–413.
[9] Widmair, J. M. and Meyer, J. C., *Macromolecules*, Vol. 14, 1981, pp. 450–453; *Rubber Chemistry and Technology*, Vol. 54, 1981, pp. 940–943.
[10] Selikhova, V. I., Zubov, Y. A., Bakeyev, F. N., and Belov, G. P., *Vysokomol. Soedin.*, Series A, 1977, pp. 19,759.
[11] Wunderlich, B., Bodily, M., and Kaplan, M. H., *Journal of Applied Physics*, Vol. 35, 1964, p. 1.
[12] Griffiths, M. D. and Maisey, L. J., *Polymer*, Vol. 17, 1976, p. 869.
[13] Tye, R. P., "Audience Discussion I: Dealing with Condensed Moisture," in *Assignment of the Glass Transition, ASTM STP 1249*, R. J. Seyler, Ed., American Society for Testing and Materials, Philadelphia, pp. 4–5 (this publication).
[14] Maurer, J. J., *Rubber Chemistry and Technology*, Vol. 38, 1965, pp. 979–990.
[15] Burfield, D. R. and Tanaka, Y., *Polymer*, Vol. 28, 1987, pp. 907–910.
[16] Maurer, J. J., in *Thermal Methods in Polymer Analysis*, S. W. Shalaby, Ed., Franklin Institute Press, Philadelphia, 1978, pp. 129–161.
[17] Maurer, J. J., in *Analytical Chemistry*, R.-S. Porter and J. F. Johnson, Eds., Vol. 1, Plenum, New York, 1968, pp. 107–118.
[18] Sircar, A. K., *Journal of Scientific and Industrial Research*, Vol. 41, 1982, pp. 536–559.
[19] Callan, J. E., Hess, W. M., and Scott, C. E., *Rubber Chemistry and Technology*, Vol. 44, 1971, pp. 814–837.
[20] Collins, E. A., Jorgensen, A. H., and Chandler, L. A., *Rubber Chemistry and Technology*, Vol. 46, 1973, pp. 1087–1102.
[21] Sircar, A. K., Galaska, M. L., Linden, S. M., et al., *Proceedings of the 22nd NATAS Conference*, pp. 319–324.

Michael J. Moscato[1] *and R. J. Seyler*[1]

Assigning the Glass Transition Temperature in Oriented Poly(ethylene terephthalate)

REFERENCE: Moscato, M. J. and Seyler, R. J., "**Assigning the Glass Transition Temperature in Oriented Poly(ethylene terephthalate),**" *Assignment of the Glass Transition, ASTM STP 1249*, R. J. Seyler, Ed., American Society for Testing and Materials, Philadelphia, 1994, pp. 239–252.

ABSTRACT: Thermoplastics like poly(ethylene terephthalate) (PET) can be melt processed into sheet and film formats for a variety of applications. Fabrication of these formats often involves application of a drawing process which can include drafting, tentering, heat setting, detentering, and heat relaxation. These various components add a thermomechanical history to the PET that will influence the glass transition. Thermomechanical analysis (TMA) in both the compressive mode and the tensile mode along with differential scanning calorimetry (DSC) techniques have been applied to a series of seven PET films generated using different drawing processes applied with an Iwamoto Biaxial Stretcher on extruded sheet from a single source. These measurements were undertaken to assign a glass transition temperature for each film. The assigned temperature ranged between 70°C and 94°C. Orientation and heat setting increased the temperature assigned as T_g while detentering both increased and lowered T_g on a direction-specific basis. Both the measurement technique and the assignment protocol employed to determine T_g contributed to differences in the assigned value. DSC results suggested the presence of two amorphous domains (mobile and constrained) in the film oriented uniaxially without constraint. TMA data demonstrated the assigned T_g may be direction specific and not a bulk property in oriented PET films.

KEYWORDS: differential scanning calorimetry (DSC), thermomechanical analysis (TMA), mobile amorphous domain, rigid amorphous domain, glass transition, glass transition temperature (T_g), poly(ethylene terephthalate) (PET), tensile mode, compressive mode, softening point (T_s), uniaxial orientation, biaxial orientation

The glass transition is not a unique first-order material property. Instead it is second-order-like with a kinetic sensitivity, making its observed value strongly dependent upon not only a material's chemistry but also its thermomechanical history and the measurement conditions employed [1–4,6]. The change in properties associated with this passage between the glassy and rubbery (or liquid) states of amorphous materials are often large, e.g., 3 orders of magnitude in elastic modulus [1,4] or 6 to 7 orders of magnitude in viscosity [2–5]. They are continuous changes which occur over a narrow temperature interval, not at a discrete temperature. In polymers this interval could be as narrow as 2° to 5° [6] or as wide as tens of degrees for plasticized resins [4,6]. In crystallizable polymers, the glass transition is further influenced by crystalline regions and how they are developed [7–9].

Frequently, values for the "glass transition temperature" of amorphous polymeric materials are reported without regard for the external influences coupled to its value. Such reporting of or use of a single number for representing the glass transition as T_g with the same certainty of conviction as any first-order thermodynamic property increases the risks

---
[1] Analytical chemist, Analytical Services, and research associate, Material Science and Engineering, respectively, Eastman Kodak Company, Rochester, NY 14653.

associated with misrepresentation. Yet strong needs remain for using a singular, but repeatable, representation of the glass transition.

To address these points, this manuscript will examine the values of the glass transition assigned from common applications of differential scanning calorimetry (DSC) and thermomechanical analysis (TMA) to film samples of a semicrystalline poly(ethylene terephthalate) (PET). These films have been prepared from a common extruded sheet using different postextrusion thermomechanical histories. The discussion that follows identifies the differences in assigned values of T_g which stem from these thermomechanical histories and from the measurement protocols used.

Experimental

The poly(ethylene terephthalate) discussed in this paper had an inherent viscosity of 0.7. The resin was extruded with a 38-mm Killion extruder into 1-mm-thick sheet. Individual 10-cm by 10-cm squares of the extruded PET were then stretched in an Iwamoto Biaxial Stretcher according to the conditions described in Table 1. Biaxial stretching was accomplished sequentially with the first stretch in the machine direction (parallel to the web direction during extrusion). All stretches were performed at a rate of 100 mm/min. Approximately 100 mm^2 were sampled from the center of each stretched sheet to provide specimens for subsequent thermal testing.

DSC

All DSC curves were obtained using a Seiko Instruments model 210 DSC that had been temperature calibrated with melting points of high-purity indium and tin according to ASTM E 967, Practice for Temperature Calibration of Differential Scanning Calorimeters and Differential Thermal Analyzers. Individual measurements were carried out on 4 to 14 mg specimens in crimped aluminum pans and heated through two 0°C to 300°C thermal cycles at a heating rate of 10°C/min. Encapsulated specimens were quench-cooled with liquid nitrogen between thermal cycles. The DSC cell was purged with a flow of dry nitrogen gas at a flow rate of 50 mL/min.

TABLE 1—*Thermomechanical history of PET samples.*

Sample	Stretch Temp, °C	Stretch Ratio $M \times T$		Heat Set Temp, °C	Detenter
Cast	—	—	—	—	—
Uniax	105	3.5	—	—	—
Uniax C	105	3.5	*	—	—
Biax	105	3.5	3.5	—	—
Biax Hs	105	3.5	3.5	109	—
Biax DtA	105	3.5	3.5	109	cond. A
Biax DtB	105	3.5	3.5	109	cond. B

All samples stretched sequentially at 100 mm/min.
*This sample was constrained to original width during stretch.

Compression TMA

All penetration and expansion curves were recorded using a Seiko Instruments model 100 TMA in the compressive mode. The sample thermocouple was calibrated with the melting points of high-purity foils of indium and tin according to ASTM E 1363, Test Method for Temperature Calibration of Thermomechanical Analyzers. A 2.7-mm-diameter quartz probe was used for both the penetration and expansion measurements. Specimens of each film or sheet ranging in thickness from 0.1 mm to 1.0 mm were cut to approximately 5-mm by 5-mm and placed on the quartz stage of the TMA. The probe was brought into contact with the specimen and an appropriate compressive force was applied through the film thickness during heating from 20°C to 110°C at 5°C/min. A purge of dry nitrogen gas was maintained through the TMA at a flow rate of 50 mL/min. Penetration results were obtained on the first heat of each specimen with an applied force of 245 mN. The expansion data were obtained on the second heating of the same specimens after rapid cooling to the original starting temperature. The applied force was reduced to 9.8 mN for recording the expansion curves.

Tensile TMA

All tensile TMA curves were recorded using the same Seiko Instruments model 100 TMA after reconfiguring with a quartz tensile probe. Specimens of thickness ranging from 0.1 mm to 1 mm were cut 7.5-mm by 3.5-mm with the long dimension parallel to either the machine direction or the transverse direction of the film sample. Each specimen was suspended within the tensile probe with a pair of stainless steel chucks under a tensile force of 49 mN. The specimens were heated from 20°C to 130°C in tension at 5°C/min. A purge of dry nitrogen gas was maintained through the TMA at a flow rate of 50 mL/min.

Discussion

Use of poly(ethylene terephthalate) or PET as a commercial thermoplastic for film, fibers, and packaging applications is leveraged by the enhancements in mechanical and optical properties that can be achieved with controlled drawing and thermal conditioning. A unique feature of semicrystalline polymers like PET is their ability to undergo strain hardening under appropriate drawing conditions above the glass transition temperature to form a crystalline morphology [7,9]. It has further been established that the arrangement of the chains in the noncrystalline regions of such oriented polymers also contributes to the ultimate properties and that there exists two distinct noncrystalline domains: a mobile amorphous phase and a constrained amorphous phase [8,10–12]. Hristov and Schultz [11] have reported the observance of two glass transitions in oriented semicrystalline PET fibers consistent with this two-noncrystalline-domain description.

DSC

Initial assessment of the PET films prepared with the seven different paths outlined in Table 1 was accomplished with DSC. The observed thermal properties are summarized in Table 2. A glass transition was readily apparent for only two of the seven specimens on the first thermal cycle while all seven exhibited melting of a crystalline phase. Figures 1 and 2 illustrate the differences in appearance of the DSC curves using the cast sheet and the uniaxially oriented samples respectively as examples. The classic heat capacity step associated with the glass transition was observed for all seven specimens in the second thermal cycle. The width of this heat capacity step ranged from 4.2°C to 7.5°C, consistent with

TABLE 2—*DSC assignments.*

Sample/Scan		T_g, °C	T_g interval, °C	T_m, °C	T_c, °C	dH, J/g
Cast	/1	73.8	70.6–77.1	250.4	155.1	32.8
	/2	77.9	74.3–80.3	253.7	141.6	42.7
Uniax	/1	—	—	250.6	—	43.6
	/2	78.5	75.3–81.2	250.4	148.4	34.8
Uniax C	/1	72.5	69.3–74.8	249.7	116.8	37.7
	/2	79.2	76.8–81.0	250.4	147.7	34.0
Biax	/1	—	—	249.7	—	39.1
	/2	78.5	74.6–80.9	251.0	147.0	34.4
Biax Hs	/1	—	—	249.7	—	42.0
	/2	78.5	74.7–81.0	251.0	148.3	34.0
Biax DtA	/1	—	—	249.7	—	42.3
	/2	77.9	74.7–80.5	250.0	145.0	35.7
Biax DtB	/1	—	—	249.9	—	41.1
	/2	78.5	74.4–81.9	250.4	147.0	33.8

All second scans were immediately preceded by a liquid nitrogen quench after initial heating to 300°C.

expectations for an unmodified resin. In all cases exhibiting a distinct glass transition, a spontaneous crystallization exotherm followed before melting.

These DSC results are typical for semicrystalline polymers like PET and illustrate an important limitation to using DSC to assign the glass transition in oriented semicrystalline polymers. Recognizing that the observed glass transition will be influenced by the thermomechanical path to which a material has been subjected and that PET is nearly half amorphous at maximum crystallinity, the PET samples considered in this paper should have given rise to different observed glass transition temperatures in the first thermal cycle of otherwise chemically equivalent specimens. The apparent absence of a glass transition for five of the seven samples during the first thermal cycle does not indicate the absence of amorphous regions in these specimens. Instead it reflects a sensitivity limitation with DSC.

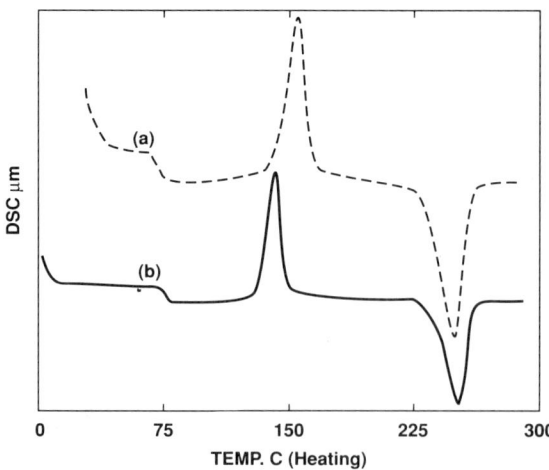

FIG. 1—*DSC curve of PET extruded sheet:* (a) *first thermal cycle;* (b) *second thermal cycle.*

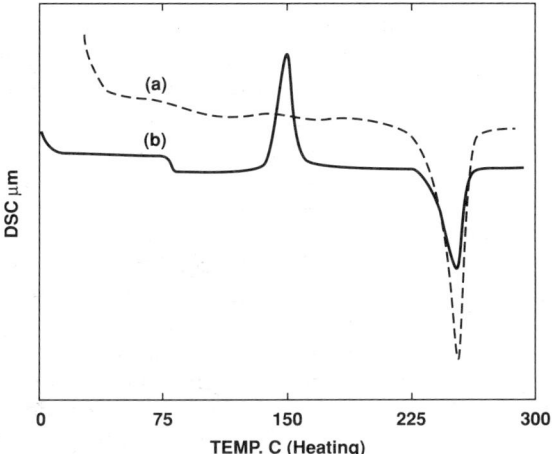

FIG. 2—*DSC curve of PET film oriented uniaxially without constraint:* (a) *first thermal cycle and* (b) *second thermal cycle.*

For a highly oriented sample like the one stretched uniaxially without constraint, two glass transitions in the first thermal cycle were possible, but not readily apparent. In Fig. 3 the first thermal cycle for the uniaxially oriented PET is redisplayed to 150°C with an expanded y-axis scale. This scale expansion reveals two consecutive heat capacity steps, 70.6°C–80.8°C and 86.6°C–102.2°C, which appear to represent the mobile and constrained amorphous domains, respectively. The breadth of each of these transitions (10.2°C and 15.6°C) is approximately twice that observed for the singular glass transition during the second thermal cycle. Similarly, the magnitude of the heat capacity increment ($dC_p = 0.075$ mJ/°C mg and $= 0.086$ mJ/°C mg) individually and combined is significantly less than the 0.376 mJ/°C mg recorded on the second thermal cycle. Hence crystallinity and orientation both lower delta

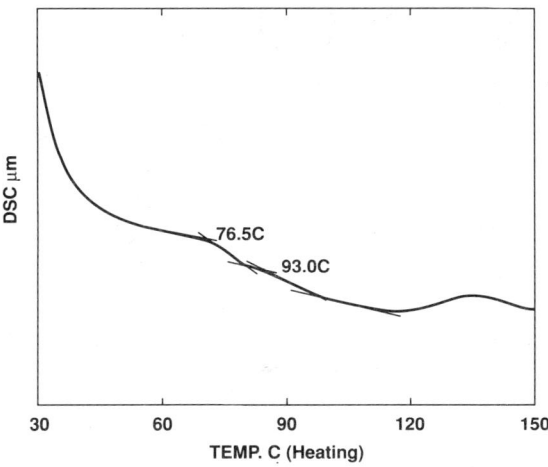

FIG. 3—*Expanded scale first thermal cycle DSC curve of PET film oriented uniaxially without constraint.*

C_p by reducing the portion of material that remains amorphous and broaden the temperature interval of the glass transition. This broadening of the amorphous transition interval in semicrystalline polymers is not completely understood but is related to mobility restrictions in the near crystal regions and the reduction in size of the amorphous domains [13]. The reduction in magnitude and the increase in transition interval combine to significantly reduce one's ability to distinguish the glass transition in "as received" specimens using DSC.

For the as-cast sheet and the constrained uniaxially oriented specimens, a distinct glass transition was easily noted in the first thermal cycle of the DSC measurements (see Fig. 1 or Table 2). This glass transition temperature, assigned as the midpoint of the heat capacity increment ($\frac{1}{2}$ dC_p), is observed to increase on the second thermal cycle. The breadth of the transition interval decreases slightly on the second thermal cycle as well. Both effects have been observed previously in PET [14] and are attributed to residual ethylene glycol monomer which plasticizes the amorphous regions. Heating to 300°C removes these volatiles eliminating the plasticization effect in the second thermal cycle.

More than two thermal cycles for a given specimen and/or imposition of a 10-min hold in the melt near 300°C before quenching were undertaken with a couple of specimens to ensure removal of process history effects in the assigned T_g from DSC measurements after the first thermal cycle. Constancy of T_g after the first thermal cycle using either approach allows acceptance of the second thermal cycle values as representative of the bulk glass transition for this PET. One can estimate the precision of the DSC method using these second cycle values from Table 2. The seven specimens yield a mean T_g = 78.4°C with a 95% confidence interval of 78.0°C to 78.8°C and a standard deviation of 0.44°C.

TMA

The diminished sensitivity of the DSC measurement in assigning the glass transition of oriented semicrystalline specimens may be overcome by using an alternative measurement. Mechanical or rheological measurements typically offer improved signal-to-noise capability in these instances. Although either static or dynamic load protocols are available, the static load approach offers the advantage of no frequency dependence. For this paper, TMA was selected. TMA may be conducted using several configurations (Fig. 4). In the compressive mode one can observe the glass transition from the z or thickness direction whereas using the tensile mode provides an opportunity to examine the glass transition from the perspective of any in-plane direction. As such, TMA makes it possible to examine direction specific influences on the glass transition resulting from the drawing operations that would not be possible with a bulk response measurement like DSC. TMA is, however, disadvantaged in that the protocols employed to assign a glass transition temperature from a TMA curve (regardless of the measurement mode used) do not address the temperature interval of the glass transition.

Compressive Mode

Operation of the TMA instrumentation in the compressive mode with a relatively large contact area probe and negligible force produces an expansion measurement. This protocol equates to that which is referred to as dilatometry and it uses the inflection in the thermal expansivity as illustrated in Fig. 5a to assign T_g. For film specimens this measurement addresses the linear thermal expansivity in the thickness direction. Increasing the applied compressive force to modest levels (100 mN to 300 mN), with or without a reduction in the probe contact area, results in a penetration measurement. More commonly practiced in polymer laboratories, this protocol determines a softening point as the extrapolated onset of

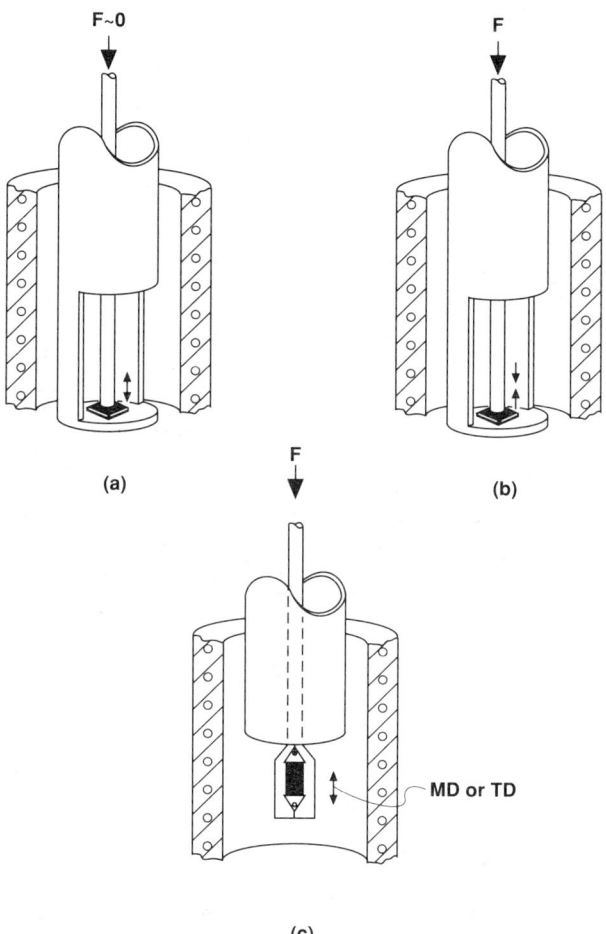

FIG. 4—*TMA test configurations:* (a) *compressive mode expansion,* (b) *compressive mode penetration, and* (c) *tensile mode.*

penetration to represent the glass transition (Fig. 5b). It is, in effect, a microrheology measurement through the film thickness and will therefore be dependent on the applied force.

Difficulties may arise from effects like residual stresses in thin films which can make expansivity measurements with TMA difficult. Such was the case with the samples studied here. This difficulty was overcome using an approach patterned after that of Turi [15]. The specimens were subjected to two thermal cycles through just the glass transition region with rapid cooling between cycles. During the first cycle a 245 mN force was applied to record T_s as the penetration mode representation of the glass transition. Because most of the specimens were quite crystalline and because the limit temperature for the TMA thermal cycle (110°C) was below that of spontaneous crystallization on heating or of melting, it was assumed that thermomechanical history contributions to T_g would not be erased with this protocol. Consequently, assigning T_g from expansion at a much reduced force of 9.8 mN during the second thermal cycle for the same specimen, should still reflect an "as prepared"

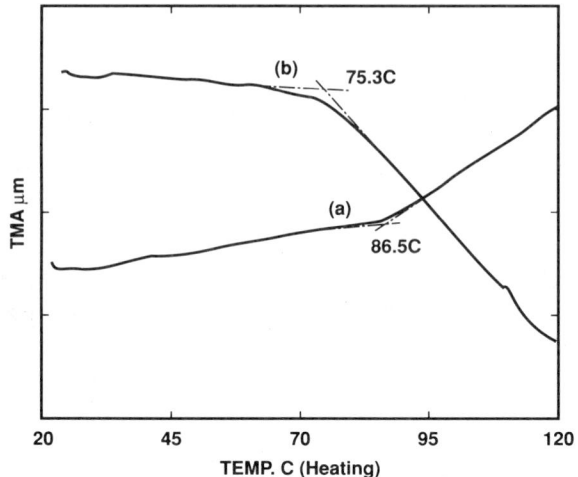

FIG. 5—*Compressive mode TMA curves of PET film biaxially oriented and heat set with detentering: (a) expansion and (b) penetration.*

film property. Values of T_g assigned for the seven samples with this protocol are provided in Table 3. Unlike DSC where little distinction in the assigned T_g was noted, a considerable range of values is assigned to both modes of compression TMA, indicative of processing contributions to the glass transition.

Considering first the expansion data, the observed differences are the result of the extent and type of crystalline morphology and orientation that was developed. The lowest value of T_g is recorded for the as-cast sheet which should have relatively little crystallinity or orientation. The sample oriented uniaxially without constraint (Uniax) yielded the highest value of T_g because of its extensive strain-induced crystallinity and consequent in-plane mechanical anisotropy. The biaxially oriented and heat-set sample (Biax Hs) was similarly high in observed T_g. Like the Uniax sample, this film should contain a considerable amount of strain-induced crystallinity but with less in-plane orientation anisotropy. It should also, by virtue of heat setting, contain some thermally induced spherulitic crystallinity. Comparison of the Biax Hs sample to the Biax sample, which underwent equivalent orientation without the subsequent heat setting, indicates a substantial increase in T_g occurred with further crystallization during heat setting. The remaining two biaxial films (Biax DtA and Biax DtB) underwent equivalent orienting and heat setting processes as the Biax Hs film but

TABLE 3—T_g *assigned from compressive mode TMA.*

Sample	Penetration T_g, °C[a]	Expansion T_g, °C[b]	Delta, °C
Cast	68.3	81.1	12.8
Uniax	70.0	93.6	23.6
Uniax C	72.8	85.5	12.7
Biax	75.9	85.7	9.8
Biax Hs	85.7	92.8	7.1
Biax DtA	73.1	87.7	14.6
Biax DtB	75.3	86.5	11.2

[a] Commonly reported as the softening point, T_s.
[b] Recorded on second heat after quench from 110°C of the initial heat.

experienced the additional process of detentering. Detentering is a controlled reduction in final extent of drawing in the second stretch (T or X) direction. Both detentering protocols used served to reduce the T_g from that of the Biax Hs conditions. The final sample Uniax C is uniaxially stretched with the width constrained to its original dimension during stretching. This is, in effect, an unequal simultaneous biaxial stretch that will result in less anisotropy than the Uniax film and less overall orientation than the biaxial films. The assigned T_g is consistent with expectations from this description.

For noncrystalline polymers like high-impact polystyrene (Fig. 6), the softening point, T_s, obtained in penetration will generally coincide with the high-temperature half of the glass transition interval measured with DSC at comparable heating rates.[2] The observed T_s = 99.1°C is within the latter half of the 91.4°C–100.5°C glass transition interval from DSC. For crystallizable polymers this relationship between T_s and the DSC glass transition interval may not hold. Consider the two films for which a distinct T_g was recorded in the first DSC thermal cycle (Cast and Uniax C). The value of T_s recorded as the "penetration T_g" in Table 3 for these two films does meet this expectation. This fit is attributable to the predominantly amorphous composition of these films. For the Uniax film the T_s occurs at the onset to the DSC recorded glass transition for the mobile amorphous domain. Thus introduction of crystallinity alters the coincidence of the TMA penetration measurement with the DSC measurement of the glass transition.

The penetration TMA measurement does appear to sense the extent of orientation and crystallization developed within each film. The Cast sheet is primarily unoriented amorphous material and exhibits the lowest T_s while the Biax Hs film has the greatest extent of orientation and thermal conditioning. The Uniax, Uniax C, and Biax films represent a progression of extent of total orientation beyond the Cast sheet and exhibit the same progression in their value of T_s. The 9.8°C increase in T_s between films Biax and Biax Hs is attributable to the effects of thermal crystal growth during heat setting which tends to "fix" the imposed orientation. Both detentering procedures significantly lowered T_s, offsetting the temperature gains in T_s achieved with heat setting.

The column labeled "delta" in Table 3 is the difference between the softening point T_s determined in penetration and the glass transition temperature T_g assigned from expansion. The softening point averaged 13.0°C lower demonstrating an aspect of the applied force dependence of the TMA measurement in the compressive mode. The substantial variability (s = 5.2) of this delta value indicates the sensitivity of this comparison to the thermal history, suggesting it be used only for estimation purposes.

Rough estimates of the precision of the compressive-mode TMA measurements for assigning the glass transition were made by examining three separate specimens from the Uniax film for both penetration and expansion. The resulting standard deviations were s = 0.1°C and s = 2.6°C for penetration and expansion, respectively. With such few samplings it is difficult to attribute any significance to this difference. Nevertheless, the results suggest the orientational anisotropy may impact expansivity measurements to a greater extent.

Tensile Mode

Operation of the TMA in the tensile mode offers a potential advantage over most other techniques for determining the glass transition when one is considering material that has been subjected to a drawing process. This advantage is the ability to examine T_g with respect to any in-plane direction. Such direction specific assessment is of value only if the imposed thermomechanical history impacts on the glass transition in a nonuniform way within the material. Although descriptions of crystalline and amorphous orientation have occurred in

[2]Seyler, R. J., unpublished work.

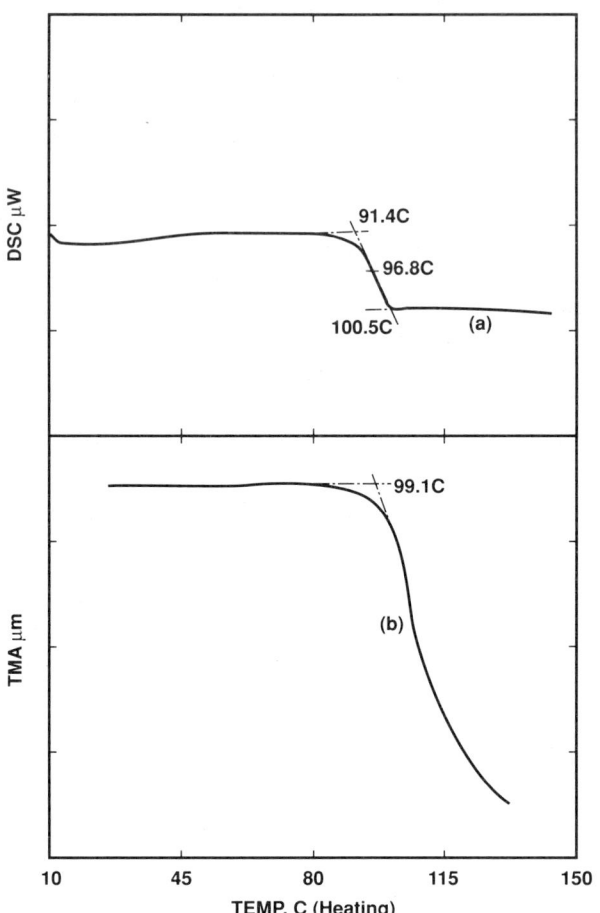

FIG. 6—*Method comparison for T_g assignment of high impact polystyrene sheet:* (a) *DSC curve and* (b) *compressive mode TMA curve in penetration.*

the literature for oriented semicrystalline polymers, the authors were not aware of any prior references to the glass transition being directionally different within a given material. Yet the existence of optical and mechanical anisotropy in certain oriented materials and the knowledge that the glass transition is process history dependent offers the potential for such to occur.

Direction specific measurements using tensile TMA may be accomplished by mounting rectangular specimens whose long axis parallels the in-plane direction of interest such that the tensile force is applied along this direction. The tensile mode could be conducted using a near-zero applied force or a more substantial force. The appearance of the tensile TMA curve and the assigned temperature will be dependent upon the level of applied force [16]. In this experiment a force of 49 mN was applied to the samples yielding a tensile stress of 0.3 MPa which is sufficient to induce elongation of the specimen during the reduction in the elastic modulus associated with the glass transition. The assigned value for T_g is the extrapolated onset to elongation as demonstrated in Fig. 7.

The tensile TMA results are summarized in Table 4. As was the case for compressive TMA, the tensile TMA assignments indicate differences between samples arising from

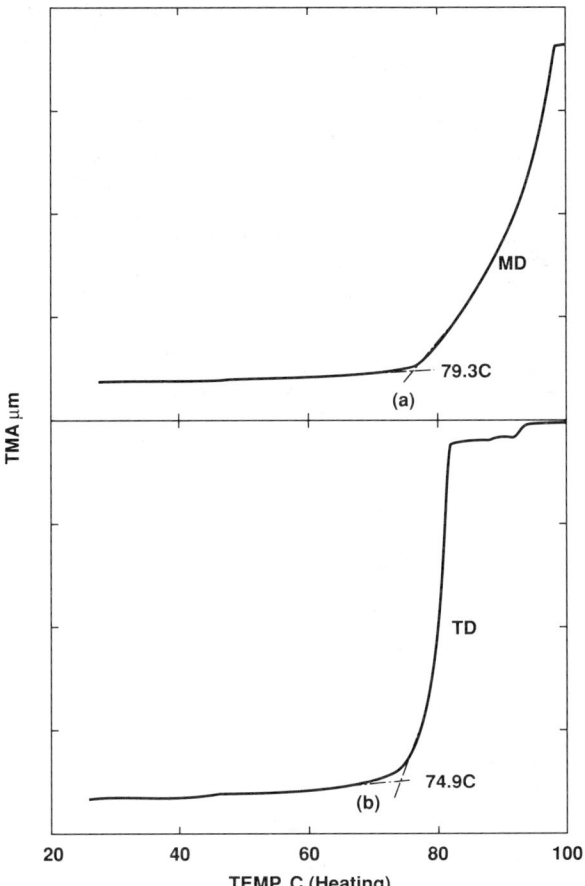

FIG. 7—*In-plane glass transition anisotropy for a PET film oriented uniaxially with constraint from tensile TMA:* (a) *machine direction and* (b) *transverse direction.*

process history contributions. Furthermore, the tensile TMA assignments of T_g differ between the machine and transverse stretch directions within a sample suggesting process contributions can influence the glass transition in a nonuniform way. These directional differences in assigned in-plane glass transition values are tabulated as "delta" in Table 4.

Examining the glass transition in the machine direction indicates the Cast and Uniax C

TABLE 4—T_g assigned from tensile mode TMA.

Sample	M direction T_g,°C	T direction T_g,°C	Delta, °C
Cast	79.9	78.7	1.2
Uniax	84.7	78.0	6.7
Uniax C	79.3	74.8	4.5
Biax	84.2	84.0	0.2
Biax Hs	83.0	85.0	−2.0
Biax DtA	85.0	81.7	3.3
Biax DtB	86.8	83.3	3.5

films have similar T_g values that are less than the remaining samples. Similarly, lesser values of T_g were assigned in the transverse direction for these two samples. These results are consistent with the DSC first-cycle results which indicated no significant crystallinity was present and process history was of minimal influence.

The Uniax film exhibited a machine direction T_g comparable to those for the biaxially oriented films indicating strain-induced crystallinity had occurred. These observations are again consistent with first thermal cycle DSC results. The significantly lower transverse-direction T_g assigned for this film is comparable to that found for the as-cast and constrained uniaxial films indicating the observed reduction in width with drawing inhibited development of any transverse orientation. The large difference in the Uniax specimen reflects the significant anisotropy associated with the strain-induced crystal morphology developed in this film by drawing.

The biaxially oriented films exhibit both machine direction and transverse direction glass transition values that are greater than those of the as-cast sheet. This is anticipated since significant orientation was imposed in both directions. The strain-induced crystallinity which resulted would be expected to elevate T_g. The delta value for the Biax film is nearly zero indicating in-plane isotropy. Addition of a heat-setting step (Biax Hs) tended to disrupt this isotropy in favor of the second stretch (transverse) direction. Detentering further promoted anisotropy (increased magnitude of delta) but favored the machine direction. This reversal in property bias arises from relaxation of the amorphous orientation along the transverse direction.

The delta value for the film oriented uniaxially with constraint is the second largest value, indicating measurable in-plane anisotropy. That it is less than that of the unconstrained case despite both machine directions being stretched 3.5× supports an earlier observation that the constraint to constant width in the transverse direction acts as a simultaneous transverse stretch of a magnitude comparable to the width reduction in the unconstrained case. Thus the constrained uniaxial drawing process is, in effect, an unbalanced simultaneous biaxial orientation process. The lack of observed crystallinity in this film as prepared despite a greater extent of total orientation when compared to the unconstrained uniaxial film suggests the simultaneous transverse component acts to lessen the "strain hardening" during drawing which develops strain-induced crystallinity.

The delta for the as-cast sheet was not zero as one might predict *a priori*. The small bias observed in the machine direction, though less than the long-term variability of this measurement, is believed real because within day variability is typically less than long-term variability and the bias can result from the imposed web tension during melt casting of the sheet. This web tension develops a residual orientation which does not create crystallinity but is sufficient to enhance T_g.

The TMA values of T_g assigned from the tensile measurements tend to be between T_s obtained with penetration and T_g obtained with expansion despite the much greater levels of effective stress sustained by specimens in the tensile mode. Testing four specimens in each direction of the film uniaxially oriented without constraint over several days yielded an average standard deviation of 1.9°C as an estimate of long-term uncertainty. This uncertainty is comparable to that estimated for compressive mode TMA measurements.

Conclusion

The glass transition of a poly(ethylene terephthalate) film has been examined using both DSC and TMA under a variety of measurement conditions on specimens sampled after exposure to seven different thermomechanical histories. These thermomechanical histories included uniaxial stretching with and without constraint, biaxial stretching with and without heat setting, and heat-set biaxial stretching with two detentering conditions. The singular values of T_g assigned to represent the glass transition of this PET ranged between 70°C and 94°C. In all cases, the postextrusion thermomechanical history had an observable influence on the assigned temperature for the glass transition. Orientation and crystallinity resulting from the process history tended to increase the temperature of the glass transition by as much as 17°C. Detentering lowered the temperature of the glass transition in the thickness and transverse directions but increased it in the machine direction.

In addition to thermomechanical history contributions to the temperature assignable as the glass transition, measurement type and measurement conditions will also influence the assigned temperature. The differences in assigned values of T_g were greater between measurements for a given thermomechanical history than between thermomechanical histories for a given measurement. Estimates of uncertainty derived from three or four specimens taken from the Uniax film indicated standard deviations of generally less than 2°C for TMA protocols and less than 1°C for DSC.

The influence of the thermomechanical history on the glass transition temperature of oriented PET was shown to be directionally nonuniform. Differences of as much as 7°C in the assigned T_g between the perpendicular in-plane draft and tenter directions from a single sample were noted using tensile mode TMA measurements. These observations further complicate the use of a single temperature to represent the glass transition.

The widely used DSC method for assigning a glass transition is of diminished utility for oriented crystalline polymers. It addresses the glass transition as a bulk property and the lesser volume of amorphous component diminishes the delta C_p step to the extent that it becomes difficult to resolve from baseline. Scale expansion of the enthalpic axis of the DSC curve can improve the ability to resolve a T_g in the first heat cycle. Removal of the thermomechanical history by quenching from the melt allows determination of the bulk polymer glass transition using DSC. For the 0.7 I.V. PET examined here, the bulk glass transition was determined to be 78.4°C with a 95% Confidence Interval of 78.0°C to 78.8°C. The breadth of the glass transition interval ranged from 4.2°C to 7.5°C.

The scale-expansion approach applied to the DSC first thermal cycle of a film oriented uniaxially without constraint identified two amorphous transitions which are believed to correspond to the mobile and constrained amorphous domains of an oriented crystallized polymer. The breadth of the transition interval for both amorphous domains was increased approximately twice that of the bulk transition as a result of the thermomechanical history contribution to T_g.

TMA offers greater practical sensitivity than DSC for assigning a singular value to glass transitions of "as prepared" crystalline polymer films. Assignment protocols applied to TMA curves, however, preclude identification of glass transition temperature interval. The TMA may be used in either the compressive mode for expansion or penetration or in the tensile mode. TMA expansion measurements on thin films may require determinations to be made on a second pass through the glass transition to avoid complications arising from residual stresses. Tensile TMA measurements are in-plane direction specific, imparting the ability to access in-plane nonuniformities.

Acknowledgments

The authors wish to thank E. K. Priebe and R. K. Schlotzhauer for preparation of the PET films used in this experiment.

References

[1] McCrum, N. G., Read, B. E., and Williams, G., *Anelastic and Dielectric Effects in Polymeric Solids*, John Wiley & Sons, New York, 1967, pp. 43–56.
[2] Sidorovich, A. V. and Kuvshinski, E. V., "Relaxation of Specific Volume and Enthalpy During Changes in Phase State of a Polymer," *Relaxation Phenomena in Polymers*, G. M. Bartenev and Y. V. Zelenev, Eds., Keter Publishing House Jerusalem Ltd., Jerusalem, 1974, pp. 53–65.
[3] Bondi, A. and Tobolsky, A. V., "Phase Transitions and Vitrification," *Polymer Science and Materials*, A. V. Tobolsky and H. F. Mark, Eds., Wiley-Interscience, New York, 1971, pp. 105–122.
[4] Aklonis, J. J., MacKnight, W. J., and Shen, M., *Introduction to Polymer Viscoelasticity*, Wiley-Interscience, New York, 1972, pp. 37–59.
[5] Doremus, R. H., *Glass Science*, John Wiley & Sons, New York, 1973, pp. 101–120.
[6] Meares, P., *Polymers-Structure and Bulk Properties*, Van Nostrand Reinhold Co., London, 1971, pp. 251–272.
[7] Misra, A. and Stein, R. S., *Journal of Polymer Science: Polymer Physics*, Vol. 17, 1979, p. 235.
[8] Ward, I. M., *Structure and Properties of Oriented Polymers*, Applied Science Publishers, London, 1975.
[9] Roland, C. M. and Sonnenschein, M. F., *Polymer Engineering Science*, Vol. 31, 1991, p. 1434.
[10] Jin, Y., Fu, Y., Mucha, M., and Wunderlich, B., "Thermal Characterization of Poly(ethylene terephthalate) Fibers," *Proceedings of the 21st NATAS Conference*, Atlanta, GA, 13–16 September 1992, pp. 683–688.
[11] Hristov, H. A. and Schultz, J. M., *Journal of Polymer Science: Polymer Physics*, Vol. 28, 1990, p. 1647.
[12] Tsou, D. L., Desai, P., Abhiraman, A. S., and Huang, T-H., *Journal of Polymer Science: Polymer Physics*, Vol. 29, 1991, p. 49.
[13] Wunderlich, B., "Determination of the History of a Solid by Thermal Analysis," *Thermal Analysis in Polymer Characterization*, E. A. Turi, Ed., Heyden, Philadelphia, 1981, pp. 11–13.
[14] Turi, E. A. and Khanna, Y. P., "Effect of Diethylene Glycol Content of Polyethylene Terephthalate on its Thermal Transitions," *Thermal Analysis in Polymer Characterization*, E. A. Turi, Ed., Heyden, Philadelphia, 1981, pp. 63–70.
[15] Turi, E., *Thermal Characterization of Polymeric Materials*, Academic Press, New York, 1981, pp. 500–502.
[16] Moscato, M. J., "The Use of Thermomechanical Analysis as a Viable Alternative for the Determination of the Tensile Heat Distortion Temperature of Polymer Films," *Materials Characterization by Thermomechanical Analysis, ASTM STP 1136*, A. T. Riga and C. M. Neag, Eds., American Society for Testing and Materials, Philadelphia, 1991, pp. 100–107.

Applications

Ernesto L. Rodriguez[1]

The Glass Transition Temperature of Glassy Polymers Using Dynamic Mechanical Analysis

REFERENCE: Rodriguez, E. L., "**The Glass Transition Temperature of Glassy Polymers Using Dynamic Mechanical Analysis,**" *Assignment of the Glass Transition, ASTM STP 1249*, R. J. Seyler, Ed., American Society for Testing and Materials, Philadelphia, 1994, pp. 255–268.

ABSTRACT: Dynamic Mechanical Analysis (DMA) is presented for four glassy polymers. Poly(vinyl acetate), poly(vinyl chloride), poly(styrene), and poly(carbonate) were studied as a function of the heating rate using ramp and step heating programs and a constant frequency of 1 Hz. The effect of frequency on the dynamic mechanical parameters was also examined from 0.01 Hz to 10 Hz. The dynamic elastic storage modulus (E'), the dynamic elastic loss modulus (E'') and the tanδ (E''/E') were affected by both the heating rate and the frequency. Apparent activation energies for the glass transition were also determined for the four polymers which were in the range from 98 of 194 kcal/mol.

KEYWORDS: activation energy, dynamic mechanical analysis (DMA), dynamic elastic storage modulus, frequency, glass transition temperature, poly(vinyl acetate), poly(vinyl chloride), poly(styrene) and poly(carbonate)

Extensive literature exists in the dynamic mechanical tests of solid polymers and polymer melts [1,2]. In the solid state, one of the particular advantages of dynamic mechanical tests lies in their sensitivity in studying glass transitions (alpha relaxations) and secondary transitions (beta, gamma, delta etc.) of polymeric materials.

The basic property that is measured in a dynamic mechanical test from low temperatures to $T_g + 50°C$ is the modulus or stiffness of the material. Either the dynamic elastic modulus (E', E'') or the dynamic shear modulus (G', G'') is determined from a dynamic mechanical test and its application is defined by the testing geometry and the instrumentation employed. Recalling that from the equation $E = 2(1 + \nu)G$, the relationship between the elastic modulus and the shear modulus G is established through the Poisson's ratio ν. For elastomers $\nu = 0.5$, then $E = 3G$, which is the relation that is commonly used in dynamic mechanical tests. A dynamic shear test (G', G'') is used in solids and melts whereas a dynamic elastic test is most commonly applied to solids up to the temperature range of $T_g + 100°C$.

For linear amorphous polymers such as poly(vinyl acetate) (PVAC), poly(vinyl chloride) (PVC), poly(styrene) (PS), and poly(carbonate) (PC) the dynamic elastic storage modulus (E') is approximately 3×10^9 Pa (3×10^{10} dynes/cm^2) in the solid or glassy state which slowly decay as the temperature is increased. In the glass transition region the modulus drops by a factor of between 10^2 to 10^3 Pa (10^3 to 10^4 dynes/cm^2), from 3×10^9 Pa (3×10^{10} dynes/cm^2) in the glassy state to $3 \times 10^7 - 3 \times 10^6$ Pa ($3 \times 10^7 - 3 \times 10^6$ dynes/cm^2) in the viscous state. This glass transition temperature usually includes a wide tempera-

[1]The BFGoodrich Company, Technical Center, Avon Lake, OH 44012.

ture range of about 10 to 30°C. The ability of a dynamic mechanical test to measure this large loss in modulus at the glass transition improves the probability of making a correct assignment of the glass transition temperature of the material. This sensitivity can be easily compared with a heat capacity measurement as a function of temperature from a differential scanning calorimetry where, in general, there is an increase of no more than 24 cal/mol through the glass transition temperature of the polymer from the glassy state to the melt [3].

As is well known, the glass transition temperature of polymers is not constant but changes with time. In a dynamic mechanical test, the glass transition is measured in seconds or minutes. If the measurements are done more rapidly so that the timescale is shortened, then the apparent T_g is raised. If the timescale is extended to hours or days, then the apparent T_g is lowered [4,5].

In this paper, the effects of heating rate and frequency on the dynamic elastic storage modulus (E'), the dynamic elastic loss modulus (E''), and tanδ (E''/E') are presented for PVAC, PVC, PS, and PC as well as their glass transition temperatures.

Experimental

The poly(vinyl acetate) (M_w = 101 600), poly(vinyl chloride (M_w = 122 000), and poly(carbonate) (M_w = 31 000) samples were obtained from Scientific Polymer Products Inc. The poly(styrene) beads (M_w = 125 000 to 250 000) were obtained from Polysciences Inc. The PVAC, PS, and PC samples were compression molded as received. The PVC sample was stabilized with two parts of organotin stabilizer, dry blended for two days, and then compression molded. Compression molded sheets of 2.54 cm by 2.54 cm (1 in. by 1 in.) and 3.175-mm-thick (0.125-in.) were prepared and specimens cut from them for dynamic mechanical testing.

The Rheometrics Solids Analyzer RSA-II was used for the dynamic mechanical tests (Rheometrics Inc., NJ). With this instrument samples are subjected to precise tension, compression, or bending and the resultant force is detected to provide viscoelastic data about the material. The deformation is controlled by a microprocessor. The instrument reports the storage (E'), loss (E''), the complex (E^*) moduli and the tanδ values of the sample. The frequency range of the instrument is from 0.0016 to 16 Hz (0.01 to 100 rad/s), with a temperature ramping capability from 0.1°C to 50°C/min. The dual cantilever testing geometry was employed. The effect of frequency, the temperature scanning rate, and amount of strain on the sample were examined.

Results and Discussion

Figures 1 to 4 show the dynamic mechanical parameters (E', E'', and tanδ) as a function of temperature for PVAC, PVC, PS, and PC, respectively, corresponding to a temperature step heating program of 1°C/min and a frequency of 1 Hz. PVAC showed a beta transition peak near -18.5°C and the alpha transition (glass transition) at 51.2°C determined from the maximum in tanδ. For PVAC the initial loss in the E' values started near 30.5°C and ended near 60°C for a loss of 1000 units in the E' values.

For PVC (Fig. 2) the beta transition covered a wide range of temperatures, from -85°C to approximately 7.4°C with a peak at -49°C. The glass transition temperature was measured at 87.3°C from the maximum in tanδ. The PVC showed an initial loss in the E' values at 64.1°C and ended near 100°C for a loss of 100 units on the dynamic elastic storage modulus. This relatively small loss in the E' values for PVC through its glass transition is due to the small degree of crystallinity present in PVC. This crystallinity raises the modulus values above T_g and then the modulus decreases with temperature until melting of the crystallites near 200°C.

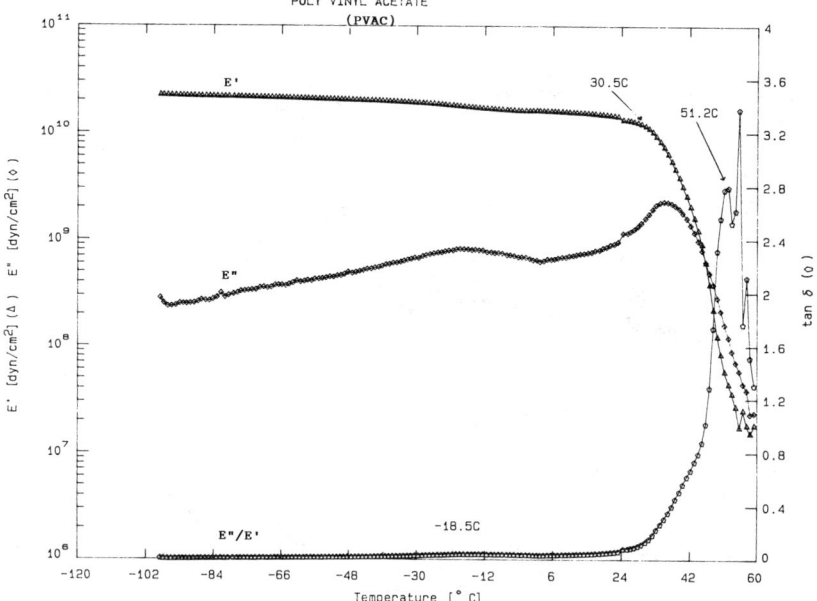

FIG. 1—*Dynamic mechanical spectrum for PVAC at 1 Hz.*

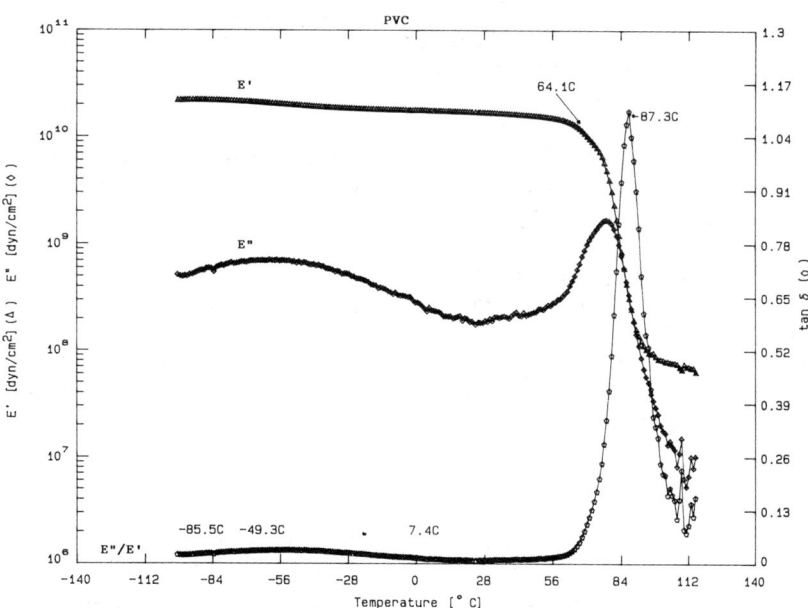

FIG. 2—*Dynamic mechanical spectrum for PVC at 1 Hz.*

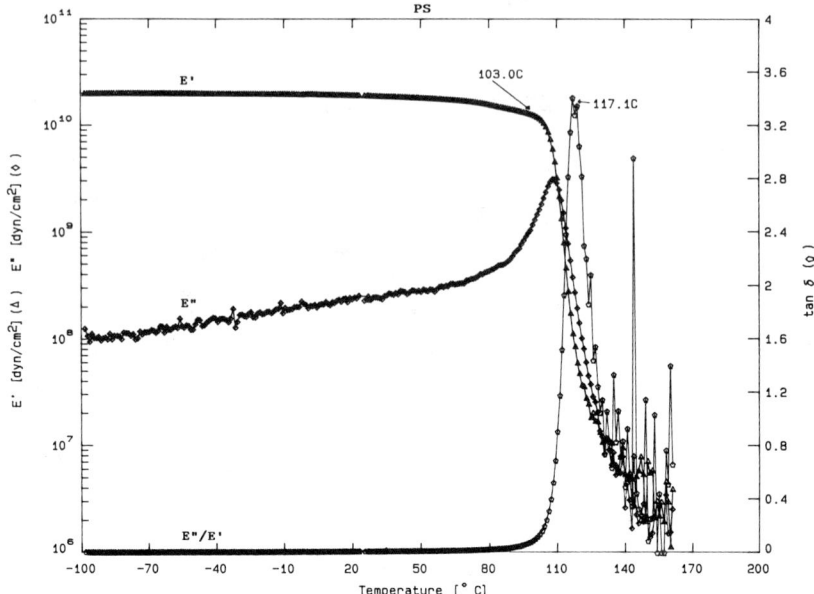

FIG. 3—*Dynamic mechanical spectrum for PS at 1 Hz.*

For PS (Fig. 3), the beta transition was very weak and detected as a small shoulder around 80 to 90°C. The glass transition was measured at 117.1°C from the maximum in tanδ. The initial loss in the dynamic storage modulus values occurred at 103°C and ended at around 140°C for a loss of 10^4 units.

For PC (Fig. 4) only half of the beta transition was detected which started well below −100.0°C and ended around −18°C. The glass transition was measured at 159.0°C from the peak in the tanδ curve. The initial loss in the dynamic elastic storage modulus occurred at 148°C and ended at near 170°C for a loss of about 1000 units in the dynamic elastic modulus.

The magnitude of the dynamic storage modulus decay through the glass transition is important because it reflects the losses in mechanical properties of the sample and influences the processing character in the viscous state.

Molecular transitions of polymers are affected by experimental variables and the techniques employed in the measurements [5–7]. The effects of frequency and heating rate on the four glassy polymers are presented next.

Effect of Heating Rate

The effect of using ramping heating rates of 1, 2, 5, 10, and 20°C/min on the dynamic elastic storage modulus is illustrated in Figs. 5 to 8 for PVAC, PVC, PS, and PC, respectively. For all these polymers the dynamic elastic storage modulus E' was shifted to a higher temperature in the glass transition region. The effect of the heating rate on the tanδ curve as a function of temperature is illustrated in Fig. 9 for PVC. In Fig. 9, changes in the shape of the curve, magnitude, and location of the tanδ peak were observed. This was also observed for PVAC and PS but a definitive trend of these changes with respect to the heating rate was not found.

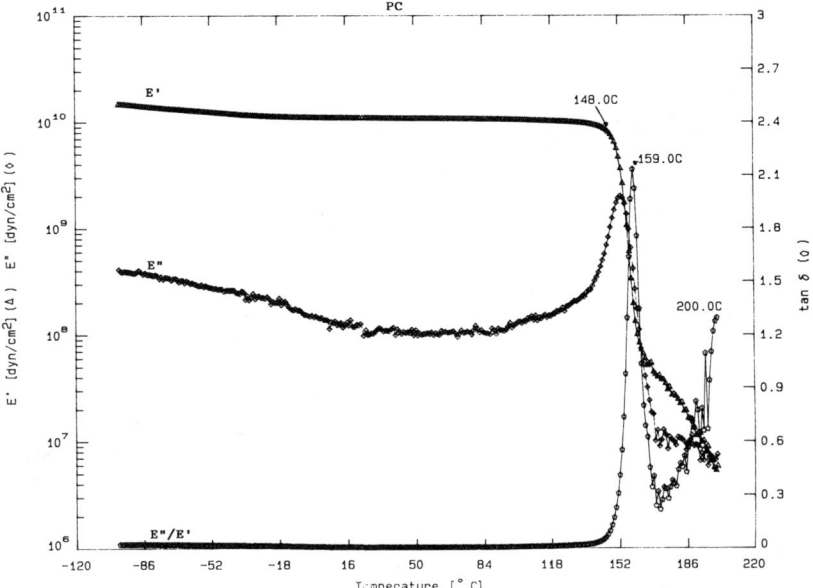

FIG. 4—*Dynamic mechanical spectrum for PC at 1 Hz.*

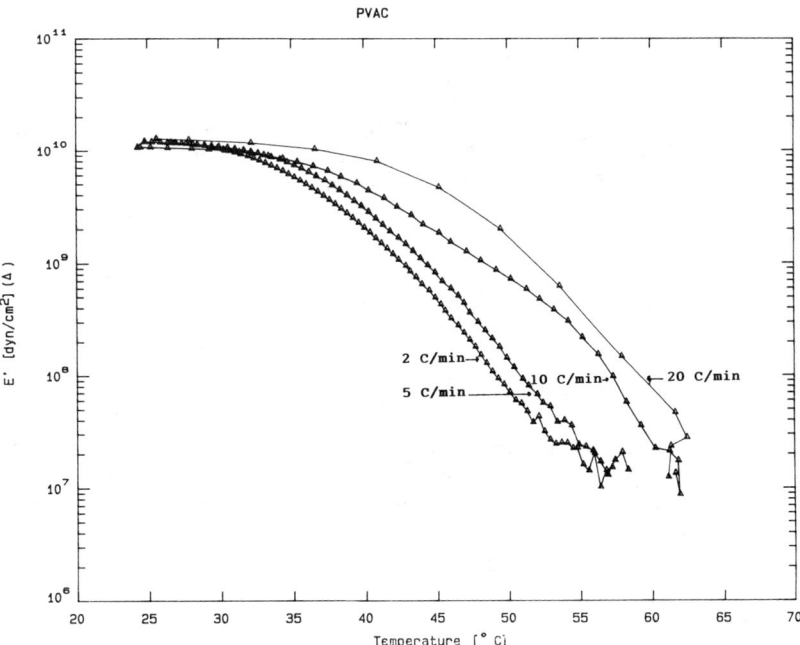

FIG. 5—*Dynamic elastic storage modulus as a function of temperature at different heating rates for PVAC.*

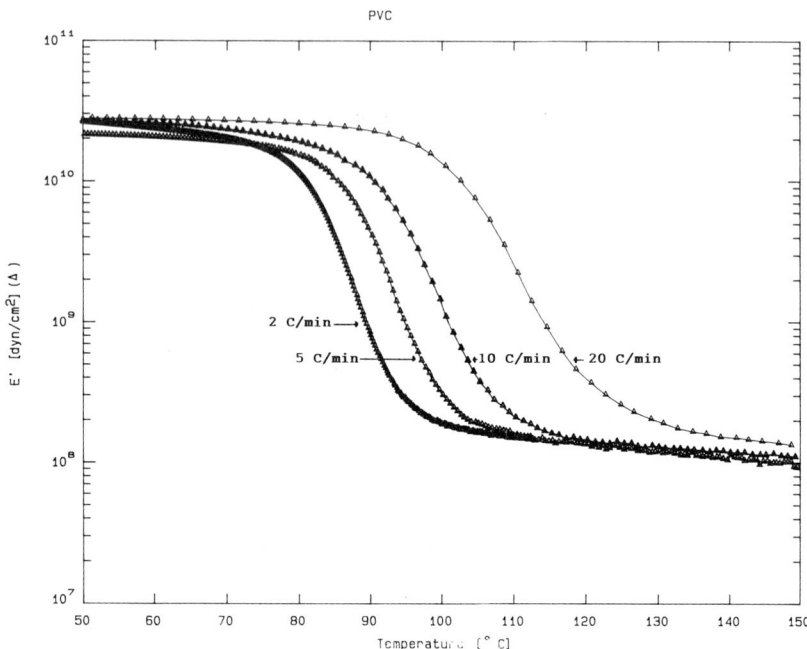

FIG. 6—*Dynamic elastic storage modulus as a function of temperature at different heating rates for PVC.*

As was illustrated in Figs. 1 through 4, the glass transition temperature covers a wide range of temperatures (30 to 40°C) for the four glassy polymers. Taking as a reference the maximum in tanδ as the glass transition temperature, the effect of the ramping heating rate is shown in Table 1. With increase of the heating rate there is an increase in the glass transition temperature for PVAC, PVC, and PS, but little change for PC.

Organic polymers are known to have low coefficients of thermal conductivity. This thermal conductivity becomes important when heating polymers at different rates and polymer samples with different dimensions. For example, a thick polymer requires more time to achieve an equilibrium temperature than a thin sample under similar heating rate. In addition, a high heating rate may result in higher thermal gradients in the polymer sample. From Table 1 the four glassy polymers behaved differently with the heating rate. The polymers with the high glass transition temperature (PS, PC) were less affected by the heating rate.

Another common heating program in a dynamic mechanical test is the step heating program where the sample is heated stepwise using a number of degrees per step along with an soak time or equilibrium time. PS was examined using this program at different °C/step, a soak time of 0.2 min, and frequency of 1 Hz. Results summarized in Table 2 show that the glass transition for PS is constant at 1, 2, and 5°C/step but increased for 10 and 20°C/step.

These results show that the dynamic mechanical parameters (E', E'', and tanδ) are a function of the heating rate for PVAC, PVC, PS, and PC and for both a ramp heating or a step driven heating program. Heating values no higher than 5°C/min or 5°C/step are recommended, with preference given to lower heating rates such as 2°C/min or 2°C/step when a better resolution is required in the glass transition temperature and for secondary and weak transitions.

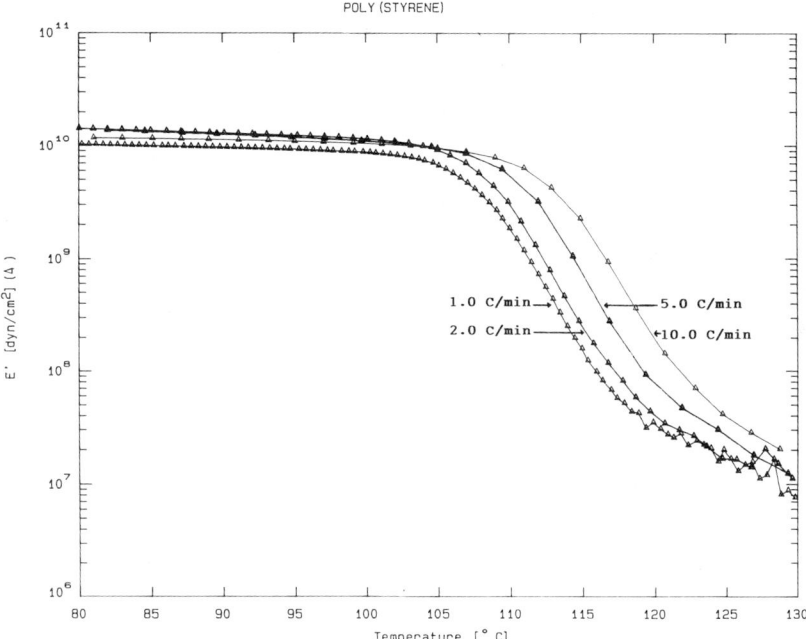

FIG. 7—*Dynamic elastic storage modulus as a function of temperature at different heating rates for PS.*

The testing geometry and the amount of strain imposed on the specimen are variables that may affect the dynamic mechanical parameters. In this study the testing geometry was maintained constant. The amount of imposed strain was varied from 0.05 to 0.5%. Results showed that there was not a significant effect on the glass transition temperature of the four polymers by varying the strain. However, for the tension or compression testing geometry the amount of strain may be important. Some studies will be done in the future to examine this effect.

Another factor that influences the dynamic properties using the dual cantilever and the flexural geometry is the surface of the specimens. It was noticed that specimens with rough surfaces and unbalanced dimensions sometimes introduce some misleading weak peak intensities in the tanδ curve that are not real.

Frequency Dependence

Figures 10 to 13 show the dynamic elastic storage modulus as a function of frequency for PVAC, PVC, PS, and PC, respectively. The E' values showed a dependence on frequency at all temperatures. From the beginning of the glass transition up to the complete loss in stiffness, the E' dependence on frequency was more accentuated. Increasing the frequency of deformation resulted in an increase in the E' values. The magnitude of the increase values in E' was different for each polymer, as shown in Figs. 10 to 13. This dependence of E' on frequency is common in viscoelastic materials and is similar to the increase in the modulus of elasticity with the strain rate.

As an example, Fig. 14 shows the tanδ curve as a function of temperature and frequency for

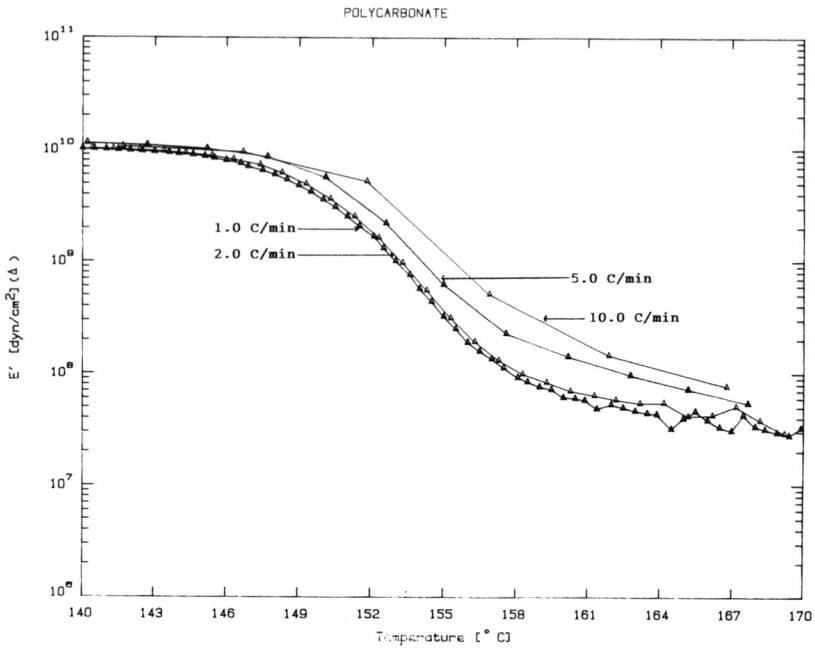

FIG. 8—*Dynamic elastic storage modulus as a function of temperature at different heating rates for PC.*

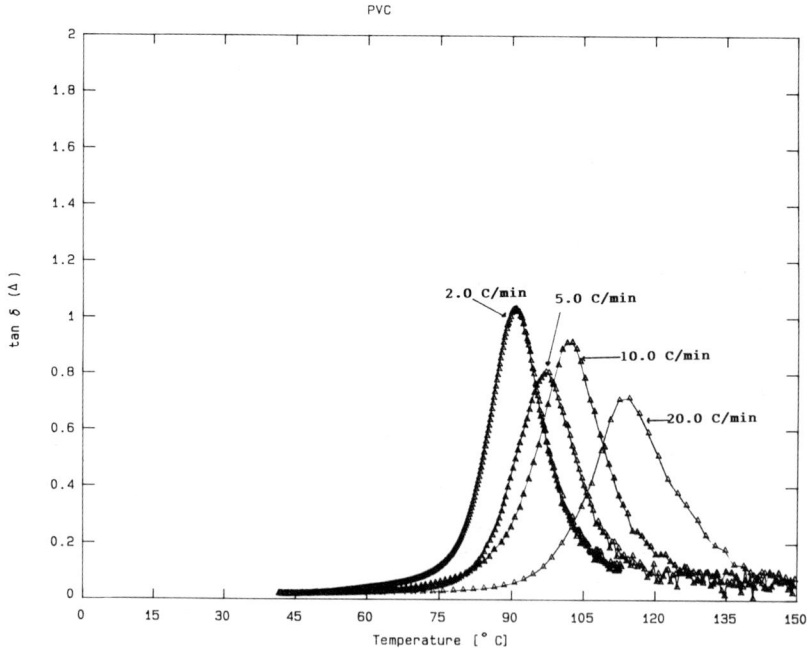

FIG. 9—*Tanδ values as a function of temperature at different heating rates for PVC.*

TABLE 1—*Effect of heating rate on the glass transition temperature.*

Ramp Heating Rate, °C/min	Glass Transition, °C			
	PVAC	PVC	PS	PC
1	—	—	117	156
2	49	91	119	156
5	52	97	119	156
10	57	103	123	157
20	60	115	126	—

TABLE 2—*Glass transitions for a temperature step heating program for poly(styrene).*

Heating Rate, °C/step	Glass Transition, °C
1	117
2	117
5	117
10	121
20	127

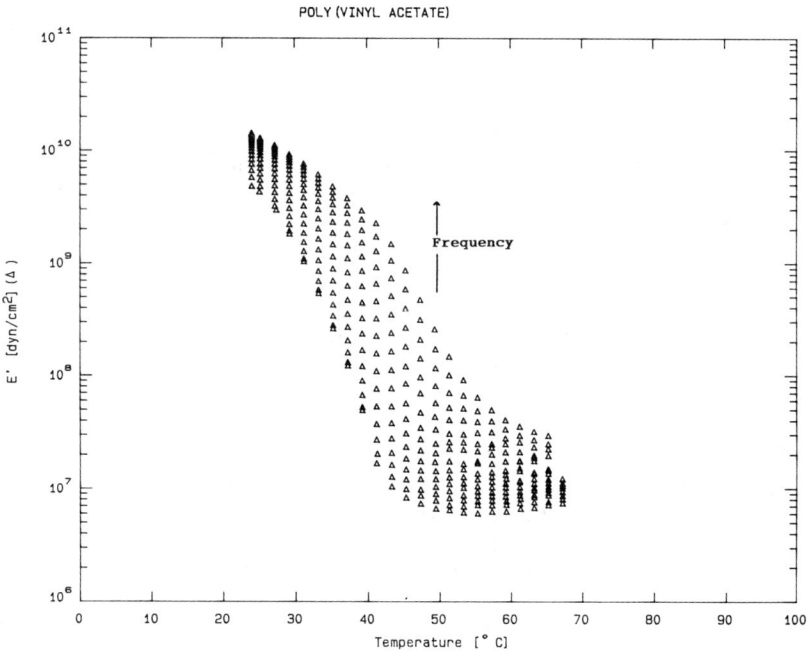

FIG. 10—*Dynamic elastic storage modulus as a function of temperature and frequency for PVAC.*

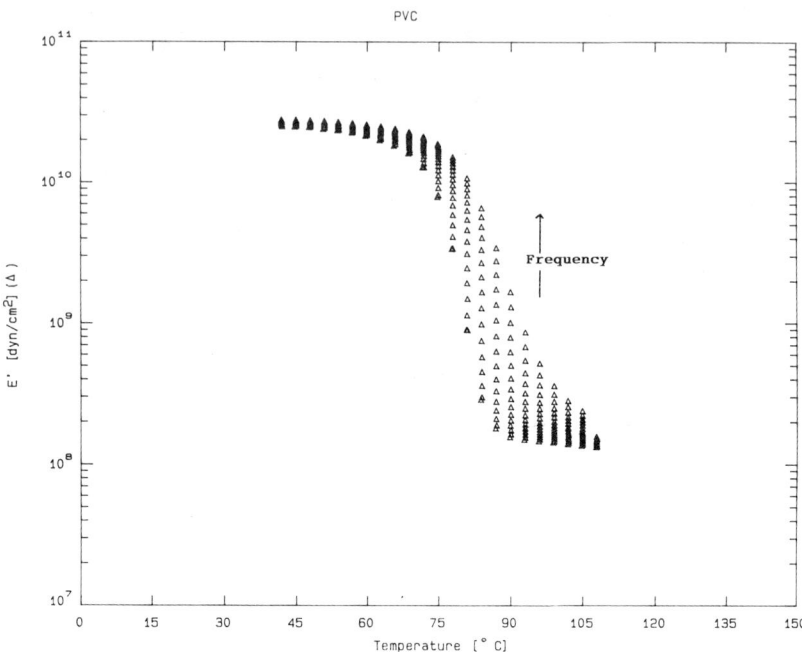

FIG. 11—*Dynamic elastic storage modulus as a function of temperature and frequency for PVC.*

PVAC. The shape of the tan δ curve is modified by frequency and the tan δ peak values or glass transition temperatures are shifted to a higher temperature with frequency. The maximum in tan δ was increased by about 5 to 7°C for a factor of 10 increase in frequency. Similar results were obtained for PVC, PS, and PC. This agreed with reported findings [5].

The tan δ peak dependence on frequency can be used to estimate the apparent activation energy for the glass transition process. This apparent activation energy defines the amount of energy required to develop molecular mobility in the polymer chains, which includes rotational carbon-carbon bonds, segmental chain bonds motion, and intermolecular separations between polymer chains (raise in the free volume). A graph of frequency versus reciprocal temperature (glass transition) is used to estimate the activation energy. Figure 15 compares the results for PVAC, PVC, PS, and PC.

The transition state theories [8] simply express that the viscoelastic behavior is related to a controlling molecular rate process with an apparent activation energy. Thus, the relationship between the frequency and the temperature at T_g is given by

$$\log f = \log f_0 - \Delta H_a/2.3\,RT$$

which states that with increasing frequency the damping maximum or T_g of a relaxation process is shifted to higher temperatures. This equation has been used by many authors to determine the activation energy of long chain motions. Using a curve fitting analysis for the curves in Fig. 15 and from the slope of the linear fit, the following apparent activation energies were determined (Table 3).

These apparent activation energy values carry an estimated error of 5 to 10% based on the glass transition temperature. They agreed with the ones reported by others [2]. For example,

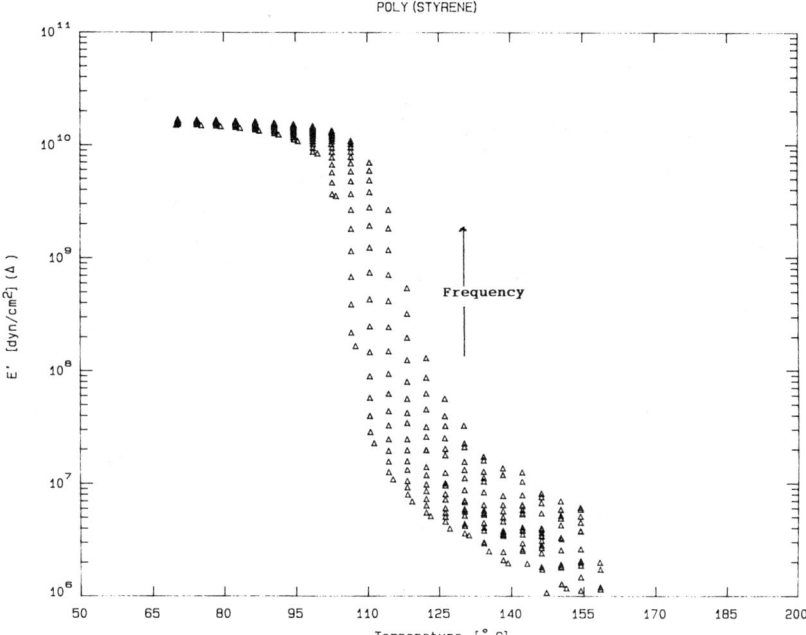

FIG. 12—*Dynamic elastic storage modulus as a function of temperature and frequency for PS.*

for PS a value of 200 Kcal/mol has been reported. For PVC the apparent activation energy reported for the alpha process is about 70.5 Kcal/mol [9]. Notice that the apparent activation energies were determined for a limited number of frequencies from 0.01 Hz to 10 Hz. The apparent energy of activation was a function of temperature, so that a universal value for the alpha or glass transition temperature cannot be given.

Finally, since the glass transition involves large segmental chain motions, the temperature at which it takes place depends on the geometry and flexibility of the polymer backbone and in the interchain molecular forces. Because many polymers have relatively high molecular weight and a wide molecular weight distribution, the low molecular weight fraction initiates the micro-Brownian motion at the beginning of the glass transition. This motion follows a molecular size with the longest polymer chains vibrating last. Thus, in a dynamic mechanical test, the shortest polymer chains will start the Brownian motion. This corresponds to the initial loss in elastic modulus. This is followed by Brownian motions corresponding to different polymer chains sizes until the largest chains are softened, which may correspond to the maximum in the tanδ curve. The rate at which this process takes place is a function of the heating rate and frequency of deformation.

Conclusions

The dynamic mechanical parameters for PVAC, PVC, PS, and PC were found to be functions of the heating rate and the type of heating program employed. Increasing the heating rate resulted in a shift of the dynamic elastic modulus to higher temperature near the glass transition and an increase in T_g to a higher value. The amount of strain imposed on the

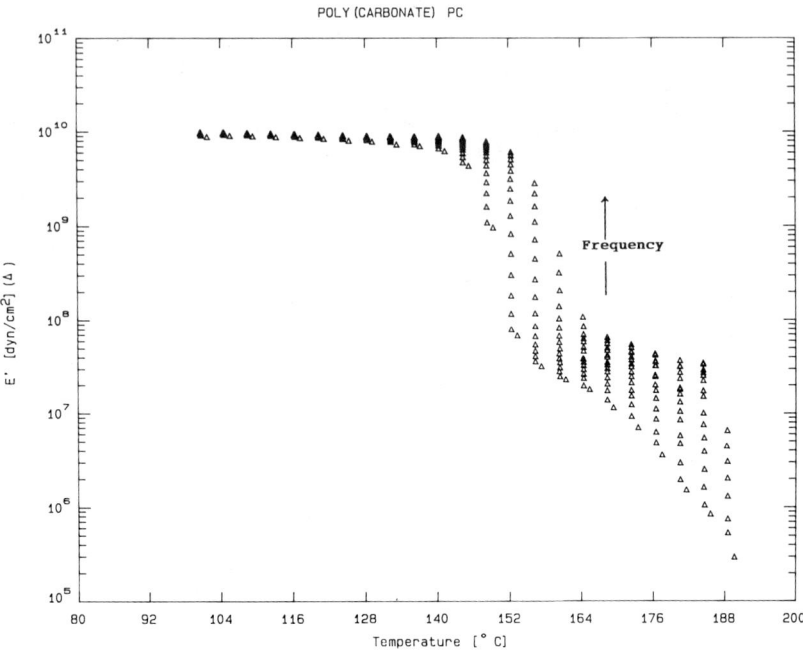

FIG. 13—*Dynamic elastic storage modulus as a function of temperature and frequency for PC.*

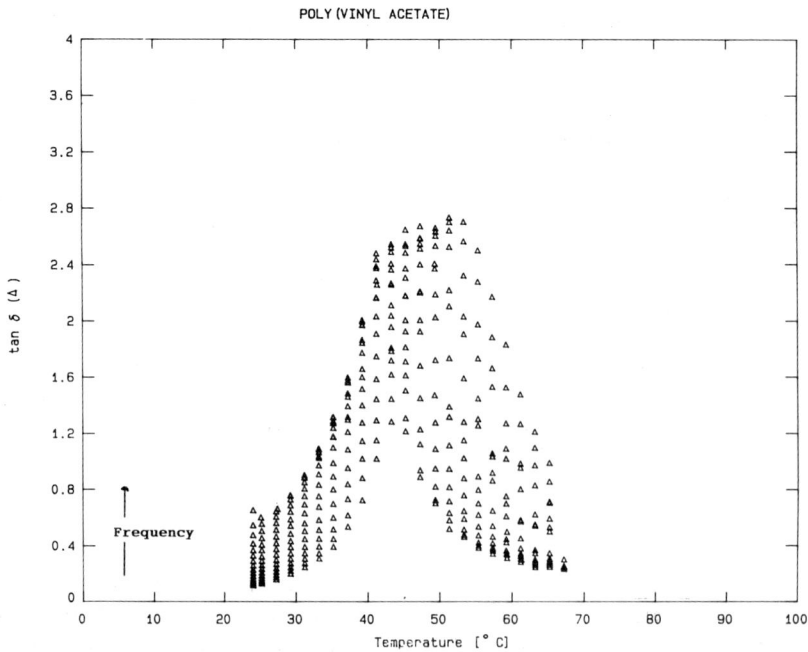

FIG. 14—*Tanδ values as a function of temperature and frequency for PVAC.*

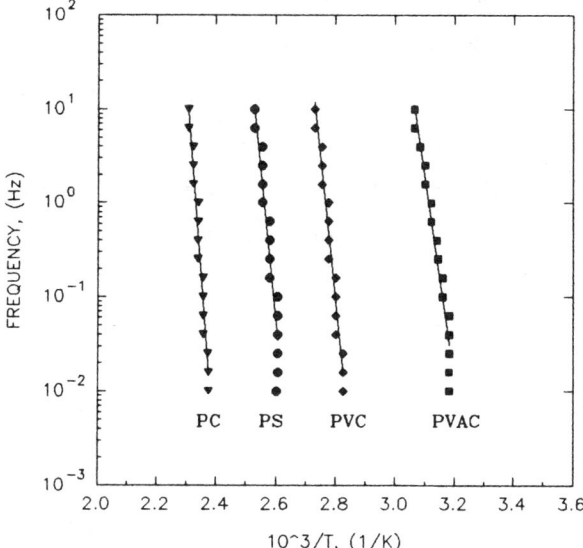

FIG. 15—*Activation diagram for PVAC, PVC, PS and PC.*

TABLE 3—*Apparent activation energies.*

Polymer	Apparent Activation Energy, Kcal/mol
PVAC	98
PVC	136
PS	194
PC	146

sample was found to have a small effect on the dynamic parameters for the dual cantilever testing geometry. However, this effect may be different for other testing geometries which await further experimentation. The effect of the frequency on the four polymers was found to be important. Both the dynamic elastic modulus and the tanδ peak are shifted to higher temperatures with increasing frequency. This behavior allowed the estimation of an apparent activation energy of the glass transition. All these findings are in accordance with the dynamic character of the glass transition temperature of polymeric materials.

Acknowledgment

This paper is the result of an independent study conducted by the author. My thanks to the BFGoodrich company for permission to publish and conduct this work.

References

[1] Ferry, J. D., *Viscoelastic Properties of Polymers,* 3rd ed., John Wiley and Sons Inc., New York, 1980.
[2] McCrum, N. G., Read, B. E., and Williams, G., *Anelastic and Dielectric Effects in Polymeric Solids,* Dover Publications Inc., New York, 1991.
[3] Wunderlich, B., *Thermal Analysis,* Academic Press Inc., San Diego, CA, 1990.

[4] Sperling, L. H., *Introduction to Physical Polymer Science*, John Wiley and Sons Inc., New York, 1986, p. 265.
[5] Nielsen, L. E., *Mechanical Properties of Polymers and Composites*, Vol. 1, Marcel Dekker Inc., New York, 1974.
[6] Ward, I. M., *Mechanical Properties of Solids Polymers*, John Wiley and Sons Inc., New York, 1983, ch. 6.
[7] *Thermal Characterization of Polymeric Materials*, E. A. Turi, Ed., Academic Press Inc., New York, 1981, ch. 1 and 2.
[8] Ward, I. M., *Mechanical Properties of Solids Polymers*, John Wiley and Sons Inc., New York, 1983, ch. 7.
[9] Bauwens-Crowet, C., Bauwens, J. C., and Homes, G., *Journal of Polymer Science*, A-2, Vol. 7, 1969, p. 735.

Ralph M. Paroli[1] and James Penn[1]

Measuring the Glass Transition Temperature of EPDM Roofing Materials: Comparison of DMA, TMA, and DSC Techniques

REFERENCE: Paroli, R. M. and Penn, J., "**Measuring the Glass Transition Temperature of EPDM Roofing Materials: Comparison of DMA, TMA, and DSC Techniques,**" *Assignment of the Glass Transition, ASTM STP 1249*, R. J. Seyler, Ed., American Society for Testing and Materials, Philadelphia, 1994, pp. 269–276.

ABSTRACT: Two ethylene-propylene-diene monomer (EPDM) roofing membranes were aged at 100°C for 7 and 28 days. The T_g of these membranes was then determined by dynamic mechanical analysis (DMA), thermomechanical analysis (TMA), and differential scanning calorimetry (DSC) and the results compared. It was found that: (1) T_g data can be obtained easily using the DMA and TMA techniques. The DSC method requires greater care due to the broad step change in the baseline which is associated with heavily plasticized materials. (2) The closest correspondence between techniques was for TMA and DSC (half-height). The latter, within experimental error, yielded the same glass transition temperature before and after heat-aging. (3) The peak maxima associated with tanδ and E'' measurements should be cited with T_g values as significant differences can exist. (4) The $T_g(E'')$ values were closer to the T_g(TMA) and T_g(DSC) data than were the T_g(tanδ) values. Data obtained at 1 Hz (or possibly less) should be used when making comparisons based on various techniques. An assessment of T_g values indicated that EPDM 112 roofing membrane is more stable than the EPDM 111 membrane. The T_g for EPDM 112 did not change significantly with heat-aging for 28 days at 130°C.

KEYWORDS: ethylene-propylene-diene monomer (EPDM), glass transition temperature (T_g), dynamic mechanical analysis (DMA), thermomechanical analysis (TMA), differential scanning calorimetry (DSC), elastomers, roofing membranes, thermal analysis

In the past, roofing membranes were mostly asphalt-based. Today, however, the choice of roofing materials is quite varied, ranging from asphalt-based or modified-asphalt to polymer-based materials such as polyvinyl chloride (PVC) and ethylene-propylene-diene monomer (EPDM). This variety has forced the international roofing industry to shift towards a multidisciplinary approach in solving roofing problems, i.e., both engineering and chemical principles are used. In 1988, a joint CIB/RILEM international roofing committee was established to investigate the applications of thermal analysis in the characterization of roofing membranes [1]. Thermal analysis, used extensively in characterizing polymeric materials, has not been widely used in the roofing industry [2–13]. Thermoanalytical techniques can be used to monitor a wide array of material characteristics. Some of the

---

[1]Research officer, Institute for Research in Construction, National Research Council of Canada, Ottawa, Ontario, Canada K1A OR6.
[2]Applications chemist, Thermal Analysis Products, Seiko Instruments USA Inc., Elk Grove Village, IL 60007.

applications include enthalpy, weight-loss, thermal stability, coefficient of thermal expansion, and the glass transition temperature.

Determination of chemical and mechanical characteristics of a material facilitate selection, application, and end-use. One important characteristic for the cold temperature performance of roofing membranes is its glass transition temperature (T_g). T_g is a property of an amorphous polymer and is defined as an assigned temperature at which the polymer changes from a rubbery to a glassy state without any change in phase, i.e., local molecular motion is hindered. Below T_g the material will be rigid and hard. Yet, above T_g the material will exhibit rubber-like characteristics such as pliability. Other properties that vary with T_g are thermal expansion coefficient, heat capacity, and dielectric constant [14–18]. Generally, the strength of polymeric materials above the glass transition temperature is lower than the strength below T_g.

Long-range translational motion of all the molecules is considered to be restrained or frozen below T_g [19]. In the case of noncrosslinked or slightly crosslinked polymers, the glass transition temperature depends on five factors: free volume, attractive forces between molecules, rotation about bonds, stiffness of chain, and chain length. Although different relationships exist between the glass-transition temperature and the melting point (T_m), T_g is always lower than T_m.

There are three thermoanalytical techniques that are commonly used to determine and monitor the changes in T_g. The traditional methods are static techniques such as differential scanning calorimetry (DSC). The drawback with DSC is that T_g is a pseudo-second-order transition and is detected by a step change in baseline (arising from a change in heat capacity). In the case of heavily plasticized materials (e.g., roofing samples) a broad transition is obtained and the step change is difficult to detect. The mechanical methods within thermal analysis include dynamic methods such as dynamic mechanical analysis (DMA) and static load techniques such as thermomechanical analysis (TMA). These techniques provide a greater change in signal arising from the T_g and therefore are easier to use for roofing materials. In this paper, a brief theory of these three thermoanalytical techniques is presented. In addition, the T_g data obtained by the various techniques on EPDM roofing materials are presented and compared.

Theory

DMA [20–24]

The DMA technique measures the stress-strain relationship for a viscoelastic material. A real and imaginary component of modulus can be obtained by resolving the stress-strain components

$$\sigma = \epsilon_0 E' \sin(\omega t) + \epsilon_0 E'' \cos(\omega t)$$

The storage modulus, E', is defined as $E' = (\sigma_0/\epsilon_0) \cos \delta$ and is a measure of recoverable strain energy in a deformed body. The loss modulus, E'', is associated with the dissipation of energy as heat due to the deformation of the material and is defined as $E'' = (\sigma_0/\epsilon_0) \sin \delta$. The ratio E''/E' yields the loss tangent or damping factor ($\tan\delta$) which is the ratio of energy lost per cycle to the maximum energy stored and therefore recovered, per cycle.

A typical dynamic mechanical analysis curve shows either E', E'', or $\tan\delta$ plotted as a function of time or temperature. In general, the most intense peak observed for either E'' or $\tan\delta$ in conjunction with a relatively pronounced drop in E' corresponds to the glass transition. However, to prove that this relaxation event is a T_g, a DMA multiplexing experiment to establish the activation energy would be required. Alternatively, other techniques such as nuclear magnetic resonance (NMR) could be used. Theoretically, the T_g is associated with large-scale molecular motion of the polymer. The T_g event may be affected by the crosslink density or degree of crystallinity and is directly related to the amorphous region within a polymer.

Care should be taken when reporting the glass transition temperature obtained by DMA. The transition temperature determined by DMA (or other dynamic techniques) is not only heating-rate dependent but also frequency dependent. In addition to heating rate and frequency, the mechanical/rheological property (E', E'', or $\tan\delta$) used to determine the T_g must also be specified. It has been found that the E'' peak maximum at 1 Hz corresponded closely with the T_g obtained from volume-temperature measurements [20].

TMA [21,25]

TMA is defined as a method for measuring the deformation of a material under a constant load as a function of temperature while the material is under a controlled temperature program (see ASTM E 473, Terminology Relating to Thermal Analysis). Essentially, the expansion or contraction of a sample with an applied force is measured as a function of time or temperature. The plot of this dimensional change versus temperature (or time) can then be used to obtain important sample parameters such as T_g, the coefficient of thermal expansion (CTE), softening temperature, and Young's modulus.

The measuring system consists of a linear voltage differential transformer (LVDT) connected to the appropriate probe. Any displacement of the probe generates a voltage which is then recorded. Various probes are available for the TMA and it is the measuring mode that generally dictates the probe to be used. The measurements can be done in either compression, expansion, penetration, flexural, or tensile mode. It is this variety of probes that allows for the measurement of a wide range of properties on samples of different configurations. The sample configuration ranges from films to fibers to powders to disks. Materials that can be studied by this technique include ceramics, polymers, and composites.

The change in linear dimension as a function of temperature can be described by the following

$$L_2 = L_1 \left(1 + \int_{T_1}^{T_2} \alpha_l \, dT\right)$$

where α_l is the coefficient of linear expansion, and L_1 and L_2 are the lengths of the specimen at temperatures (or time) T_1 and T_2, respectively. If the difference between T_2 and T_1 is relatively small, then the equation can be represented by

$$L_2 - L_1 = L_1 \alpha_l (T_2 - T_1)$$

or it can be rewritten as

$$\alpha_l = \frac{1}{L_1} \frac{\Delta L}{\Delta T}$$

Therefore, the slope of the curve of length versus temperature yields $\alpha_l L_1$ and the coefficient of linear thermal expansion is obtained by dividing by L_1.

There are some drawbacks with TMA. Proper calibration is required to obtain reliable and

DSC

DSC is a simple and yet powerful thermoanalytical technique. It is widely used in providing valuable information on chemical and physical properties of materials. The DSC technique measures the amount of energy (or heat) absorbed or released as the material is heated, cooled, or held at an isothermal temperature. The plot obtained is known as a DSC thermal curve and shows the amount of heat evolved or absorbed as a function of temperature or time. This technique yields thermodynamic data such as enthalpy and specific heat, as well as kinetic data.

The shape and appearance of DSC curves can give a clue as to the type of transition taking place. Generally, distinct peaks are obtained by the DSC technique. These correspond to first-order transitions such as melting or solid-solid phase transformations. These peaks can be integrated and a value for enthalpy (ΔH) can be determined. In the case of second-order transitions such as the glass transition, a stepwise increase in heat capacity is observed. This is also detected in the DSC by a step change in baseline slope.

Calibration of a DSC instrument is needed to determine the T_g with accuracy. Normally, the melting points of pure standards are used. Generally, it is best to use a standard that will have a transition in the same temperature range as the samples being studied.

The glass transition temperature may be determined by taking the middle of the change in baseline (half-height method). This method requires establishing a tangent line. This step is facilitated by using the first derivative of the heat-flow signal. T_g values obtained by DSC are generally different from those obtained by dynamic techniques such as DMA since DSC is a static technique.

Experimental

Two commercially available, nonreinforced EPDM roofing membrane samples (111 and 112) were obtained. One specimen was used as a control. Another specimen was placed in an air-circulating oven preheated to 130°C and heat-aged for 7 and 28 days.

The Seiko DMS110 was used to characterize the rheological properties of the specimens. The instrument was operated in the 8 mm dual cantilever mode of deformation. The sample size was 30-mm-long by 3-mm-wide with a thickness of 1 mm. The temperature range was from −160°C to 100°C and a heating/cooling rate of 1°C per min was used. The parameters also included a constant amplitude of deformation of 30 μm at frequencies of 1, 2, 5, 10, and 20 Hz.

A small specimen (1 mm by 7 mm by 7 mm) was placed upon a quartz platform inside a Seiko TMA/SS120C thermomechanical analyzer. The quartz expansion probe was positioned onto the sample under a compressive force of 1 g. The analysis was recorded from −100°C to 20°C at a rate of 2°C per min under static air environment. The temperature calibration of the TMA/SS was achieved by using the melting points of indium and mercury.

The DSC experiment (using a Seiko DSC210 module) was conducted by placing 15 to 20 mg of roofing material in an aluminum pan and a corresponding amount of inert material in a reference pan to counterbalance the specific heat capacity of the sample. The sample was tested between −150°C and 100°C at a cooling/heating rate of 20°C per min under a static air environment. The temperature calibration of the instrument was achieved by using the melting points of indium and tin.

Results and Discussion

Three rheological properties are recorded in a DMA experiment, storage modulus (E'), loss modulus (E''), and tanδ (E''/E'). The glass transition temperature was determined using the E'' and tanδ peak maxima. The two methods yield the same trend for both samples, i.e., the T_g shifts to higher temperatures with heat-aging. Furthermore, an increase of the E'' values above T_g in the rubber plateau region is also noted. The DMA data are summarized in Tables 1 to 4.

As expected, the T_g values shifted to higher temperatures as a function of heat-aging time and increased frequency. The EPDM 111 (as received) has a T_g (E'') at $-64.6°C$ at 1 Hz while at 20 Hz it is shifted to $-58.4°C$. The T_g (E'') was found to be $-48.2°C$ at 1 Hz and

TABLE 1—T_g determined by the DMA E'' peak maximum for EPDM 111.

	Frequency, Hz				
	1	2	5	10	20
As-received	−64.6	−63.3	−61.8	−60.7	−58.4
7 days	−50.5	−49.5	−47.3	−45.6	−44.5
28 days	−48.2	−46.4	−44.6	−42.8	−41.6

TABLE 2—T_g (in °C) determined by the DMA E'' peak maximum for EPDM 112.

	Frequency, Hz				
	1	2	5	10	20
As-received	−43.7	−42.8	−41.3	−40.0	−38.3
7 days	−43.1	−41.5	−40.2	−38.9	−37.3
28 days	−42.9	−41.3	−39.9	−38.5	−36.9

TABLE 3—T_g (in °C) determined by the DMA tanδ peak maximum for EPDM 111.

	Frequency, Hz				
	1	2	5	10	20
As-received	−52.9	−50.2	−48.0	−47.6	−45.4
7 days	−42.7	−41.1	−40.0	−38.4	−36.8
28 days	−40.5	−39.5	−37.3	−36.2	−34.9

TABLE 4—T_g (in °C) determined by the DMA tanδ peak maximum for EPDM 112.

	Frequency, Hz				
	1	2	5	10	20
As-received	−34.9	−33.8	−32.2	−30.8	−29.2
7 days	−32.8	−31.0	−29.1	−27.2	−25.6
28 days	−32.6	−30.9	−28.9	−27.0	−25.3

−41.6°C at 20 Hz, after oven aging for 28 days at 130°C. This is a significant difference since the ΔT_g is ~16°C. Sample EPDM 112 underwent much smaller changes (due to heat-aging) compared to EPDM 111. The T_g (E'') of the as-received specimen was −43.7°C at 1 Hz and −38.3°C at 20 Hz. The glass transition temperature increased to −42.9°C and −36.9°C for 1 Hz and 20 Hz, respectively, after 28 days at 130°C. It can be seen that EPDM 112 undergoes smaller changes than EPDM 111. Similar observations were noted for the T_g (tanδ) except that the peak maximum occurred at much higher temperatures. For example, the tanδ glass transition temperature for the as-received specimen EPDM 111 was determined to be −52.9°C at 1 Hz and −45.4°C at 20 Hz. This is approximately 12°C higher than the equivalent $T_g(E'')$.

The TMA results, summarized in Table 5, follow a similar trend. The T_g for the as-received specimen EPDM 111 was −66.2°C and −52.8°C after 28 days of heat-aging. In contrast, the glass transition temperature of EPDM 112 was −54.6°C and −52.4°C for the unaged and 28 day heat-aged specimens, respectively. These results were compared to the DMA data taken at 1 Hz and the following was observed. Firstly, the T_g for EPDM 111 shifts by 16.4°C (DMA E''), 12.4°C (DMA tanδ) and 13.4°C (TMA) after 28 days of heat-aging. Secondly, the $T_g(E'')$ for EPDM 111 was much closer to the T_g determined by TMA than that determined by the DMA-tanδ peak maximum. The results were −64.6°C (DMA E''), −52.9°C (DMA tanδ), and −66.2°C (TMA) for the as received specimens. After 28 days of aging, the difference in values obtained with the techniques was much greater −48.2°C (DMA E''), −40.5°C (DMA tanδ), and −52.8°C (TMA) but the T_g(TMA) is still closer to the DMA $T_g(E'')$ than the DMA tanδ glass transition temperature. Similar results were obtained for the EPDM 112 sample except difference in T_g values between the techniques was much greater (by as much as 20°C between the TMA and DMA-tanδ T_g). The coefficient of linear thermal expansion (CTE) above the T_g decreased with increased aging. The T_g for EPDM 112 did not shift significantly with aging (the values also match those obtained by DSC); however, the CTE values did decrease with increased heat-aging.

The DSC results are summarized in Table 6. The average onset of the T_g of the as-received EPDM 111 specimen was −78.5°C; the average T_g determined by the half-height

TABLE 5—*TMA data.*

	EPDM 111		EPDM 112	
	T_g(°C)	CTE Above T_g ($\times 10^{-6}$ per °C)	T_g(°C)	CTE Above T_g ($\times 10^{-6}$ per °C)
As-received	−66.2	342	−54.6	319
7 days	−59.2	235	−53.4	300
28 days	−52.8	202	−52.4	284

TABLE 6—*DSC data.*

	EPDM 111		EPDM 112	
	Onset of T_g (°C)	T_g (°C)	Onset of T_g (°C)	T_g (°C)
As-received	−78.5	−66.0	−61.6	−54.0
7 days	−70.4	−58.9	−60.8	−53.5
28 days	−66.6	−52.8	−60.6	−52.9

technique was $-66.0°C$. Heat-aging for 7 and 28 days at 130°C caused the T_g to shift to higher temperatures as was the case for the previous techniques. The DSC technique also confirmed that for the EPDM 112 sample the T_g remains relatively constant with heat-aging. It was observed that the T_g determined by TMA matched the T_g measured by DSC using the half-height technique. The average T_g(TMA) for the as-received EPDM 111 specimen was $-66.2°C$ while the average T_g (DSC half-height method) was $-66.0°C$. A similar relationship was found for the EPDM 112 sample. Although the T_g(DSC) results matched those obtained by TMA, they were more difficult to determine. This was due to the broad step change in the baseline. This broad change is typical of heavily plasticized materials such as roofing membranes.

Conclusions

The following conclusions can be drawn from the results obtained by DMA, TMA, and DSC:

- The T_g data for the EPDM roofing materials used in this study were obtained more easily using the DMA and TMA techniques. In this case, due to the high amounts of plasticizer, the DSC method required greater care. The step change in baseline for the EPDM material was broad and weak (due to plasticizer) which created some difficulty in detecting the T_g.
- The peak maxima used to determine T_g must be reported since significant differences can exist between $\tan\delta$ and E'' peak maxima. Furthermore, if both peak maxima were recorded then they should be reported.
- The closest correspondence of T_g data was for TMA and DSC (half-height) techniques. Both techniques, within experimental error, yielded the same glass transition temperature before and after heat-aging.
- The $T_g(E'')$ values from DMA were closer to the T_g(TMA) and T_g(DSC) data than were the $T_g(\tan\delta)$ values. It is suggested that, where applicable, T_g data obtained from dynamic measurements be measured at 1 Hz (or possibly less), thus allowing the T_g to be compared more readily with T_g obtained from nondynamic techniques.
- T_g can be used to monitor changes in EPDM roofing membranes. In this case, EPDM 112 roofing membrane was more stable than the EPDM 111 membrane. The T_g for EPDM 112 did not change significantly with heat-aging for 28 days at 130°C.

Acknowledgments

The authors are grateful to Ms. Ana H. Delgado, Mr. Ryoichi Kinoshita, and Mr. Nobuaki Okubo for experimental assistance.

References

[1] "Performance Testing of Roofing Membrane Materials," Recommendations of the Conseil International du Batiment pour la Recherche l'Etude et la Documentation (CIB) W.83 and Réunion Internationale des Laboratoires d'Essai et de Recherche sur les Matériaux et les Constructions (RILEM) 75-SLR Joint Committee on Elastomeric, Thermoplastic and Modified Bitumen Roofing, RILEM, Paris, France, November 1988.
[2] Cash, C. G., "Thermal Evaluation of One-Ply Sheet Roofing," *Single Ply Roofing Technology, ASTM STP 790*, 1982, pp. 55–64.
[3] Backenstow, D. and Flueler, P., "Thermal Analysis for Characterization," *Proceedings, 9th Conference on Roofing Technology*, National Roofing Contractors Association, Rosemont, IL, April 1987, pp. 54–68.
[4] Gaddy, G. D., Rossiter, W. J. Jr., and Eby, R. K., "The Application of Thermal Analysis to the

Characterization of EPDM Roofing Membrane Materials," *Roofing Research and Standards Development (2nd Volume), ASTM STP 1088*, American Society for Testing and Materials, Philadelphia, 1990, pp. 37–52.

[5] Dutt, O., Paroli, R. M., Mailvaganam, N. P., and Turenne, R. G., "Glass-Transitions in Polymeric Roofing Membranes," *Proceedings, 1991 International Symposium on Roofing Technology*, F. Kocich, Ed., 1991, pp. 495–501.

[6] Paroli, R. M. and Dutt, O., "Dynamic Mechanical Analysis Studies of Reinforced Polyvinyl Chloride (PVC) Roofing Membranes," *Polymeric Materials Science and Engineering*, Vol. 65, 1991, pp. 362–363.

[7] Paroli, R. M., Dutt, O., Delgado, A. H., and Mech, M., "The Characterization of EPDM Roofing Membranes by Thermogravimetry and Dynamic Mechanical Analysis," *Thermochimica Acta*, Vol. 182, 1991, pp. 303–317.

[8] Gaddy, G. D., Rossiter, W. J. Jr., and Eby, R. K., "The Application of Thermal Analysis to the Characterization of EPDM Roofing Membrane Materials After Exposure in Service," *Proceedings, 1991 International Symposium on Roofing Technology*, F. Kocich, Ed., 1991, pp. 502–507.

[9] Gaddy, G. D., Rossiter, W. J., Jr., and Eby, R. K., "The Use of TMA to Characterize EPDM Roofing Membrane Materials," *Materials Characterization by Thermomechanical Analysis, ASTM STP 1136*, A. T. Riga and C. M. Neag, Eds., American Society for Testing and Materials, Philadelphia, 1991, pp. 168–175.

[10] Paroli, R. M., Dutt, O., Delgado, A. H., and Stenman, H. K., "Ranking Polyvinyl Chloride (PVC) Roofing Membranes Using Thermal Analysis," *Journal of Materials in Civil Engineering*, Vol. 5, 1993, pp. 83–95.

[11] Delgado, A. H., Paroli, R. M., and Dutt, O., "Thermal Analysis in the Roofing Area: Applications and Correlation with Mechanical Properties," CIB World Building Congress, Montreal, Canada, 18–22 May 1992.

[12] Paroli, R. M., Dutt, O., and Delgado, A. H., "The Characterization of Roofing Membranes Using Thermal Analysis and Mechanical Testing," CIB World Building Congress, Montreal, Canada, 18–22 May 1992.

[13] Penn J. J. and Paroli, R. M., "Evaluating the Effects of Aging on the Thermal Properties of EPDM Roofing Materials," 21st North American Thermal Analysis Society (NATAS) Conference Atlanta, GA, 13–18 September 1992, pp. 612–617.

[14] Feldman, D., *Polymeric Building Materials*, Elsevier Science Publishers Ltd., New York, 1989.

[15] Billmeyer, F. W., *Textbook of Polymer Science*, 3rd Ed., John Wiley and Sons, New York, 1984.

[16] Young, R. J., *Introduction to Polymers*, Chapman and Hall, New York, 1981.

[17] Bikales, N. M., *Mechanical Properties of Polymers*, John Wiley and Sons, New York, 1971.

[18] Flory, P. J., *Principles of Polymer Chemistry*, Cornell University Press, 1967.

[19] Rosen, S. L., *Fundamental Principles of Polymeric Materials*, Barnes and Noble, Inc., 1971.

[20] Murayama, T., *Dynamic Mechanical Analysis of Polymeric Material*, Elsevier Scientific Publishing Company, New York, 1978.

[21] Wendlandt, W. W., *Chemical Analysis 19: Thermal Analysis*, 3rd Edition, John Wiley and Sons, 1986.

[22] Campbell, D. and White, J. R., *Polymer Characterization: Physical Techniques*, Chapman and Hall, New York, 1989.

[23] Crompton, T. R., *Analysis of Polymers: An Introduction*, Pergamon Press, 1989.

[24] Skrovanek, D. J. and Schoff, C. K., *Progress in Organic Coatings*, Vol. 16, 1988, pp. 135–163.

[25] *Materials Characterization by Thermomechanical Analysis, ASTM STP 1136*, A. T. Riga and C. M. Neag, Eds., American Society for Testing and Materials, Philadelphia, 1991.

Joanna L. Jankowsky,[1] Denise G. Wong,[1] Michael F. DiBerardino,[1] and Roland C. Cochran[1]

Evaluation of Upper Use Temperature of Toughened Epoxy Composites

REFERENCE: Jankowsky, J. L., Wong, D. G., DiBerardino, M. F., and Cochran, R. C., "**Evaluation of Upper Use Temperature of Toughened Epoxy Composites,**" *Assignment of the Glass Transition, ASTM STP 1249*, R. J. Seyler, Ed., American Society for Testing and Materials, Philadelphia, 1994, pp. 277–292.

ABSTRACT: Traditionally, the glass transition temperature (T_g), as determined by thermal analysis, has been used to calculate the material use threshold. The main drawback to the use of T_g as a measure of upper use temperatures is that it results in use temperatures that are significantly higher than the use temperatures developed through mechanical testing. This study compared operating limits obtained from mechanical tests to glass transition temperatures obtained by several methods of thermal analysis.

A toughened epoxy composite, IM7/977-3 from ICI Fiberite, was tested in this study, along with the corresponding neat resin. The mechanical tests performed were interlaminar shear, resin compression, and material operating limit (MOL) testing (a modified open-hole compression test). Specimens were tested over a range of temperatures to determine the upper use temperature.

Thermal analysis testing included differential scanning calorimetry (DSC) and thermal mechanical analysis (TMA). Two types of dynamic mechanical spectroscopy (DMS) tests were performed: dynamic mechanical analysis (DMA) and Rheometrics dynamic analysis (RDA). The effects of moisture, heating rates, and test frequency on the measured T_g were examined.

The DMS tests were found to give more reproducible results than either the DSC or TMA. Higher heating rates and test frequencies resulted in higher measured T_g's. Shift factors were developed to allow comparison between data obtained at various heating rates and tests frequencies. Results from the mechanical testing corresponded more closely to the storage modulus measurements of the DMS tests than the values obtained from the tan delta curves.

KEYWORDS: carbon/epoxy composites, upper use temperature, glass transition temperature (T_g), thermal analysis, frequency effects, fiber orientation effects, thermal lag, mechanical testing

The use of organic matrix composites in the primary structure of military and commercial aircraft is increasing due to weight and life cycle cost savings offered by composite materials. Because of the advantages offered by composite materials, structural applications are expanding to include higher temperature structures. Areas of the wing and fuselage subjected to elevated temperatures caused by aerodynamic heating and areas near the engine are now potential candidates for composite application. Organic matrix resins, such as epoxies and bismaleimides, soften as the temperature is increased, at some point passing through a glass transition temperature (T_g). Above the T_g, the resin will exhibit a significant decrease in strength and stiffness due to increased polymer chain mobility. This loss of

[1] Chemist, materials engineer, chemical engineer, and materials engineer, respectively, Naval Air Warfare Center, Aircraft Division, Warminster, PA 18974-0591.

strength is detrimental to structural integrity and limits the use of composite materials in many elevated temperature aircraft structures.

In simple one-component amorphous systems, the glass transition occurs over a well-defined temperature range [1,2]. Newly developed matrix resins for composites are much more complex than the original single component resins developed in the 1970s. The new systems contain rubber and thermoplastic tougheners which have significantly different thermal and mechanical properties from the base resin. The tougheners form second and third phases in the polymer resin, each of which has its own T_g. This often results in the T_g of the matrix resin extending over a wide range of temperatures, reducing the upper use temperature of the composite material.

As the resin systems become more complex and as composite structures are required to operate in higher temperature areas, the maximum service temperature of the composite becomes an important material parameter. The upper use temperature for composite materials is often defined as 28°C (50°F) below the T_g at equilibrium moisture content [3]. This arbitrary value of 28°C was selected to ensure that the material would retain a significant degree of stiffness and also allow for large amounts of variability in the measurement of the T_g. The effect of moisture on the T_g is an important aspect in selection of a material for use in naval aircraft due to the constant high humidity environment in which they operate. As polymer matrix resins become more complex, their ability to retain property performance at temperatures approaching T_g differs from older, simpler resin systems. Thus the conventional rule may be overly conservative or, conversely, not restrictive enough, and screening materials by this convention may no longer be appropriate.

The glass transition can be measured by several different analytical techniques [1,2,4]. Each of these methods has several variables that will affect the results and all of the test techniques have some latitude for determining specific test procedures. Thermal analyses most often used include static techniques such as Differential Scanning Calorimetry (DSC) and Thermomechanical Analysis (TMA) and dynamic techniques such as Dynamic Mechanical Spectroscopy (DMS). DSC measures the difference in the specific heat of a material as it goes through the glass to rubber transition. This method is popular due to the very small specimen required (<20 mg). TMA is used to measure the change in the thermal expansion of a material as it goes through the transition temperature. The T_g is observed as a change in the slope of the linear expansion of a material as it is heated at a constant rate. TMA also requires a small specimen and is relatively easy to prepare. DMS measurements are divided into two types: those that deform the material at a constant frequency, such as the Rheometrics Dynamic Analyzer (RDA), and those that deform the material at the resonant frequency of the material, such as the DuPont Dynamic Mechanical Analyzer (DMA).

Some recent work at the Naval Air Warfare Center, McDonnell Douglas, Northrop, and ICI Fiberite [5] was initiated in an effort to determine how the various methods of thermal analysis correspond to one another and identify the effects of specific machine and material variables. The work reported here covers T_g determination by DSC, TMA, and DMS. The dynamic mechanical spectroscopy measurements were made using DMA and RDA test techniques to evaluate specimen variables of fiber orientation and moisture content. This section also evaluates machine variables of oscillation frequency, heating rate, and temperature lag between the specimen and the furnace. The thermal lag study established the difference between the actual specimen temperature and the recorded specimen temperature for each thermal dynamic test.

Mechanical testing was conducted to determine upper use temperature and these results were correlated with the glass transition temperatures measured by thermal analysis techniques. Neat resin compression, composite interlaminar shear, and open hole compression tests were performed over a range of temperatures. While these mechanical tests more closely simulate actual use conditions, the cost and time required to perform these tests make them less desirable than thermal analysis as materials screening tests.

Experimental

Materials

The polymer and polymer-composite system examined in this study were the ICI Fiberite 977-3 epoxy resin and IM7/977-3 carbon/epoxy composite system [6]. Composites samples were prepared from 16- and 24-ply 0°, 16-ply ±45°, and 24-ply quasi-isotropic laminates. The laminates were cured for 6 h at 180°C using the manufacturer's suggested cure cycle. Composite DMS specimens were cut with the fibers oriented 0°, 90°, and ±45° to the major specimen dimension. A schematic diagram of the composite DMS specimens is shown in Fig. 1. The neat resin samples were degassed in a vacuum oven at 150°C for 30 min and subsequently cured for 6 h at 180°C.

All thermal and mechanical specimens were dried at 121°C under vacuum until weight loss did not exceed 0.05% for three consecutive readings prior to testing, with the exception of specimens used for the heating rate evaluation. Specimens for the heating rate study were exposed to ambient laboratory conditions prior to testing; these specimens will be referred to as ambient specimens. Weight loss measurements on specimens used for later tests showed that the ambient specimens used in the heating rate studies may have contained up to 0.5% moisture by weight.

Moisturized IM7/977-3 composite specimens were environmentally exposed at 60°C, 95% relative humidity for over 60 days, until weight increase did not exceed 0.05% for three consecutive readings. The average weight increase due to moisture in these specimens was 1.2%.

Differential Scanning Calorimetry

All DSC experiments were performed on a Perkin-Elmer DSC7 with a Perkin-Elmer 7700 Professional Computer controller. Additional DSC scans were performed on a DuPont DSC 912 controlled by a DuPont 1090 Thermal Analyzer. A nitrogen purge was used in both instruments at 25 cc/min. DSC experiments were run on ambient and wet composite samples and dried neat resin samples at 5, 10, and 20°C/min between 30 and 425°C. The glass transition was determined from the heat flow curve by the intersection of the tangents to the baseline and inflection. DSC was also used to determine the residual exotherm.

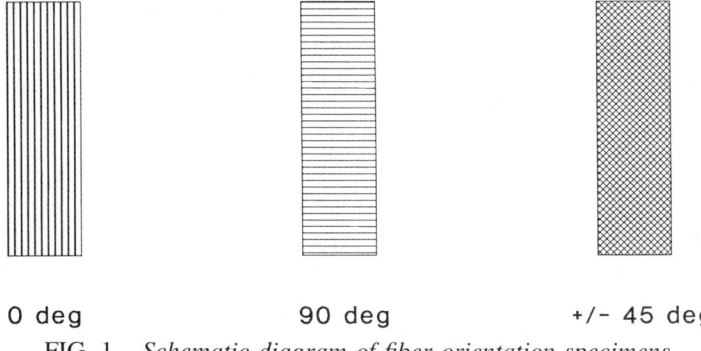

FIG. 1—*Schematic diagram of fiber orientation specimens.*

Thermomechanical Analysis

A Perkin-Elmer TMA7 connected to a Perkin-Elmer 7700 Professional Computer controller was used for thermomechanical experiments. TMA experiments were run in the expansion mode with a quartz probe and platform under a compressive force of 10 g and a nitrogen purge at 25 cc/min. Ambient and moisturized composite specimens were tested at 5, 10, and 20°C/min using a straight temperature ramp from 30 to 300°C. Additional tests were also performed by heating the samples to just above the expected glass transition temperature, holding for 15 min, quench-cooling or slow-cooling to the start temperature, and then reheating to 300°C at 5, 10, and 20°C/min.

Dynamic Mechanical Spectroscopy

Dynamic mechanical spectroscopy was performed on IM7/977-3 composite and 977-3 neat resin specimens using both the DuPont Dynamic Mechanical Analyzer and the Rheometrics Dynamic Analyzer II.

Dynamic Mechanical Analysis

A DuPont DMA 982 connected to a DuPont 1090 Thermal Analyzer was used for dynamic mechanical testing. Specimens were tested under a forced constant amplitude-resonance flexural oscillation mode. Vertical clamp test fixtures were used to hold specimens approximately 50 mm by 10 mm by 2 mm. Clamping distance was varied between 9 and 30 mm to bring the initial resonance frequency into a range of 20 to 60 Hz. Specimens were tested at an oscillation amplitude of 0.2 mm with a nitrogen purge rate of 90 cc/min. Glass transition temperatures were determined from both the E' and tan delta curves. The T_g was taken as the intersection of the tangents to the breakpoint of the log E' curve and the major peak of the tan delta curve.

Rheometric Dynamic Analysis

All rheological experiments were performed on a Rheometrics RDAII controlled by a dedicated IBM computer workstation. Specimens were tested under a forced constant amplitude-fixed frequency, torsional oscillation mode. Rectangular torsion test fixtures were used to hold specimens measuring approximately 50 mm by 12 mm by 2 mm. All tests were run in air, without additional purge. In order to ensure that the tests were performed in the linear viscoelastic region, strain and frequency sweeps were conducted prior to testing. All experiments were performed at 0.05 or 0.1% strain; both are within the linear viscoelastic region for the material tested. Glass transition temperatures were determined from both the G' and tan delta curves. The T_g was taken as the intersection of the tangents to the breakpoint of the log G' curve and the major peak of the tan delta curve.

Dynamic Mechanical Experiments

The effect of test heating rate on the measured glass transition was examined by performing DMA and RDA tests at 5, 10, and 20°C/min. Tests were performed on 0° ambient and moisturized composite specimens. The RDA experiments were performed at 10 rad/sec. The effects of sample thermal lag on DMS tests were performed as a function of furnace heat up rate. Both DMA and RDA tests were performed with composite samples containing an embedded thermocouple (see Fig. 2). A separate thermocouple was placed near the oven control thermocouple to monitor the oven temperature. A thermocouple reader with a built-in chart recorder was used to measure both specimen and oven temperature as a function of

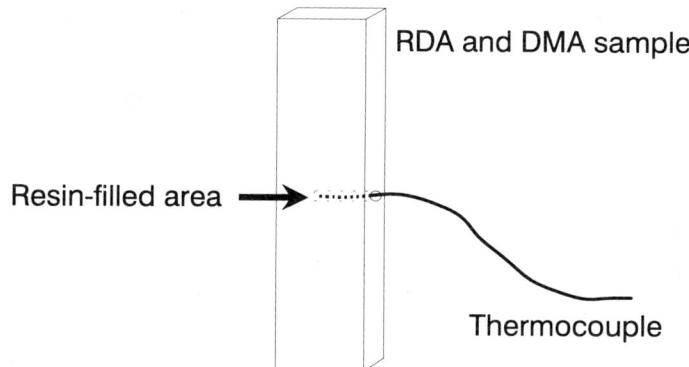

FIG. 2—*Diagram of a thermocoupled DMS specimen.*

time. All specimens were tested at 5, 10, and 20°C/min. RDA specimens were tested at 10 rad/sec.

In order to determine the effect of test frequency on the measured T_g, RDA experiments were performed at 0.1, 0.5, 1, 10, and 100 rad/sec. Tests were conducted on 0° dried composite samples heated at 10°C/min. The G' and tan delta T_g's were reported.

The effect of fibers and fiber orientation on the measured glass transition temperature was investigated by DMA and RDA tests on 0°, 90°, ±45° and neat resin samples. Tests were performed on dried specimens at 10°C/min. The RDA experiments were performed at 10 rad/sec. The modulus breakpoint and tan delta T_g's were reported.

Mechanical Testing

Mechanical tests were performed as a function of temperature. The upper use temperature was evaluated by two methods: the tangent intercept and the temperature at which the strength drops to 50% of the dry room temperature value.

Neat Resin Compression

Compression tests were performed on neat resin specimens (0.508 cm by 0.508 cm by 1.016 cm) according to ASTM D695, Test Method for Compressive Properties of Rigid Plastics. All resin specimens were dried prior to testing. Tests were performed at 24°C, 93°C, 121°C, 149°C, 163°C, 177°C, 191°C, 204°C, 218°C, and 232°C. Four specimens were tested at each temperature.

Interlaminar Shear (ILS) Tests

ILS tests, conducted on 16-ply 0° composite laminates, were based on a modified Suppliers of Advanced Composite Materials Association (SACMA) Recommended Test Method for Apparent Interlaminar Shear Strength of Oriented Fiber-Resin Composites by the Short Beam Shear Method (SRM-8). The specimens were 0.208-cm-thick and 0.635-cm-wide. The specimen length was 3.429 cm with a span-to-thickness ratio of 10:1. This is longer than recommended by SACMA. In previous work performed at the Naval Air Warfare Center, it was found that using a span-to-thickness ratio of 4:1 (as recommended by SACMA) did not produce the appropriate interlaminar failure mode for this particular carbon/epoxy system. A 4:1 ratio produced a crushing type of failure rather than a shear-like one, with small multiple cracks produced under the loading pin. A study was conducted to determine a better span-to-thickness ratio. This study showed that a 10:1 ratio produced acceptable failure modes. The specimens were tested at 24°C, 66°C, 93°C, 121°C, 149°C, 177°C, and 204°C. Five specimens were tested at each temperature.

Material Operating Limit (MOL) Test

The MOL test is a modified open hole compression test designed by Northrop [7]. The specimens are 7.62 cm by 2.54 cm with a 0.635-cm-diameter hole in the center. MOL tests were performed on moisturized 24-ply quasi-isotropic composite laminates. The compression tests were conducted using a special fixture designed to eliminate specimen buckling. The load is applied at a cross-head speed of 0.127 cm/min until failure. The tests were performed at 82°C, 93°C, 104°C, 127°C, 149°C, 163°C, and 177°C. At least four specimens were tested at every temperature.

Results and Discussion

Differential Scanning Calorimetry

Differential scanning calorimetry traces of the IM7/977-3 composite showed no discernible infection indicative of the glass transition temperature on either the Perkin-Elmer or the DuPont instruments. The initial concern was that the amount of resin present, less than 30% of the total specimen weight, was not sufficient to show the small changes in heat capacity that occur near the glass transition. Larger specimens (approximately 20 mg) were then used to ensure an adequate amount of resin was present during the DSC tests. These larger specimens still failed to show consistent T_g's, indicating that some other factor may be responsible for the inability of the DSC equipment to determine T_g.

Samples from the neat 977-3 resin gave relatively clear, reproducible glass transitions on the DuPont DSC scans although not on the Perkin-Elmer. Scans at 5, 10, and 20°C/min gave glass transition temperatures of 177, 179, and 185°C, respectively. Both instruments did reveal residual exotherms in the composite and in the neat resin plaques, even after completing the manufacturer's recommended cure cycle. The magnitude of this exotherm was monitored by DSC to ensure that the cure state of the material was not advanced by the method used to dry the specimens at 121°C. Indeed, this exotherm did not change considerably after the dry conditioning.

The fact that neat resin samples showed a T_g and the composite samples did not would indicate that the presence of the fibers may interfere with the measurement of the transition. Although a number of other researchers have found difficulty with this method [8], some researchers have successfully used DSC to determine T_g for composites [9,10]. It is unclear at the present time why DSC tests failed to indicate a glass transition for this particular composite system. Factors other than the presence of the fibers may have contributed to the lack of a clear transition, such as the onset of the residual exotherm in the temperature range where T_g is expected and the two-phase nature of the resin system in which the transitions of the two phases occur within 50°C of one another. Conflicting data within the literature and problems mentioned above present some concerns about the use of DSC to determine the glass transition temperature of polymer composite materials as a measure of upper use temperature.

Thermomechanical Analysis

The data from thermomechanical analyses of composite and neat resin samples were inconsistent at best. The TMA traces did not often display a simple, single change in slope at the glass transition. Rather, they curved gradually into the final slope reached at the final test temperature. No definitive intersection of tangents could be assessed; measured T_g could vary by up to 30°C on a single experiment depending on the choice of tangents. The two phase construction of 977-3 resin made this a nonideal system for thermal testing. The transitions in each phase occurred too close together in time and temperature for the baseline to be reestablished before the onset of the second slope deviation. The TMA traces were also frequently marred by a small expansion in the region of the glass transition. Two temperature programs were attempted at each of the three heating rates to alleviate what we believed was the result of residual stresses in the material. Neither the straight ramp nor the addition of an isothermal hold and quench cool neutralized this effect. As a result of these unresolved problems, TMA was not an effective tool for the determination of the glass transition in this resin system.

Dynamic Mechanical Spectroscopy

Heating Rate Experiments—The effect of oven heating rate on the T_g measured by DMA is shown in Table 1 for both ambient and moisturized conditions. The measured glass transition for the IM7/977-3 composite samples increased dramatically as the heating rate was raised from 5 to 20°C/min. Conditioning the composite specimens to attain equilibrium moisture content markedly decreased the T_g measured by DMA. The E' break point decreased an average of 47°C for each of the three heating rates tested; the tan delta peak temperature decreased an average of 43°C.

The effect of heating rate on T_g measured by RDA was similar to that described for DMA as seen in Table 1. The T_g determined for moisturized specimens decreased by an average of 65°C for the G' break point and 60°C for the tan delta peak temperature from the ambient values. The increase in the measured T_g as determined by RDA, however, was affected to a lesser extent than the T_g measured by the DMA.

The difference in transition temperature measured at each heating rate can be attributed to the lag between the oven temperature and the specimen equilibrium temperature. The results of the thermal lag experiments for DMA and RDA are shown in Fig. 3. As seen in this figure, the temperature difference between the specimen and furnace is greater for the DuPont machine than for the Rheometrics machine. This is particularly evident at the higher heating rates. The difference in thermal lag is due to the oven configuration and the location of the specimen in relation to the heating element. Using temperature offset factors, delta T,

TABLE 1—*Effect of oven heating rate on T_g.*

Heating Rate, C/min	Specimen Condition	DMA		RDA	
		T_g, E'(°C)	T_g, tan delta (°C)	T_g, G'(°C)	T_g, tan delta (°C)
5	ambient	220	255	184	226
10	ambient	233	263	198	238
20	ambient	251	283	214	259
5	wet	171	207	133	182
10	wet	186	221	135	182
20	wet	206	244	147	195

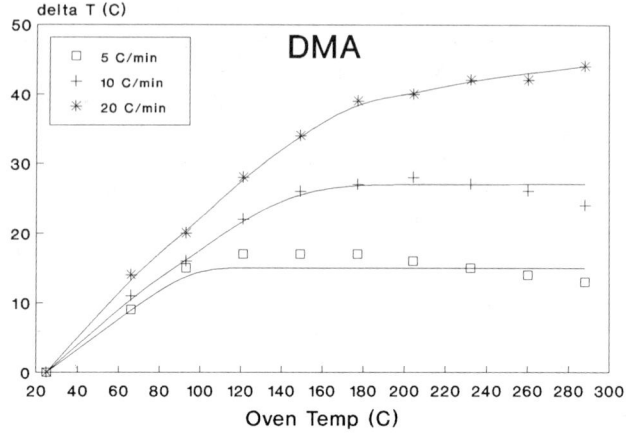

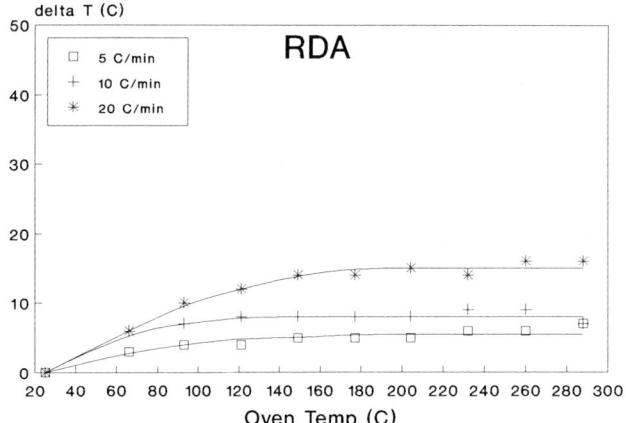

FIG. 3—*Thermal lag as a function of heating rate for DMA and RDA.*

determined from Fig. 3, the T_g's determined from tests with different heating rates can be corrected to an equilibrium specimen temperature. Using a delta T of 15°C, 27°C, and 42°C for DMA heating rates of 5, 10, and 20°C/min, respectively, the calculated ambient T_g's for all three rates is between 205 and 209°C for E' break point and between 236 and 240°C for tan delta. Similar corrections can be made for the RDA data using the appropriate delta T factors from Fig. 3. Table 2 shows the calculated T_g's shifted to the specimen equilibrium temperature for both ambient and moisturized specimens. These data show good agreement on a glass transition temperature for all heating rates. It is important to note that even after shifting the measured T_g to account for the thermal lag, the highest heating rate (20°C/min) still produces a slightly higher T_g.

Increasing the oven heating rate on both machines showed similar trends for the moisturized specimens. The glass transition temperatures of wet specimens increased with the oven heating rate similar to ambient specimen results. Our initial concern was that heating

TABLE 2—*Heating rate values adjusted for thermal lag.*

Heating Rate, C/min	Specimen Condition	DMA		RDA	
		T_g, E' (°C)	T_g, tan delta (°C)	T_g, G' (°C)	T_g, tan delta (°C)
5	ambient	205	240	178	220
10	ambient	206	236	190	230
20	ambient	209	241	199	244
5	wet	156	192	128	177
10	wet	159	194	127	174
20	wet	166	204	133	181

moisturized specimens at slow rates would actually dry the specimen during the experiment and lessen the measured effects of the moisture. The experimental results show that this was not the case at the heating rates tested. The data show that the decrease in glass transition due to moisturization is relatively constant for all three heating rates and the increase in wet T_g with increasing heating rate is due to the thermal lag between the specimen and the furnace. However, this may not have been the case if much slower heating rates (e.g., 1°C/min) were used. The delta T shift factors from Fig. 3 provide a reasonable measure of the equilibrium glass transition for moisturized specimens without the drying effects that may occur at very slow heat-up rates.

Frequency Experiments—The effects of test frequency on the measured glass transition temperature is shown in Fig. 4. Tests were performed over three decades, from 0.1 to 100 rad/sec. As expected [1,4,11,12], the measured glass transition temperature increased as the testing frequency was raised. The plot of log frequency versus measured glass transition temperature reveals a linear relationship of increasing T_g with increasing frequency. Regression of the data in Fig. 4 yields a 7.5°C change in tan delta peak temperature per decade of frequency and a 5°C change in G' break point. This difference in shift factors is

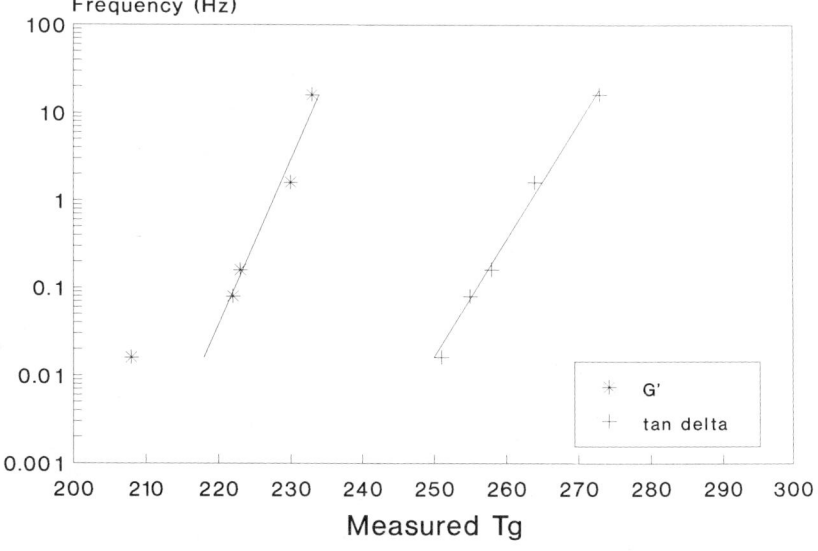

FIG. 4—*Frequency versus measured T_g.*

most likely due to the difficulty associated with determining the tangents prior to and after the breakpoint in the modulus curve. Glass transition temperatures measured from the 0.1 rad/sec were not used in either the G' or the tan delta regression due to the much longer sampling interval associated with this frequency. The sampling time for 0.1 rad/sec, approximately 90 sec, is very long considering the heating rate, 10°C/min, so the specimen temperature changes considerably during the sampling interval. These factors resulted in too few data points to obtain a smooth curve. Sparse data at the transition reduced the confidence in the measured transition temperature at this frequency.

Examination of the tan delta thermal curves (Fig. 5) reveals a noticeable difference in the shape of the curves as the frequency is changed. At lower frequencies there is an increase in definition between the two phases of the 977-3 resin. At high frequencies, the tan delta peak of the toughener appears as a shoulder on the larger, higher temperature epoxy peak. As the frequency is decreased, the distance between the two peaks increases, and the toughener peak begins to emerge as a separate event. It would appear from these data that the toughener component of the matrix resin is more sensitive to the changes in test frequency than the epoxy phase, although it would be difficult to quantify this effect due to the absence of a distinct toughener peak at higher frequencies without the use of curve-fitting analysis [*13*].

Fiber Orientation—It is well documented that the addition of fibers and other reinforcing matter to polymeric materials often results in a higher T_g than unreinforced materials

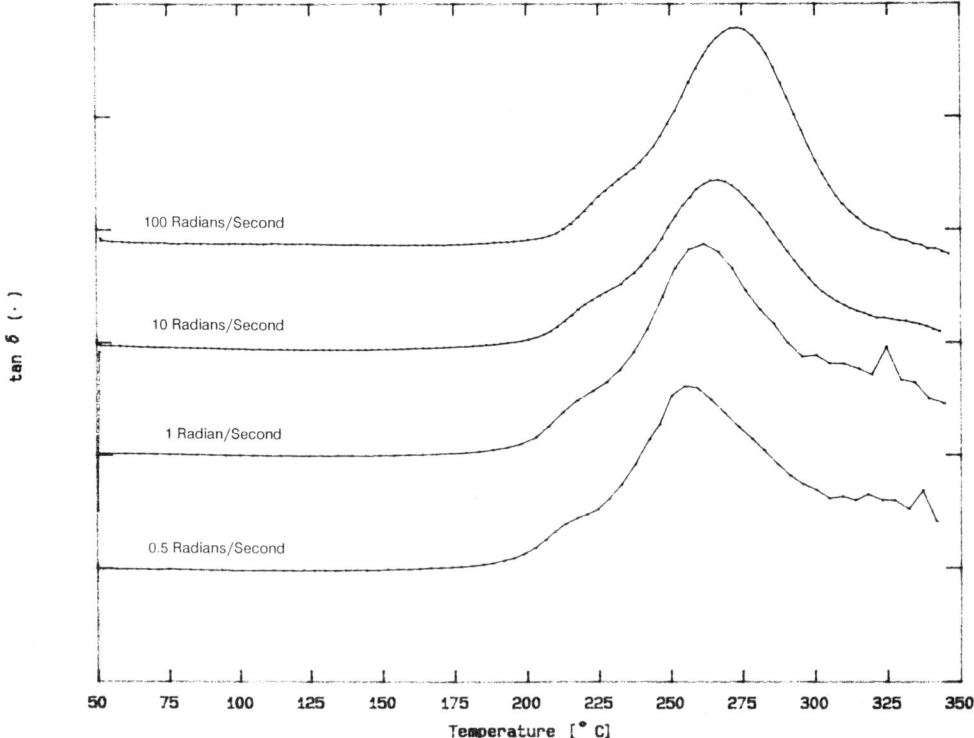

FIG. 5—*RDA thermal curves showing effects of frequency on* tan *delta* T_g (tan *delta curves shifted for clarity*).

[2,4,14]. Three different fiber orientations were tested to examine the effect of carbon fiber reinforcement on the measured glass transition of 977-3 resin. The results of the fiber orientation experiments by DMA and RDA are shown in Figs. 6 and 7. Figure 6 shows the storage modulus curves of the three fiber orientations determined by DMA. As seen in this figure, the amount the measured modulus was affected depended on the orientation of the fiber direction in relation to the deformation mode of the testing equipment. For the DMA machine, which deforms the specimen longitudinally in a flexural/bending mode, the 0° specimen had the highest modulus, followed by the ±45° and the 90° specimen. Figure 7 shows the storage modulus versus temperature plots for RDA composite and neat resin tests. This figure shows how the measured modulus was affected by the orientation of the fibers. Because the RDA deforms the specimen in a torsional manner, the ±45° orientation provided the most resistance to deformation, followed by the 0°, the 90°, and the neat resin samples.

Transition temperatures determined by DMA for composite samples containing IM7 fibers were consistently higher than those measured for the 977-3 neat resin as seen in Table 3. T_g's for composite specimens increased with the amount of reinforcement provided by the fibers. Thus the modulus curves in Fig. 6 are not only shifted to a higher modulus by the fibers, but they are also shifted to higher temperatures. The glass transition temperatures determined by RDA were also affected by the presence of reinforcing fibers. Contrary to the incremental increase in DMA results, the RDA transitions fell into two distinct groups. The ±45° and the 0° specimens displayed a G' T_g 15°C and tan delta 10°C higher than the 90° and the neat resin specimens.

The difference in measured T_g between the DMA and RDA for the fiber orientation tests was further investigated. The different fiber orientations for the DMA specimens affected the initial resonant frequency as well as the frequency at the transition temperature. As seen above, changes in frequency will change the measured T_g. The data in Table 3 were then

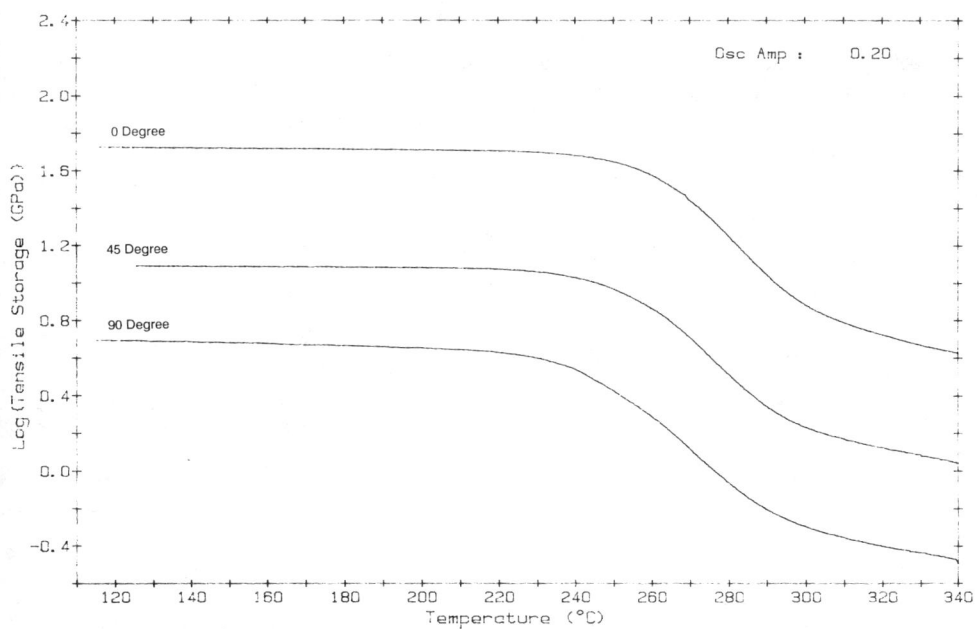

FIG. 6—*DMA thermal curves showing effects of fiber orientation on storage modulus.*

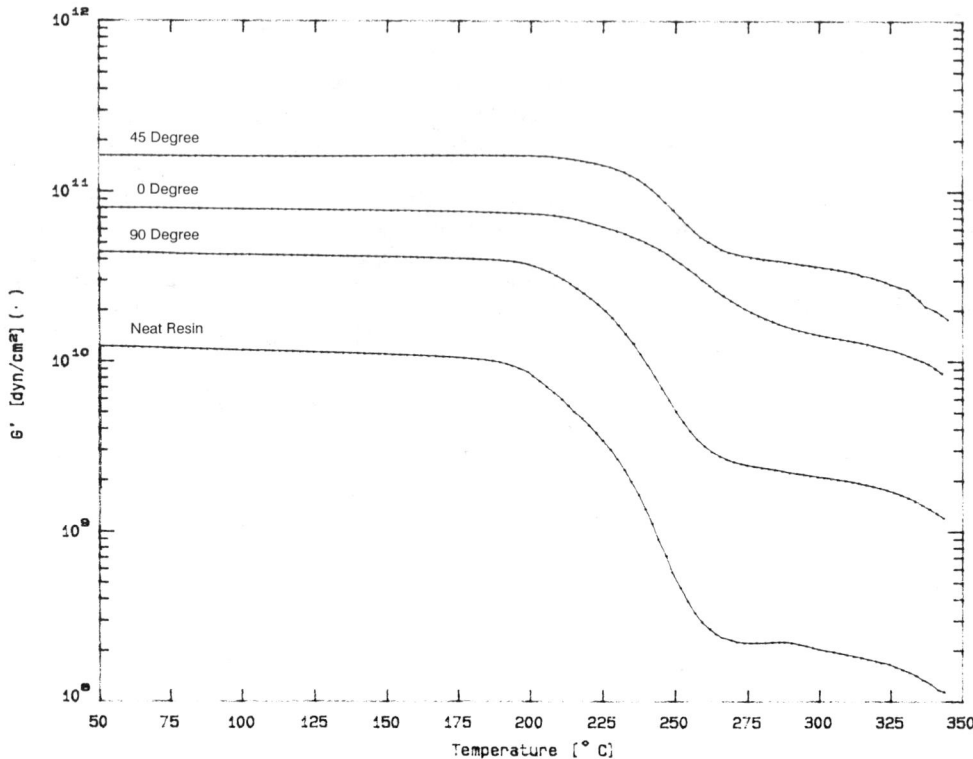

FIG. 7—*RDA thermal curves showing effects of fiber orientation on storage modulus.*

corrected to an equilibrium specimen temperature at a frequency of 1 Hz. These data are presented in Table 4. The corrected values show good agreement between the T_g's determined from the DMA and RDA. The 90° specimens show a slight 5°C increase in T_g over the neat resin, and the 0° and ±45° specimens display a T_g elevated by 20°C. Such increases in the T_g due to inclusion of fibers have been attributed to filler surface area effects, restriction of polymer molecular motion, packing density changes of polymer chains, and modification of conformation and orientation of polymer chain segments as a result of

TABLE 3—*Effects of fiber orientation on T_g.*

	DMA		RDA	
	T_g, E'(°C)	T_g, tan delta (°C)	T_g, G'(°C)	T_g, tan delta (°C)
0°	258	294	230	264
45°	253	286	230	258
90°	242	279	217	249
Neat Resin	233	261	212	251

Test frequency of 10 rad/s on RDA.
Tested at material's resonant frequency on DMA.
Heating rate of 10 C/min.

TABLE 4—*Fiber orientation effects on T_g adjusted for thermal lag and frequency.*

	DMA		RDA	
	T_g, E'(°C)	T_g, tan delta (°C)	T_g, G'(°C)	T_g, tan delta (°C)
0°	218.7	256.6	220.5	254.5
45°	215.3	250.0	220.5	248.5
90°	204.3	243.0	207.5	239.5
Neat Resin	197.5	227.9	202.5	241.5

Values shifted for thermal lag.
Values adjusted to correspond to a test frequency of 1 Hz.

strong interfacial adhesion [2,4,14]. The dramatic increase in T_g due to fibers oriented to provide reinforcement in the deformation direction is most likely due to the increase in modulus as the composite temperature approaches T_g [4].

Mechanical Testing

Neat Resin Compression Tests—The results of the neat resin compression testing are shown in Fig. 8. The compressive strength shows a linear decrease as the temperature increases from room temperature to 204°C. Between 204°C and 218°C a sharp drop in compressive strength from 54 MPa to 20 MPa occurs. Above this temperature, the strength again appears to decrease in a linear fashion. Using a technique similar to that used on the storage modulus curves, a dry T_g would be determined as approximately 204°C. At approximately 162°C, however, the resin compressive strength has been decreased to T_{50}, defined as 50% of the room temperature value. From these data an upper use temperature based on

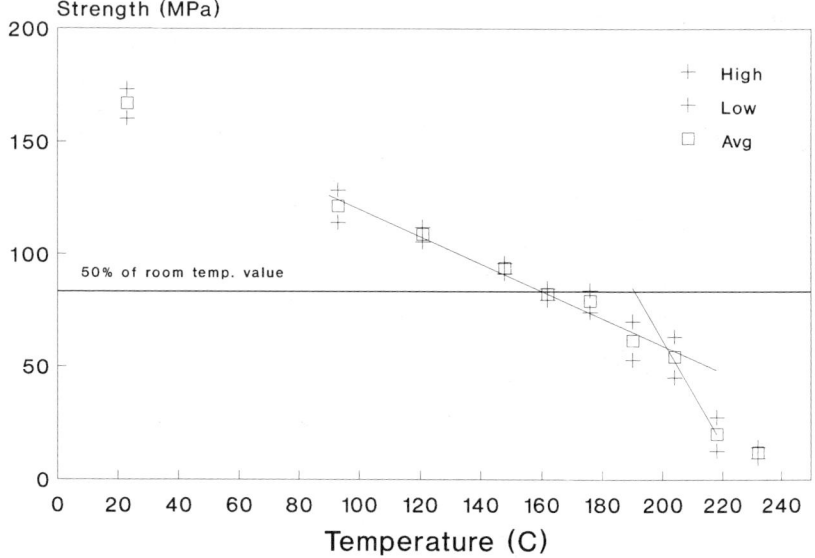

FIG. 8—*Neat resin compression strength as a function of temperature.*

50% retention of room temperature dry strength would be 162°C and an upper use temperature based on T_g would be 204°C.

Interlaminar Shear Testing—Figure 9 shows the data generated from the interlaminar shear testing. These results show a similar trend to the neat resin compression data. The strength decreases in a linear fashion from room temperature to approximately 204°C. Between 204°C and 218°C the strength drops rapidly from 33 MPa to 23 MPa. At 177°C, the interlaminar shear strength is 38 MPa, 45% of the room temperature strength of 84 MPa. Using this method, the dry upper use temperature obtained from this test is again significantly less than the upper use temperature based on T_g. The fibers in the specimen may have helped maintain the strength of the composite as the temperature was increased; however, when T_g was reached the strength was severely degraded.

Material Operating Limit Testing—The MOL test results are shown in Fig. 10. The two points for the dry specimen tests at room temperature and 100°C show a similar trend to the neat resin and interlaminar shear results (a 30% decrease over this temperature range). The wet test results show a slight plateau between 82°C and 104°C followed by a nonlinear decrease to 175°C. There is no obvious knee in the curve and construction of tangents to the curve is difficult. An upper use temperature from the tangent is between 125°C and 150°C. An upper use temperature based on T_{50} is approximately 135°C.

The test results for MOL are not as well defined as the other test methods. This may be the result of the difficulty and scatter associated with compression testing or of the moisture content in the specimens. The MOL test, although designed to be a simple reliable test to determine material operating limit, does not appear to provide any more accurate upper use temperature data than any of the other mechanical or thermal techniques.

Mechanical Testing Summary

Comparison of the upper use temperatures determined by the mechanical testing with the T_g's determined through thermal analysis revealed some useful relationships. The corrected

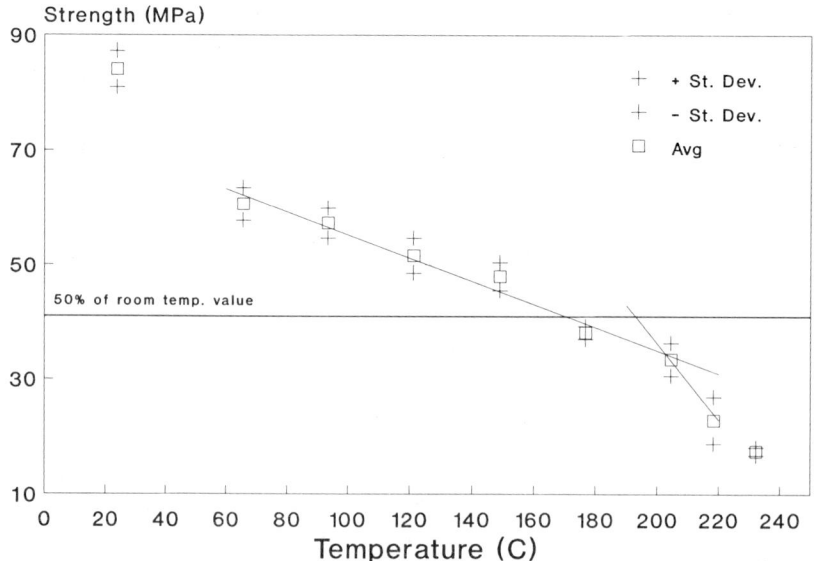

FIG. 9—*Interlaminar shear strength as a function of temperature.*

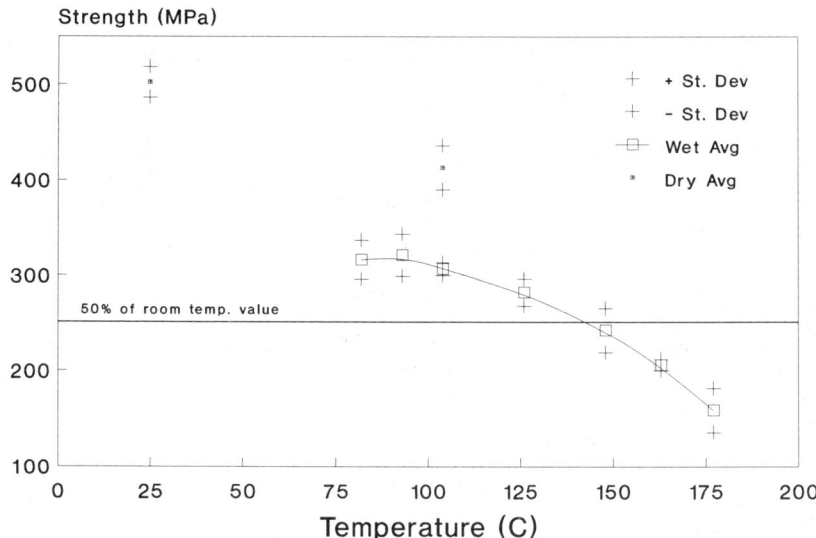

FIG. 10—*Wet MOL strength as a function of temperature.*

break point T_g's measured for the 90° and the neat resin specimens (Table 4) correlate well with the upper use temperature determined from the tangent intercept of mechanical test results. If the T_g minus 28°C rule is applied, these transition temperatures approximately match the T_{50} upper use temperatures. The rule of conversion, though, may not be conservative enough for the higher T_g 0° and 45° samples. Care must be taken when converting to upper use temperature for such samples, and a larger shift allowance will be required to equate T_g to T_{50}.

Conclusions

Based on the results of this study, the following conclusions can be made for evaluation of the upper use temperature of toughened epoxy composite materials:

1. DSC and TMA were not effective tools for determining upper use limits by glass transition temperature for composite systems.
2. DMS experiments performed at different heating rates can be accurately corrected to account for the inherent thermal lag of the equipment.
3. DMS can be performed on moisturized specimens at rates above 5°C/min without drying the sample.
4. The presence of reinforcing fibers in a polymer matrix will raise the T_g; significant differences in T_g are only observed if the fibers are oriented to provide resistance to the testing deformation.
5. After application of appropriate thermal and frequency shifts factors, the T_g data from different DMS equipment show good agreement.
6. The glass transition data corrected for thermal lag at 1 Hz correlate well with upper use temperature determined by mechanical testing.

References

[1] Sperling, L. H., *Introduction to Physical Polymer Science*, Wiley, New York, 1992.
[2] *Thermal Characterization of Polymeric Materials*, E. A. Turi, Ed., Academic, New York, 1981.
[3] MIL-SPEC SD-565-3, "Detail Specification for F/A-18E/F Aircraft Weapon System," Para 3.4.1.2.1.
[4] Nielsen, L. E., *Mechanical Properties of Polymers and Composites*, Marcel Dekker, Inc., New York, 1974.
[5] Thomas, J. A., "Round Robin Glass Transition Temperature Test Summary," *McDonnell Douglas Aerospace Interim Report No. H12-380-0406*, 1992.
[6] Almen, G., Mackenzie, P., Malhotra, V., and Maskell, R., *SAMPE International Symposium 35*, 1990, p. 419.
[7] Dyere, T. and Whitehead, R., "Use of Materials Operating Limit Tests of Composite Materials," *Northrop Report No. 87-80*, Vol. 4, 1987.
[8] Martin, R. H., Siochi, E. J., and Gates, T. S., *ACS Technical Conference 7*, 1992, p. 207.
[9] Williams, R. J. J., Benavente, M. A., Ruseckaite, et al., *Polymer Engineering Science*, Vol. 30, 1990, pp. 1140–1145.
[10] McGrath, J. E., Grubbs, H., Rogers, M. E., et al., *SAMPE Technical Conference 23*, 1991, p. 119.
[11] Ward, I. M., *Mechanical Properties of Solid Polymers*, Wiley, New York, 1983.
[12] Jones, D. I. G., *Journal of Sound Vibration*, Vol. 140, 1990, p. 85.
[13] Rotter, G. and Ishida, H., *Macromolecules*, Vol. 25, 1992, p. 2170.
[14] Tomkinson, G. D. and Vallance, M. A., *Polymer Composites*, Vol. 8, 1987, pp. 237–243.

Manoj K. Gupta[1]

Glass Transition Measurements on Automotive Coatings by DSC, DMA, and TMA

REFERENCE: Gupta, M. K., "**Glass Transition Measurements on Automotive Coatings by DSC, DMA, and TMA,**" *Assignment of the Glass Transition, ASTM STP 1249*, R. J. Seyler, Ed., American Society for Testing and Materials, Philadelphia, 1994, pp. 293–301.

ABSTRACT: The glass transition temperature (T_g) is a very useful parameter that characterizes the temperature dependence of the physical and mechanical properties of coatings. Differential scanning calorimetry (DSC), thermomechanical analysis (TMA), and dynamical mechanical analysis (DMA) were used to determine the T_g of automotive coatings.

DMA analyses were done on coatings deposited on a steel substrate in the fixed frequency mode, at frequencies of 0.01, 0.1, 1.0, and 10 Hz. Specimens were heated in either step mode (2°C) or ramp mode (5°C/min). TMA analyses were done either on coatings deposited on steel substrate or on free films at 5°C/min. DSC analyses were done at 5°C/min either on free films or on specimens that were sanded out from steel panel. Indium metal was used for the temperature calibration of DSC, TMA, and DMA. For one-coat films, the correlation between the temperature of midpoint in DSC, the onset temperature of penetration by TMA, and the onset temperature of loss modulus by DMA at 0.1 Hz was good. However, for multicoat films the T_g's for individual layers were visible only if their T_g's were far apart. It was possible in some cases to remove the top layer by scraping and analyzing the residual layers. The use of DSC, DMA, and TMA to determine the T_g's of different automotive coatings on metal substrate has been demonstrated, which is very useful in research and development, and problem solving.

KEYWORDS: differential scanning calorimetry (DSC), thermal mechanical analysis (TMA), dynamical mechanical analysis (DMA), glass transition temperature, automotive coatings, metal substrate

The physical and mechanical properties of coatings are greatly influenced by environmental factors such as light, heat, and moisture [1]. Since organic coatings are viscoelastic in nature, the T_g, the temperature at which a coating changes from a glassy and hard material to a viscoelastic one, is a very important parameter which predicts the performance of a coating with variation in temperature [2,3]. Thermal analysis techniques such as differential scanning calorimetry (DSC), dynamic mechanical analysis (DMA), and thermal mechanical analysis (TMA) have been used to measure the T_g of organic coatings [4]. However, these techniques often give different values for the T_g. Even for a given technique, the T_g value depends on factors such as the heating rate, heating mode, frequency of measurement, data analysis method, and substrate behavior, etc. Therefore an objective of this paper is also to define the experimental conditions under which the T_g values for coatings on metal substrates, as determined by DSC, DMA, and TMA, are in agreement.

Figure 1 shows the three layers of organic coatings on a typical automobile. The e-coat is used for corrosion protection and is electrodeposited onto both sides of the substrate. The

---

[1]Senior research associate, BASF Corporation, Southfield, MI 48086.

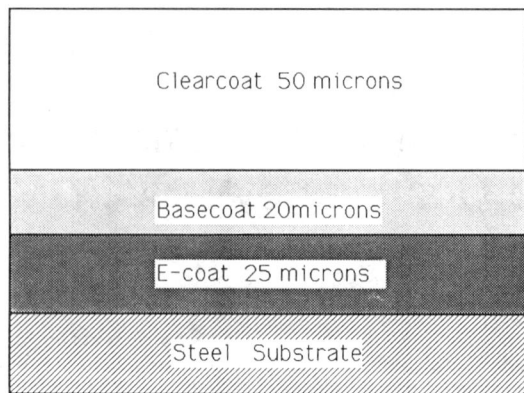

FIG. 1—*Organic coatings layers on a typical automobile.*

basecoat contains pigments and provides color. The clearcoat is transparent and provides protection against ultraviolet (UV) light. In this paper two types of e-coats on metal substrate were analyzed by DSC, DMA, and TMA. The DMA and TMA analyses of some multicoat systems were also included.

Experimental Methodology

TA Instruments' 9000 thermal analysis system equipped with the 910 Differential Scanning Calorimeter, 943 Thermal Mechanical Analyzer, and 983 Dynamic Mechanical Analyzer were used. DSC specimens were prepared either from the free films or by sanding the coating off by Emery paper from the panel and sealing a 10 to 15 mg specimen in the aluminum pan. The specimen was heated at 5°C/min from ambient to 175°C, then cooled to −50°C and then scanned again at 5°C/min. The midpoint temperature from the second DSC scan was assigned to the T_g. The TMA analysis was done either on free films or on coatings on metal substrates in the penetration mode (probe diameter ~1 mm). Specimens of about 11 mm diameter were punched out from the panel for TMA analysis. The specimens were heated at 5°C/min under 5 g load. The onset temperature for the penetration was assigned to the T_g.

The DMA was run in the fixed frequency mode at 0.01 Hz, 0.1 Hz, 1.0 Hz, and 10.0 Hz. The specimen dimensions were on the order of 50 mm by 13 mm by 1 mm (the substrate is about 0.8 mm thick and the thickness of coating is comparatively small). Only limited measurements were done at 0.01 Hz because of the long duration of the experiment (12 h) and the extreme sensitivity to the vibrations. Two different heating programs were used for the DMA analyses:

1. The scanning mode viz. scanning at 5°C/min to 225°C. The measurements were possible only at 1 Hz in this mode because the equilibration time is too long at 0.1 Hz and therefore the instrument could not scan at 5°C/min at 0.1 Hz.

2. The increment mode viz. taking a measurement at fixed frequency and then increasing temperature by 2°C, equilibrium for 3 cycles and then taking the next measurement. The process was repeated until the temperature reached 175°C.

Instead of using the conventional method of T_g determination by taking the maximum in tan delta or the maximum in loss modulus, the onset temperature for the loss modulus was taken as the T_g. However, it was observed that in our specimens, the maxima in loss modulus was about 20°C higher than the onset temperature for the loss modulus.

The DSC module was calibrated with ice and indium using ASTM E 967, Practice for Temperature Calibration of Differential Scanning Calorimeters and Thermal Analyzers. The TMA was calibrated using indium and ice in the penetration mode, following ASTM E 1363, Test Method for Temperature Calibration of Thermomechanical Analyzers. The temperature calibration of the DMA was done with indium metal strip. Polycarbonate and ABS standards (specimens available through TA Instruments) were also run on the DMA periodically to check temperature accuracy. Nevertheless a small specimen of indium was incorporated in all DSC, DMA, and TMA specimens and any shift in indium reference temperature was incorporated in the final calculations.

The two e-coat specimens used in this study were electrodeposited on 10 cm by 30 cm galvanized steel panels and cured at 177°C for 30 min. The solvent-borne white basecoat and clearcoat were sprayed on top of the e-coated panels. These topcoats were cured together at 135°C for 30 min.

Results

Table 1 summarizes the thermal analysis results for the two e-coats. The results are the average of at least two, but mostly three or four specimens. Although the instrument modules were calibrated via the software, a small specimen of indium was used in all analyses and any difference between the measured and the literature values for indium were incorporated in the calculations. The computer calculated values for transitions on the thermal curves do not represent the correction due to the differences in indium melting point.

TMA Analysis

Figure 2 shows the TMA thermal curve for the e-coat. The thickness of coating was about 25 μm (on each side) whereas the steel substrate was about 800 μm (0.8 mm) thick. The TMA analysis of the substrate showed a continuous downward movement of the probe, however, without any transition. This made the thermal analysis of coatings on metal substrates very challenging and difficult. For the specimens used in this study, a 5 g load at a

TABLE 1—*Glass transition temperatures (°C) of e-coats by DSC, DMA, and TMA.*

	E-coat 1	E-coat 2
DSC Midpoint	76.4	83.6
(5°C/min scanning)		
TMA Onset Point	76.9	83.1
(5°C/min scanning)		
DMA Loss Modulus Onset[a]	76.6	81.9
(2°C increments)		
DMA Loss Modulus Maxima[a]	95.8	102.2
(2°C increments)		
DMA Loss Modulus Onset[b]	78.2	83.7
(5°C/min scanning)		
DMA Loss Modulus Maxima[b]	99.9	103.9
(5°C/min scanning)		

[a] DMA data in increment mode at 0.1 Hz.
[b] DMA data in scanning mode at 1.0 Hz.

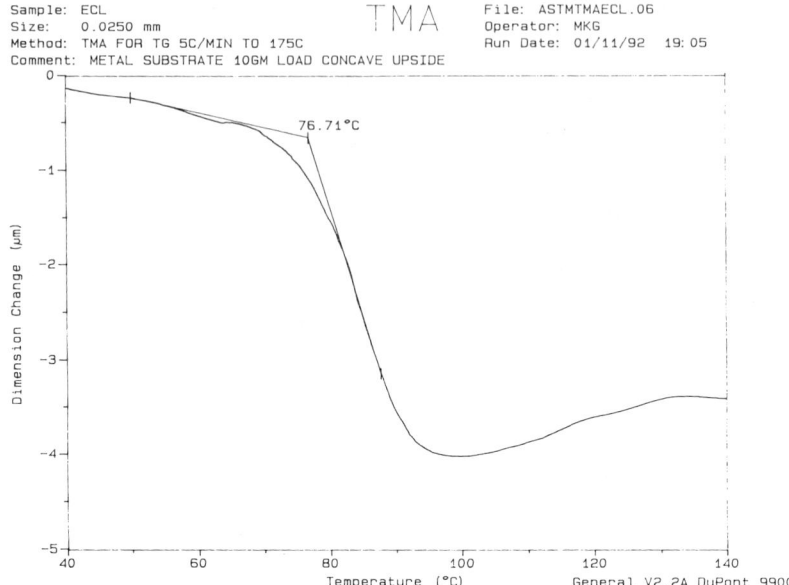

FIG. 2—*TMA thermal curve of the e-coat 1 in the penetration mode.*

heating rate of 5°C/min was found to be the optimum. A thin foil of indium was put underneath the specimen as the internal standard. The positioning of the probe was also found to affect the nature of the thermal curve and temperature readout. For one-coat films, the determination of the onset point is not too complex. However, for multicoat films, the determination of onset points is often very difficult due to the overlapping transitions.

Figures 3 and 4 show TMA thermal curves of multicoat films. Figure 3 represents an e-coat, basecoat, and clearcoat system whereas Fig. 4 shows the above system without clearcoat (removed by scrapping off by a razor blade). It is evident that the clearcoat transition is absent in Fig. 4 and the onset points for the basecoat and the e-coat were calculated with higher precision. The T_g's for e-coat, basecoat, and clearcoat in Fig. 3 were comparable to the T_g's of these layers measured individually. In some cases, the transitions for the basecoat and the clearcoat or for the e-coat and the clearcoat strongly overlapped and scrapping was necessary before separate transitions could be observed.

DSC Analysis

Figure 5 shows the DSC thermal curve for an e-coat. The specimen was prepared by sanding off the e-coat from the panel. The temperature of the midpoint in the transition was taken as the T_g. The maximum of the first derivative of the heat flow with temperature was also used to calculate the T_g. The difference between T_g's measured by the midpoint and by the first derivative was often within 1°C. In this author's opinion the derivative method is superior because the calculations used to determine the T_g are operator-independent. The midpoint method requires the selection of tangents for the curve, which is subjective and operator-dependent.

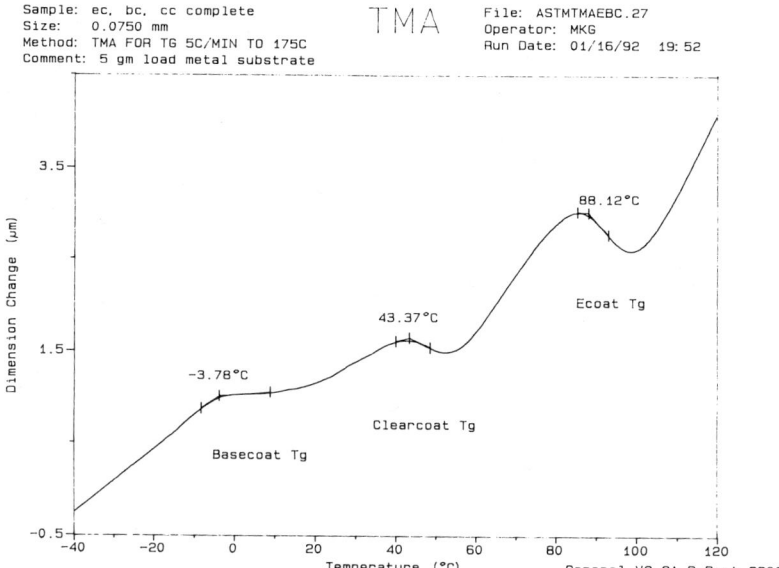

FIG. 3—*TMA thermal curve of a complete automobile paint system.*

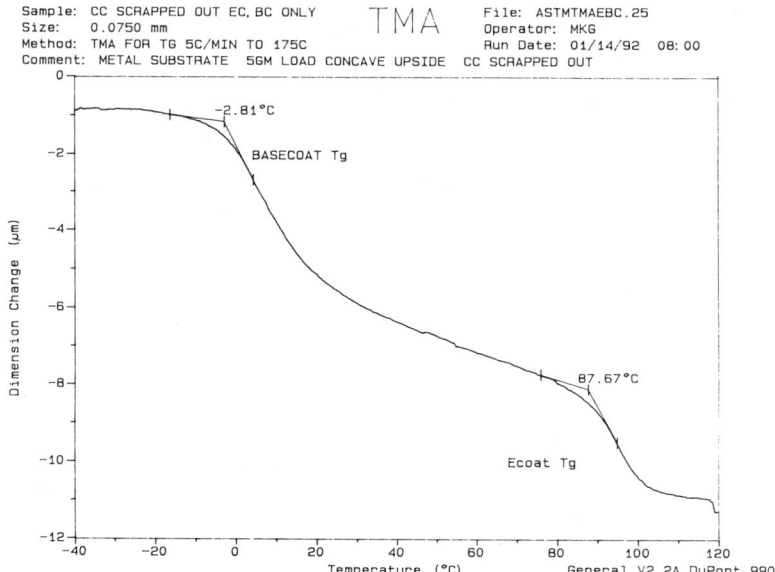

FIG. 4—*TMA thermal curve of an automobile paint system after scraping off clearcoat.*

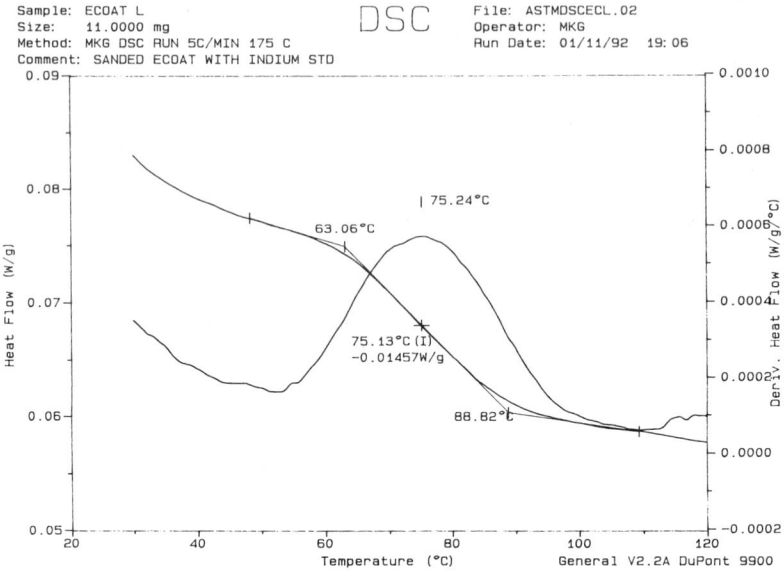

FIG. 5—*DSC thermal curve of the e-coat 1.*

DMA Analysis

Figures 6 and 7 are the thermal curves for the two e-coats used in this study. Figure 6 shows the storage and loss modulus of the e-coat as a function of temperature at 0.1 Hz, 1.0 Hz, and 10.0 Hz. The absolute values for the modulus of the specimen could not be

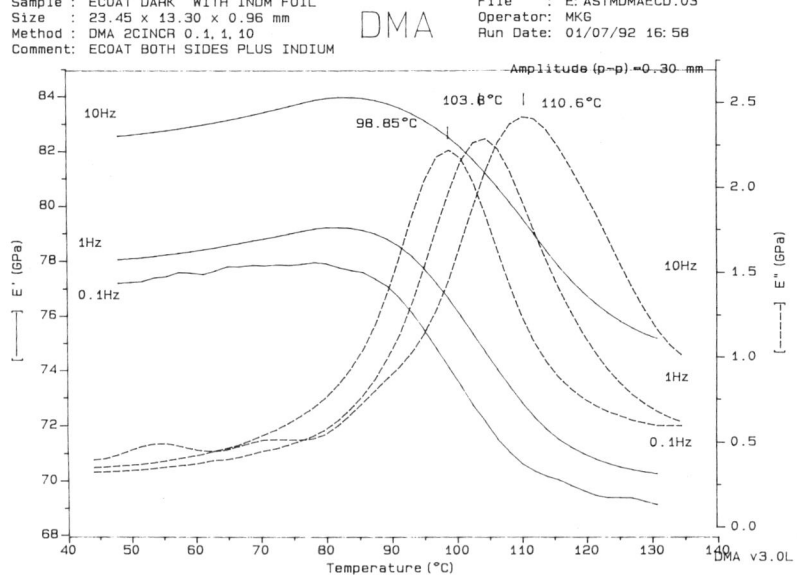

FIG. 6—*Storage and loss modulus of the e-coat 2 at 0.1, 1.0, and 10.0 Hz.*

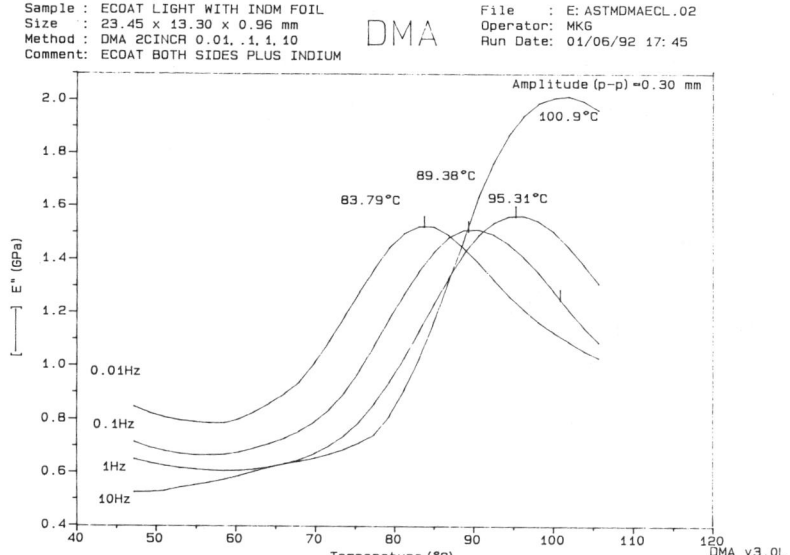

FIG. 7—*Loss modulus of e-coat 1 at 0.01, 0.1, 1.0, and 10.0 Hz.*

calculated due to the steel substrate and only the transition temperatures were estimated. The onset temperature for the storage modulus was significantly higher than the T_g obtained by TMA or DSC. It is suspected that the large storage modulus of the substrate modulates the storage modulus of the system and it is therefore not possible to get information about the coating layers.

Figure 7 shows the loss modulus as a function of temperature for the second e-coat at 0.01, 0.1, 1.0, and 10.0 Hz. As expected, the maximum in loss modulus depends on the frequency. Figure 8 shows a plot of maxima in loss modulus as a function of the frequency. The DMA measurement at 0.01 Hz took almost 12 h and therefore no efforts were made to make measurements at 0.001 Hz or lower frequency. However the interpolation of the curve in Fig. 8 predicts that at 0.0001 Hz, the loss maximum will be at 75.43°C, which is comparable to the T_g values obtained by DSC or TMA.

Figure 9 shows the DMA thermal curve of a complete automotive paint system (e-coat, basecoat, and clearcoat). The analysis was done in the scanning mode at 5°C/min and 1 Hz, and therefore the T_g values were shifted upwards. The storage modulus does not show transitions for three layers; however, the loss maxima are visible for all three layers. In fact, after correcting for the temperature difference with indium and frequency dependence, the values are in fair agreement with those obtained by TMA. It was observed for numerous systems on the metal substrate that the maximum in loss modulus was always about 20°C higher than the onset temperature in loss modulus. This shift would be very helpful for multicoat systems, where onset points could not be easily calculated.

Conclusions

DSC, DMA, and TMA have been used for the measurement of T_g for automotive coatings on the metal substrate. It was found that despite the very thin specimens (<50 μm), TMA was very successful for the measurement of T_g on one-coat as well as multicoat systems

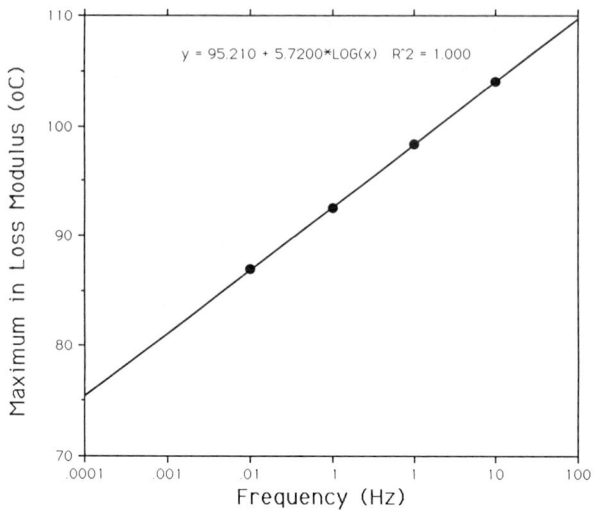

FIG. 8—*Loss modulus maxima versus frequency for e-coat 1.*

DMA was also very successful on metal substrates and could complement the TMA analysis. The T_g determination by DSC at 5°C/min (midpoint or the maximum in first derivative), TMA at 5°C/min (onset point in the penetration mode using a 1 mm diameter probe on metal substrate) and DMA at 0.1 Hz in 2°C increments (onset temperature in the loss modulus or the temperature of maximum loss modulus offset by 20°C) were in good agreement. It is strongly believed that the thermal characterization of automotive coatings

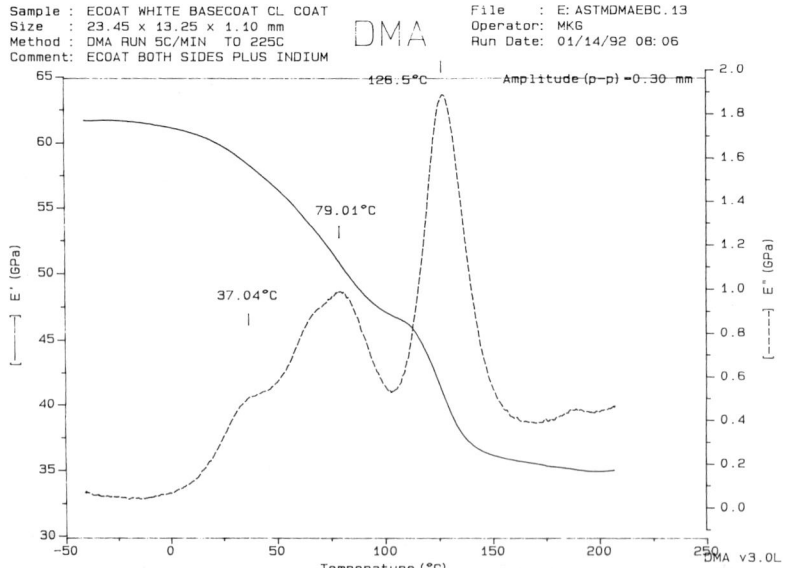

FIG. 9—*DMA thermal curve of an automobile paint system.*

on the metal substrate, in particular on multicoat systems, is very useful for research and development, and problem solving.

Acknowledgments

The author would like to thank BASF Corporation for supporting this work and permitting publication of these results. The author would also like to thank Dr. Sohail Akhter and Dr. Thomas G. Savino for their helpful comments.

References

[1] Bauer, D. R., Mielewski, D. F., and Gerlock, J. L., "Photooxidation Kinetics in Crosslinked Polymer Coatings," *Polymer Degradation and Stability*, Vol. 38, 1992, pp. 57–67 and references therein.
[2] Reichert, K. W. and Doennerbrink, G.,"Characterization of Coatings by Thermomechanical Analysis," *Organic Coatings*, Vol. 8, 1986, pp. 383–395.
[3] Zosel, A., "Mechanical Behavior of Coating Films," *Progress in Organic Coatings*, Vol. 8, 1980, pp. 47–79.
[4] Skrovanek, D. J. and Schoff, C. K., "Thermal Mechanical Analysis of Organic Coatings," *Progress in Organic Coatings*, Vol. 16, 1988, pp. 135–163.

Discussion

Written discussion (anonymous discusser)—My questions stem from a Rheometrics Applications bulletin that suggests, and has been confirmed to me by their technical personnel, that the magnitude of the glass transition is dependent upon the type and thickness of the substrate, coating, and adhesive forces at each interface.

For two urethane coatings on aluminum substrate, if the composition of urethanes are identical, what effect on the T_g as measured by DMA (Rheometric RSA-3 point bending mode) will be detected if a pigment or carbon black is added to one?

For a urethane coating on aluminum, what effect on the T_g as measured by DMA (Rheometric RSA-3 point bending) will be detected if either the thickness of the coating or of the substrate is increased?

Do the adhesive forces at the adhesive/substrate interface in a coated sample influence the glass transition temperature? If so, how is this correlated to the idea of the glass transition being a bulk property of the polymer?

Your comments or further references would be greatly appreciated.

M. K. Gupta (author's closure)—Aluminum substrate (0.8 mm thick) shows significant change in modulus in the temperature range of these experiments. This makes the analysis of transitions more complicated. Furthermore, I do not have access to a RSA-3 instrument, so I cannot comment on the specific changes. However, for moderate concentration of pigments or increase in thickness, T_g should not change, although absolute modulus will change.

Rickey J. Seyler[1]

Closing Discussion: Highlights and the Challenges that Remain

As editor, I will now attempt to summarize for you what I found to be some of the key concepts from these well prepared presentations and the numerous well founded audience discussions. It was particularly gratifying to experience the cross-fertilization of ideas offered by the physics/inorganic and the polymer/organic communities. For instance, the structural relaxation process introduced in Prof. Moynihan's lecture appears to substantiate the thermodynamic and kinetic aspects of the glass transition offered by Prof. Wunderlich. This leads us to the observation that:

the glass transition is the phenomenological passage between a glassy solid and a "melt" encountered through the promotion of or suppression of relaxation processes active in the time scales of the measurement used.

Conceptually then, the glass transition is a passage between short range vibrational processes in the glassy solid and long-range translational and rotational processes in the rubber/liquid state within a finite temperature interval. This may be observed by studying optical, electrical, mechanical, or thermal properties as a function of temperature. We are reminded that the glass transition is not reversible, but rather is bidirectional with hysteresis. The extent of this hysteresis is dependent upon the time scale of the measurement employed.

Despite the recognition that the glass transition requires a finite temperature interval over which to occur, there is a strong practical need to represent this interval with a singular temperature referred to as T_g. Confusion and discrepancies may, however, result from such practices because the temperature reported is somewhat arbitrary and strongly influenced by a host of extrinsic factors. Sound arguments of merit for assigning T_g as the onset, midpoint, endpoint, peak, or fictive temperature have been presented. Of these temperatures, the fictive temperature, although derived rather than directly measured, promises to be least impacted by many of the extrinsic factors influencing the temperatures where the glass transition is observed.

A number of recurring points about T_g were cited:

- A consistently assigned T_g will still vary because of differences in **heating rate** or **frequency of an oscillating load.** Frequency shifts are time-temperature superposable and are of the order of 6° to 8°C per decade. The frequency most often mentioned for T_g measurements was 1 Hz although comparisons of dynamic mechanical data with DSC data suggest that frequencies between 0.01 and 0.1 Hz most closely match the DSC midpoint value recorded at less than 20 K/min.

[1]Eastman Kodak Company, Rochester, NY 14650-2158; chairman of the symposium and editor of this STP.

- **Aging effects** (enthalpic relaxation, densification, or free volume reduction) can influence the observed temperature, signal appearance (curve shape), or the ability to detect the glass transition, or a combination thereof.
- **Composition** (both chemical and physical) dependence of T_g and the temperature interval of the glass transition has been noted.
- Depending upon circumstances, certain **measurement techniques** may offer relatively better **sensitivity** to observe the glass transition. There does not appear to be a universally superior measurement method.
- Many of the techniques used to monitor the glass transition measure the specimen temperature indirectly. In order to achieve a reasonable **certainty of observed temperatures,** temperature calibration of the instrumentation at the measurement conditions to be used for measuring the glass transition is required.
- **Diluent/filler** presence in specimens usually affects the glass transition. **Moisture** is particularly troublesome in glass transition measurements, especially when subambient temperatures are required as part of the measurement protocol. Unless water levels are of interest, every effort should be made to prevent moisture gain in a specimen prior to examining the glass transition.
- T_g may not necessarily represent the bulk material. Assigned values may differ within localized areas of a sample and may even be directionally different in oriented or structural forms.
- **Discrepancies or confusion** can and do exist in the open literature regarding T_g's of materials. These are particularly problematic in digests, handbooks, or reviews where the measurement and material descriptions are typically not included.
- Measurement of **thin films** properties including the glass transition remains an analytical challenge. Acoustic wave microsensors were shown to be suitable for such purposes.
- Manufacturing and engineering applications of amorphous and semicrystalline materials require T_g to predict use temperature limits or to establish process conditions. These application needs underscore the need for a uniform, consistent approach to assigning the glass transition temperature, T_g.

What is the challenge we of the technical community face regarding the glass transition? Consider observance of the glass transition as a process:

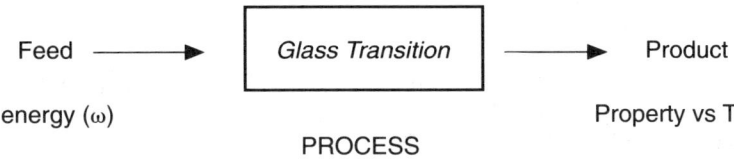

Energy is input to a material specimen as the process "feed." The product of the process is a signal of the specimen response to the energy input. At some level of input the specimen undergoes a change in its response to the "feed" which signifies the occurrence of the glass transition. This change is neither discontinuous nor instantaneous. Two needs exist regarding this process: process understanding which would lead to a precise definition of the glass transition temperature and process representation which provides for assessment of the practical use of glasses as engineering materials. Both needs require a **unique identification** of the glass transition process. Providing this unique identification of the glass transition which exists over a finite, but real temperature interval over which large changes occur is the **measurement challenge** facing us.

A recommendation has been offered that the glass transition be measured on cooling from the "melt" as the reference state. In practice most measurements of the glass transition are accomplished on heating because of generally better temperature control in equipment with heating, often, employing multiple temperature excursions through the glass transition to impose a reference thermal history. Although the recommended approach gives a more material specific assessment for academic purposes, the practiced approach may be more beneficial in applications where thermal history contributions are important or need to be understood.

In attempting to formulate a measurement concept for a unique identification of the glass transition, I am reminded of thermodynamics problems. It was always necessary to define the limits of the SYSTEM within which the problem was formulated and subsequently solved. The same holds for "assigning the glass transition." Any measurement of T_g is a unique but meaningless outcome without a complete description of the system from which the unique identifier has been derived. This is illustrated in the schematic below where the system which yields a T_g is that encompassed by the large box. It includes not only the property whose change is used to identify the glass transition but also conditions of the test, specimen, and atmosphere.

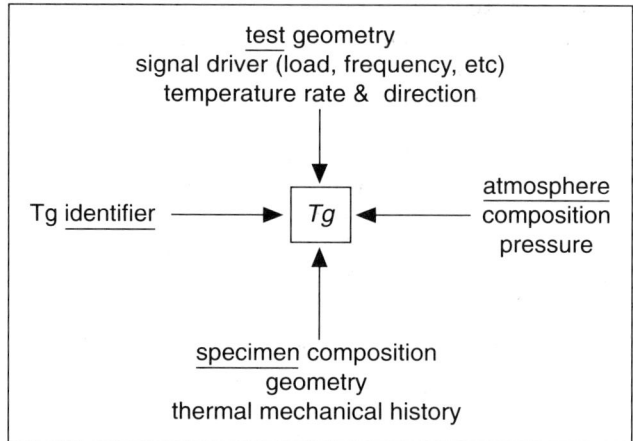

Our goal must be to meet the measurement challenge for unique identification of the glass transition by achieving appropriate measurement protocols which will lead to reliable, reproducible ASSIGNMENTS of the glass transition as a singular temperature and to accurately communicate this assignment not simply as T_g but as

$$T_g + S Y S T E M!$$

Author Index

B

Bair, H.E., 50
Bayer, G., 174
Bidstrup, S.A., 108
Bloyer, D., 185

C

Cassel, B., 202
Chang, S.-S., 120
Chartoff, R.P., 88, 226
Cho, J., 185
Cochran, R.C., 277

D-E

Day, D.R., 108
DiBerardino, M.F., 277
Earnest, C.M., 75

G-J

Grate, J.W., 153
Gupta, M.K., 293
Handa, Y.P., 165
Jankowsky, J.L., 277

K-M

Kincs, J., 185
Martin, S.W., 185
Moscato, M.J., 239
Moynihan, C.T., 32

O-R

O'Neill, M.L., 165
Paroli, R.M., 269
Penn, J., 269
Riga, A.T., 202
Rodriguez, E.L., 255

S-T

Saffell, J.R., 137
Seyler, R.J., 1, 13, 239, 302
Sircar, A., 88, 226
Tye, R.P., 4, 6

W

Weidemann, H.G., 174
Weiss, R.A., 214
Weissman, P.T., 88
Widmann, G., 174
Wong, D.G., 277
Wunderlich, B., 17

Subject Index

A

Acoustic wave microsensors, 153, 154(figs), 155(table)
Activation energy, glassy polymers, 255
Aging, physical, nature of glass transition, 17
Alkali borate, 185
Alkali thioborate, 185
Amorphous materials, 1(overview)
Amorphous phase, glass transition, 17
Analytical techniques, glass transition in polymers, 75, 174, 202, 211(table)
Assignment of glass transition temperatures, 13-16, 75, 137
ASTM Standards
 D 83, 13
 D 695, 281
 D 3418, 13, 228
 D 4065, 13, 90
 D 4092, 13, 90
 E 37, 13
 E 375, 13
 E 831, 76
 E 1142, 13
 E 1356, 1, 13, 76, 228
 E 1363, 81, 82
 E 1545, 1
Automotive coatings, 293, 295(table)

B

Biaxial orientation, poly(ethylene terephthalmate)(PET), 239
Blocks, 50
Borate, inorganic glasses, thermal properties, 185

C

Calorimetric studies, glass transition phenomena, 120
Calorimetry, glass transition of ionomers, 214
Carbon/epoxy composites, 277, 283(table), 285(table), 288-289(tables)
Chalcogenide, inorganic glasses, thermal properties, 185
Coatings, automotive, 293, 295(table)
Composition and behavior of polymeric materials, glass transition measurement, 50
Compressive mode, poly(ethylene tereph- thalate)(PET), 239
Condensed moisture, discussion, 4
Cooling exotherm, 137
Cooling rate, 137
Copolyester, 202
Crosslinking, 88
Crystallinity, 88
Cure, elastomer systems, 226

D

Dielectric properties, 108
Dielectric relaxation, 32
Dielectric thermal analysis for glass transition temperatures, 108
Differential scanning calorimetry (DSC), assignment of glass transition temperatures
 automotive coatings, 293, 295(table)
 elastomer systems, 226
 epoxy composites, 277, 283(table), 285(table), 288-289(tables)
 instrumentation, 175(fig)
 inorganic glasses, 185
 liquid crystal polymer, 202
 measurement, 50, 75, 120, 174, 226
 oriented poly(ethylene terephthalate), 239
 overview) 1
 polymers, 174
 structural relaxation process, 32
 thermal curves, 137
DSC. *See* Differential scanning calorimetry.
Dynamic elastic storage modulus, 255
Dynamic mechanical analysis for T_g
 automotive coatings, 293, 295(table)

determination in polymers, 88
elastomer systems, 226
glassy polymers, 255
liquid crystal, 202
toughened epoxy composites, 277, 283(table), 285(table), 288-289(tables)

E

Elastomers, measurement of glass transition temperature, 226, 230(table), 232(table)
Epoxies, 108
Epoxy composites, 277, 283(table), 285(table), 288-289(tables)
Ethylene-propylene-diene monomer (EPDM), measuring glass transition temperature, 269, 273-274(tables)

F

Fiber orientation effects, toughened epoxy composites, 277, 283(table), 285(table), 288-289(tables)
Fictive temperature, 32, 50, 120, 137
Flexural plate wave (FPW), 153
Frequency effects, 255, 277, 283(table), 285(table), 288-289(tables)

G

Gases, high pressure, polystyrene plasticization, 165, 170(table)
Germanate, inorganic glasses, thermal properties, 185
Glass transition, 17, 50, 239
Glass transition region, time dependencies, 32
Glass transition phenomena, calorimetric studies, 120
Glass transition temperature, T_g
  acoustic wave microsensors, 153
  assignment of values, 13-16
  assignment of values, using thermo-mechanical analysis, 75, 137
  automotive coatings, 293, 295(table)
  calorimetric studies, 120
  definitions, discussion, 13-16, 32,
  depression, 165
  determination by dynamic mechanical methods, 88
  determination by thermal analysis, 17

dielectric analysis, 108
differential scanning calorimetry, 137, 174
DSC thermal curves, 137
effects of moisture, discussion, 4
epoxy composites, 277, 283(table), 285(table), 288-289(tables)
ethylene-propylene-diene-monomer (EPDM), comparison of measurement techniques, 269, 273-274(tables)
glassy polymers, 255
inorganic glasses, 185
ionomers, 214, 216(table)
liquid crystal polymer, 202
measurements by DSC, 50, 174
measurements, comparative, 174, 234-236(tables), 269, 273-274(tables), 273-274(tables), 293, 295(table)
measurements, discussion, 6, summary, 302
measurements in elastomer systems, 226, 234-236(tables)
oriented poly(ethylene terephthalate), 239
phenomenology of structural relaxation process, 32
poly(ethylene terephthalate), 239
polymeric materials, 50, 137, 174
polystyrene plasticization, 165
stress relaxation, 32
structural relaxation, 32, 44(table)
temperature dependence, 32
thermal curves (DSC), 137
thermomechanical, 174
thermo-optical, 174
thin polymer films, 153
toughened epoxy composites, 277, 283(table), 285(table), 288-289(tables)
Glasses, inorganic, transition and heat capacities, 185
Glassy polymers, 255
Grafts, 50

H

Halide, inorganic glasses, thermal properties, 185
Heat capacity
  carbon, 127(figs)
  inorganic glasses, 185
  polyethylene, 29(fig)
Heat/cool rate (DSC), 137
Heat-flow calorimeter, 165

SUBJECT INDEX 309

Heating rate (DSC), 137
High-pressure calorimetry, 165, 167(fig)
Hot state, 174
Hysteresis, nature of glass transition, 17, 23(fig)

I

Inorganic glasses, thermal properties, 185, 197(table)
Instrumental factors, Tg, 88
Instruments, thermomechanical analyzer, 77-78(figs)
Ionic domains, 214
Ionomers, glass transition, 214, 216(table)

L

Linear thermodilatometry, 75
Linear viscoelastic test methods, 88
Liquid crystal polymer, glass transition, 202
Liquid, nature of glass transition, 17
Loss factor, 108, 113(fig), 115(fig)

M

Measurement of glass transition temperature
  comparative methods, 174, 234-236(tables), 269, 273-274(tables), 273-274(tables), 293, 295(table)
  dielectric analysis, 108
  differential scanning calorimetry, 174
  thermomechanical analysis, 174
  thermo-optical analysis, 174
Measurement of temperature, 6, 137
Mechanical relaxation, glass transition of ionomers, 214
Mechanical testing, toughened epoxy composites, 277, 283(table), 285(table), 288-289(tables)
Mesophase
  liquid crystal polymer, 202
  nature of glass transition, 17
Metal substrate, 293, 295(table)
Microsensors, acoustic wave, in thin polymer films, 153
Microphase separation, nature of glass transition, 17, 214
Mobile amourphous phase, glass transition, 17, 239

Moisture, T_g determination in polymers, 88
Moisture, condensed, discussion, 4

N

Nanophase, nature of glass transition, 17

O

Onset temperature, 137
Operational definition, 17

P

Peak temperature, 137
Penetrometry, 75
Permittivity, 108, 113(fig), 115(fig)
Phosphate, inorganic glasses, thermal properties, 185
Physical aging, 50
Plasticization effect of dissolved gas, 165
Plasticizers, 50, 165
Polycarbonate, 255
Poly(ethylene terephthalate)(PET), glass transition temperatures, 174, 180(table), 239, 240(table), 242(table), 246(table), 249(table)
Polymer-based materials, glass transition temperatures
  acoustic wave microsensors, 153
  blends, glass transition measurements by DSC, 50
  dielectric analysis, 108
  differential scanning calorimetry, 50, 174
  DSC thermal curves, 137
  elastomer systems, 226
  glass transition, discussion, 4
  glassy polymers, 255
  heat capacity, 29(fig)
  ionomers, 214
  measurement systems, comparison, 179(table)
  mechanical methods for T_g determination, 88
  polystyrene plasticization, 165
  T_g measurements by DSC, 50, 88
  temperature measurement, discussion, 6
  thermodynamic transition, 50
  thermomechanical analysis, 174
  thermo-optical analysis, 174
  thin films, 153
  transition behaviors, 153

Polymer gas interactions, 165
Polymer plasticization, 165
Polystyrene(PS), 165, 174, 255
Poly(vinyl acetate), 255
Polyvinylchloride, 108, 255

R

Relaxation properties, thin polymer films, 153
Residual entropies of glasses, 120, 124(table)
Rigid amorphous phase, glass transition, 17, 50, 239
Roofing materials, mechanical testing, 269

S

Salt groups, in relation to glass transition of ionomers, 214
Semicrystalline materials, 1(overview)
SH-APM. See Shear horizontal acoustic plate mode.
Shear horizontal acoustic plate mode (SH-APM), 153
Silicate glasses, thermal properties, 185
Simultaneous differential scanning calorimetry, 174
Softening temperature ($_s$), 75, 239
Solid, nature of glass transition, 17
Strain, nature of glass transition, 17, 24(fig)
Stress relaxation, 32
Structural relaxation process, 32
Surface acoustic wave (SAW), 153

T

T_g criterion, instrumental factors, 88
T_g, glass transition temperature, 32
T_s, softening temperature, 75
Temperature calibration, 75
Temperature dependence
of macroscopic properties in glass transition region, 32
Temperature measurement
assigned glass transition
discussion, 6
Tensile mode, poly(ethylene terephthalate), (PET), 239
Thermal analytical techniques
liquid crystal polymer, 202, 211(table)
nature of glass transition, 17
overview, 1

T_g determination in polymers, 88, 202
Thermal analysis, toughened epoxy composites, 277, 283(table), 285(table), 288-289(tables)
Thermal curves (DSC), 137
Thermal/mechanical history, Tg in polymers, 88
Thermal properties, inorganic glasses, 185
Thermoanalytical methods
discussion, 4
Thermodilatometry, 75
Thermodynamic transition, time dependent, 50
Thermomechanical analysis (TMA)
assignment of glass transition temperatures, 75
automotive coatings, 293, 295(table)
elastomer systems, 226, 234(table)
epoxy composites, 277, 283(table), 285(table), 288-289(tables)
in polymers, comparison of results, 174
liquid crystal polymer, 202
oriented poly(ethylene terephthalate), 239
Thermo-optical analysis (TOA), 174, 175(fig)
Thermoplastics, 88
Thermosets, 88
Thickness/shear mode (TSM), 153
Thin polymer films, 153
Thioborates, inorganic glasses, thermal properties, 185
TMA. See Thermomechanical analysis.
TOA. See Thermo-optical analysis.
Transition behavior, thin polymer films, 153
TSM. See Thickness-shear mode.

U

Ultrasonic, 153
Uniaxial orientation, poly(ethelene terephthalate)(PET), 239
Upper use temperature, 277, 283(table), 285(table), 288-289(tables)

V

Viscoelastic test methods, 88
Viscosity, 32
Vitreous state, 120
Vulcanization, elastomer systems, 226